S0-BQM-861

From Government to E-Governance:

Public Administration in the Digital Age

Muhammad Muinul Islam
Jahangirnagar University, Bangladesh

Mohammad Ehsan
University of Dhaka, Bangladesh

Managing Director: Lindsay Johnston
Senior Editorial Director: Heather A. Probst
Book Production Manager: Sean Woznicki
Development Manager: Joel Gamon
Development Editor: Christine Smith
Assistant Acquisitions Editor: Kayla Wolfe
Typesetter: Lisandro Gonzalez
Cover Design: Nick Newcomer

Published in the United States of America by
Information Science Reference (an imprint of IGI Global)
701 E. Chocolate Avenue
Hershey PA 17033
Tel: 717-533-8845
Fax: 717-533-8661
E-mail: cust@igi-global.com
Web site: http://www.igi-global.com

Library of Congress Cataloging-in-Publication Data

From government to e-governance: public administration in the digital age / Muhammad Muinul Islam and Mohammad Ehsan, editors.
p. cm.
Includes bibliographical references and index.
Summary: "The book explores the nature, structure, experiences and processes of public administration in past civilizations and its changing trends and characteristics with an evolutionary perspective"--Provided by publisher.
ISBN 978-1-4666-1909-8 (hardcover) -- ISBN 978-1-4666-1910-4 (ebook) -- ISBN 978-1-4666-1911-1 (print & perpetual access) 1. Public administration--Technological innovations. 2. Internet in public administration. 3. Public administration--History. I. Islam, Muhammad Muinul, 1980- II. Ehsan, Mohammad.
JF1525.A8F77 2012
352.3'802854678--dc23

2012005069

British Cataloguing in Publication Data
A Cataloguing in Publication record for this book is available from the British Library.

All work contributed to this book is new, previously-unpublished material. The views expressed in this book are those of the authors, but not necessarily of the publisher.

Dedicated to the memories of Professor Peter Aucoin, who was an exceptionally inspiring human being, and Professor Gregory B. Lewis for illustrating how to think differently with statistics.

Editorial Advisory Board

List of Reviewers

Table of Contents

Preface xiv

Acknowledgment xvi

Chapter 1
From e-Government to e-Governance: A Holistic Perspective on the Role of ICTs 1
Wolter Lemstra, Delft University of Technology, The Netherlands

Chapter 2
E-Government: Some Factors for a Conceptual Model 25
Mehdi Sagheb-Tehrani, Bemidji State University, USA

Chapter 3
Understanding E-Governance: A Theoretical Approach 38
Muhammad Muinul Islam, Jahangirnagar University, Bangladesh
Mohammad Ehsan, University of Dhaka, Bangladesh & Dalhousie University, Canada

Chapter 4
E-Government, M-Government, L-Government: Exploring Future ICT Applications in Public Administration 50
Alberto Asquer, University of Cagliari, Italy

Chapter 5
Challenges of Implementing E-Governance in a Politically Driven Environment 64
Manish Pokharel, Korea Aerospace University, South Korea
Jong Sou Park, Korea Aerospace University, South Korea

Chapter 6
Marketing E-Government to Citizens 75
María de Miguel Molina, Universitat Politècnica de València (UPV), Spain
Carlos Ripoll Soler, Universitat Politècnica de València (UPV), Spain

Chapter 7
Alignment of IT Projects and Investments to Community Values 93
Marcus Vogt, Heilbronn University, Germany
Kieth Hales, Bond University, Australia

Chapter 8
From Town Hall to the Virtual Community: Engaging the Public with Web 2.0
and Social Media Applications 114
Kathryn Kloby, Monmouth University, USA
Leila Sadeghi, Kean University, USA

Chapter 9
Key Success Domains for Business-IT Alignment in Cross-Governmental Partnerships 131
Roberto Santana Tapia, Ministry of Security and Justice, The Netherlands
Pascal van Eck, University of Twente, The Netherlands
Maya Daneva, University of Twente, The Netherlands
Roel J. Wieringa, University of Twente, The Netherlands

Chapter 10
Information Technology Product Quality, Impact on E-Governance, Measurement,
and Evaluation 162
Jiri Vanicek, Czech University of Life Science, Czech Republic
Ivan Vrana, Czech University of Life Science, Czech Republic
Zdeněk Struska, Czech University of Life Sciences, Czech Republic

Chapter 11
From E-Government to E-Governance in Europe 195
Rebecca Levy Orelli, University of Bologna, Italy
Emanuele Padovani, University of Bologna, Italy
Carlotta del Sordo, University of Bologna, Italy

Chapter 12
Core Values: e-Government Implementation and Its Progress in Brunei 207
Kim Cheng Patrick Low, Universiti Brunei Darussalam, Brunei
Mohammad Habibur Rahman, Universiti Brunei Darussalam, Brunei
Mohammad Nabil Almunawar, Universiti Brunei Darussalam, Brunei
Fadzliwati Mohiddin, Universiti Brunei Darussalam, Brunei
Sik Liong Ang, Universiti Brunei Darussalam, Brunei

Chapter 13
E-Government in Bangladesh: Prospects and Challenges 217
Noore Alam Siddiquee, Flinders University, Australia
Md. Gofran Faroqi, Ministry of Public Administration, Bangladesh

Chapter 14
Combating Corruption through e-Governance in India 233
Durga Shanker Mishra, State Government of Uttar Pradesh, India

Chapter 15
Grey Hair, Grey Matter, and ICT Policy in the Global South: The Ghana Case 245
Lloyd G. A. Amoah, Ashesi University, Ghana

Chapter 16
A Positive Hegemony? Arguing for a Universal Knowledge Regime led by an
e-Governance 'Savvy' Global Knowledge Enterprise! 261
Amlan Bhusan, Bhusan Taylor Ltd, UK

Chapter 17
Successfully Applying "e" to Governance 271
Evangelia Mantzari, Athens University of Economics and Business, Greece
Evanthia Hatzipanagiotou, Ministry of Finance, Greece

Compilation of References 289

About the Contributors 315

Index 323

Detailed Table of Contents

Preface xiv

Acknowledgment xvi

Chapter 1
From e-Government to e-Governance: A Holistic Perspective on the Role of ICTs 1
Wolter Lemstra, Delft University of Technology, The Netherlands

In this chapter, the author elaborates on the concepts of e-Government and e-Governance and place these concepts in the broader context of the introduction and diffusion of information and communication technologies. The question that is being addressed is "what should an effective and pro-active e-Government do?" For that purpose, the ICTs are not only considered general purpose technologies, but recognized as the driving force of an unfolding technological revolution. As such, a recurrent pattern in techno-economic and socio-economic development can be discerned that goes back to the First Industrial Revolution in Britain around the late 1700s and early 1800s. It is argued that appreciating the fifth instance of this pattern and recognizing the challenges each transition invokes can inform the policy formation process and make policy action more effective. In the broader context of roles that governments may assume, a stepwise approach is introduced to address the many challenges the diffusion of ICTs is bringing about in the economic and social realm, aimed at reaping the benefits implied in the new techno-economic paradigm.

Chapter 2
E-Government: Some Factors for a Conceptual Model 25
Mehdi Sagheb-Tehrani, Bemidji State University, USA

Some state, national, and local governments around the world have long played active roles in the use of Information Technologies (IT) to stimulate economic development. Electronic government utilizes information technology to provide the citizenry access to a wide range of public services. Governments in many countries around the world indeed perceive IT as a way to improve the quality of life of their citizens. Today, governments at all levels respond to millions of citizen demands electronically. Many public organizations are implementing Electronic Government (e-Government) projects. There is a need to put forward a conceptual model focusing on steps towards implementing more successful e-Government projects. This chapter argues that several key success factors are appropriate for e-Government project implementation. About twelve e-Government websites are examined upon those key success factors. This chapter puts forward a conceptual model for a better implementation of electronic government.

Chapter 3
Understanding E-Governance: A Theoretical Approach..38
Muhammad Muinul Islam, Jahangirnagar University, Bangladesh
Mohammad Ehsan, University of Dhaka, Bangladesh & Dalhousie University, Canada

Another new paradigm shift is in the offing and slowly becoming distinct from the amorphous sphere of public administration. It is the ICT-blessed governance, or e-Governance. The adoption of ICTs and the new approach to management in symbiosis are e-Governance. E-governance speaks of a new way and style in every beat and pulse of the system of public administration. It brings about changes in the structure and functions of public services, ushering transformation through effectively engaging the government, businesses, and citizens—all stakeholders. It not only ensures efficiency in public service delivery but also offers unlimited potential to combat corruption and many other bureau-pathologies in the public administration system. Based on secondary sources, this chapter offers brief theoretical discussions of e-governance, including its emergence, types of service delivery, transformation stages, and relevant other issues.

Chapter 4
E-Government, M-Government, L-Government: Exploring Future ICT Applications
in Public Administration 50
Alberto Asquer, University of Cagliari, Italy

The development of Information and Communication Technologies (ICT) brings about considerable changes in the ways public administration provides information and delivers services to citizens, businesses, and other public administration systems. This chapter reviews the application of ICT in the provision of public administration services. e-Government tools have been introduced in various countries in the world and enabled the strengthening of existing public administration services and the activation of innovative ones. m-Government tools, which are related to the emergence and diffusion of Internet mobile technology and devices, allow both overcoming infrastructure deficits and providing innovative services, which are particularly sensitive to users' context conditions. Finally, l-Government tools—i.e., ubiquitous, seamless, user-centric, and automated application of Internet technology to public administration services—have the potential to further redefine the terms of access of users to public administration services and to enhance the ties among citizens, businesses, and the government.

Chapter 5
Challenges of Implementing E-Governance in a Politically Driven Environment...............................64
Manish Pokharel, Korea Aerospace University, South Korea
Jong Sou Park, Korea Aerospace University, South Korea

The authors in this chapter underline the significance of a wider environment in which e-governance has to flourish. Even though each and every country in the world has programs to promote e-governance systems to provide maximum benefits to the citizens, and huge budgets and efforts are employed in developing master plans for e-governance in the developing countries, the systems often get terminated. There are many reasons for this termination; hence, the chapter has identified some of the key issues surrounding it.

Chapter 6
Marketing E-Government to Citizens 75
María de Miguel Molina, Universitat Politècnica de València (UPV), Spain
Carlos Ripoll Soler, Universitat Politècnica de València (UPV), Spain

This chapter explores the different literature that analyses the application of marketing strategies in e-Government to help government managers promote citizen communication and participation. On occasion, the application of marketing strategies to encourage citizens to use G2C (Government to Citizen) tools is poor. Providing citizens with information is the first step whilst the final step of governments should be to reach citizen participation in decisions. To achieve this final step, governments will need to carry out a network analysis in order to find out more about their citizens and their needs. They will then need to communicate effectively how they can cover these needs through e-Government. Based on several experiences, the authors aim to study the gaps that governments at any administrative level should cover to enhance their communication with citizens and so increase the use of e-Government. In this chapter, the authors propose the use of Web 2.0 tools as a new way of communication.

Chapter 7
Alignment of IT Projects and Investments to Community Values 93
Marcus Vogt, Heilbronn University, Germany
Kieth Hales, Bond University, Australia

Information Technology (IT) investments and IT management have become increasingly important for an organization's success. The principles of IT Governance, the use of IT Governance Frameworks, the application of IT Portfolio Management, and IT Value Management are proven methodologies to lead private companies to a successful and efficient implementation of their ITT investments. However, due to the structure of public organizations, they seem to have their limits. This chapter shows the issues of IT Value Management in public organizations and discusses the usability of classic IT alignment methods.

Chapter 8
From Town Hall to the Virtual Community: Engaging the Public with Web 2.0 and Social Media Applications 114
Kathryn Kloby, Monmouth University, USA
Leila Sadeghi, Kean University, USA

Engaging the public is a vital component of the public policy process. Traditional strategies for civic engagement include town hall meetings as well as citizen surveys, 311 call systems, and more interactive meetings for public deliberation. Each of these approaches has their limitations, leading many to consider new ways of engaging the public and the role that technology can play in the process. The authors focus on a discussion of the traditional citizen engagement approaches that are widely used by government to communicate with and interact with the public. Focusing on new interactive media, they discuss what is meant by "Web 2.0" and present the capabilities and potential applications of social media in the public sector. Highlighting government programs that utilize these technologies and interviewing subject-matter experts on this new form of communication, the authors present some of the adoption concerns and implementation strategies that public administrators should consider as they adopt Web 2.0 technologies. They conclude with a discussion of the potential that these new civic engagement techniques can offer the public sector as strategies to communicate, interact, and engage the public.

Chapter 9
Key Success Domains for Business-IT Alignment in Cross-Governmental Partnerships 131
Roberto Santana Tapia, Ministry of Security and Justice, The Netherlands
Pascal van Eck, University of Twente, The Netherlands
Maya Daneva, University of Twente, The Netherlands
Roel J. Wieringa, University of Twente, The Netherlands

Business-IT alignment is a crucial concept in the understanding of how profit-and-loss organizations use Information Technology (IT) to support their business requirements. This alignment concept becomes tangled when it is addressed in a socio-political context with non-financial goals and political agendas between independent organizations, i.e., in governmental settings. Collaborative problem-solving and coordination mechanisms are enabling government agencies to deal with such a complex alignment. In this chapter, the authors propose to consider four key domains for successful business-IT alignment in cross-governmental partnerships: partnering structure, IS architecture, process architecture, and coordination. Their choice of domains is based on three case studies carried out in cross-governmental partnerships, in Mexico, The Netherlands, and Canada, respectively. The business-IT alignment domains presented in this chapter can guide cross-governmental partnerships in their efforts to achieve alignment. Those domains are still open to further empirical confirmation or refutation. Although much more research is required on this important topic for governments, the authors hope that their study contributes to the pool of knowledge in this relevant research stream.

Chapter 10
Information Technology Product Quality, Impact on E-Governance, Measurement, and Evaluation 162
Jiri Vanicek, Czech University of Life Science, Czech Republic
Ivan Vrana, Czech University of Life Science, Czech Republic
Zdeněk Struska, Czech University of Life Sciences, Czech Republic

The chapter deals with the problems of information technologies' quality aspects and their importance for the state and public administration, e-governance, and e-democracy. The concept of quality is defined, and it is demonstrated that for quality evaluation, the clarification of needs and specification of exact requirements represents a key problem. The chapter focuses on general quality management and specifically on quality of ICT-solutions. The ISO/IEC software product quality and data quality is thereafter briefly presented. The principle of quality evaluation based on the quality decomposition into several characteristics is also described along with the principles of measurement theory. Finally, this is supplemented by selected approaches for expert rating of trends for ICT-solutions having a direct impact on long-term quality prediction.

Chapter 11
From E-Government to E-Governance in Europe 195
Rebecca Levy Orelli, University of Bologna, Italy
Emanuele Padovani, University of Bologna, Italy
Carlotta del Sordo, University of Bologna, Italy

The influence of e-government on the modernization and growth of public sector initiatives in Europe has been deeply claimed. Little is known, however, about how the so-called shift from e-government to e-governance takes place in European governments. This chapter presents a view of both challenges and advantages of implementing e-governance strategies, by examining how closely and critically intertwined e-government and e-governance are in European countries.

Chapter 12
Core Values: e-Government Implementation and Its Progress in Brunei 207
Kim Cheng Patrick Low, Universiti Brunei Darussalam, Brunei
Mohammad Habibur Rahman, Universiti Brunei Darussalam, Brunei
Mohammad Nabil Almunawar, Universiti Brunei Darussalam, Brunei

Fadzliwati Mohiddin, Universiti Brunei Darussalam, Brunei
Sik Liong Ang, Universiti Brunei Darussalam, Brunei

In this chapter, e-Government and national cultures of the island republic of Singapore and the Sultanate of Negara Brunei Darussalam (henceforth Brunei), both small countries, are examined. The authors discuss the salient core values in the two national cultures that enable e-Government to be successfully implemented or at least have the right ingredients to be successful.

Chapter 13
E-Government in Bangladesh: Prospects and Challenges 217
Noore Alam Siddiquee, Flinders University, Australia
Md. Gofran Faroqi, Ministry of Public Administration, Bangladesh

In this chapter, the authors delineate the overall policy and institutional framework of e-government from the perspective of Bangladesh. Recognizing the current government's attempt at branding the country as "Digital Bangladesh," the authors explore major e-government programs and initiatives in operation. Most importantly, they eloquently elaborate on the constraints and challenges facing Bangladesh in its pursuit of electronic governance and also shed light on the way forward.

Chapter 14
Combating Corruption through e-Governance in India 233
Durga Shanker Mishra, State Government of Uttar Pradesh, India

Studies have shown a prevalence of high level of corruption in the Indian Administrative System, which adversely affects the day-to-day lives of common citizens. This chapter examines the role of e-governance in combating corruption in delivering public services. Through a literature review assessing the outcomes of a few e-governance initiatives related to improving service delivery in different parts of India, this chapter argues that even though technology assists in instituting a transparent, accountable, consistent, reliable, and efficient system for delivery services, it cannot overcome corruption by itself. It will require political will, focused administrative strategy, business process reengineering for simplifying and opening up the system, and persistent efforts to ensure that corruption entrepreneurs do not subvert the gains of the technology.

Chapter 15
Grey Hair, Grey Matter, and ICT Policy in the Global South: The Ghana Case 245
Lloyd G. A. Amoah, Ashesi University, Ghana

By exploring the case of Ghana, this chapter examines the much vaunted linkages among good governance, ICTs, and development in developing societies. Though some significant ICT-related infrastructural development projects have been undertaken in Africa, the empirics indicate that the region, compared to other regions, such as Asia, has yet to see the magic results. To the author, Ghana is not an exception. Using an e-government project at the presidency in Ghana as a case study, this chapter attempts to understand why the vast potential benefits of ICTs have not been realized in countries like Ghana. The argument put forward by the author is that e-government and by extension ICT policy outcomes in developing polities must be understood as partly a reflection of the world view of policy elites, which is at best generally antagonistic, ambivalent, and even apprehensive of the very notion of a cyber society. The chapter concludes with recommendations relevant to Ghana and other developing polities.

Chapter 16
A Positive Hegemony? Arguing for a Universal Knowledge Regime led by an e-Governance 'Savvy' Global Knowledge Enterprise! 261
Amlan Bhusan, Bhusan Taylor Ltd, UK

Amlan Bhusan raises important questions in "A Positive Hegemony? Arguing for a Universal Knowledge Regime led by an e-Governance 'Savvy' Global Knowledge Enterprise!" To him, there is a growing academic consciousness, regarding the use of e-governance, to deliver social goods in a better way. This voice advocates that more needs to be done by public institutions, governments, and more importantly, the academia, to develop e-governance as an enabler for social efficiency. Such developments would help reach debates and discussions on this area to the grassroots of the policy system. His chapter is neither a commentary of the application of e-governance to deliver social change nor a study of how different governments have handled this area around the world. Rather, it is a practicing consultant's views of the power of e-governance to refine public choice and social decision making and how this process was enriched by a more vigorous role of the academia. Taking specific examples from the education sector, particularly universities, this chapter is a comment on some of the ways in which e-governance 'can' be handled across the education system and how lessons from the developed countries can be used to inspire similar revolutionary changes to the status quo in the developing world. His objective is to promote a greater role for the academia in the public policy making process. The idea is to support a more constructive engagement of the academia with the more vulnerable parts of the social system. Above all, he argues for the benefits of spreading the values of information democracy, right to access to information, among the people. He envisages that the power of a more vocal and active academia would be profound in how it could positively affect the information apartheid affecting many large sections of the developing world. He proposes greater research and development on the means of engaging with e-governance and to establish the mechanisms to enhance, converge, simplify, homogenize, and structuralize the knowledge and information enterprise of the global political and social systems.

Chapter 17
Successfully Applying "e" to Governance 271
Evangelia Mantzari, Athens University of Economics and Business, Greece
Evanthia Hatzipanagiotou, Ministry of Finance, Greece

The challenge of better public service offering and the expectations of modern citizens and businesses, as well as the poor past practices of public organizations, bring forward the need to design and implement new systems. These systems are based on current information and communication technologies and are points of reference on the path towards e-government "enlightenment." However, the transition from the traditional processes to the modern ones can be long and strenuous, if relevant projects are not carefully implemented. Therefore, in order to successfully apply electronic practices and methods to public systems of governance, a step-wise approach needs to be formulated starting from traditional standards, leading to transitional procedures, and finally achieving simplification and increased quality of public service by exploiting previous experiences, overcoming past limitations, and applying the lessons learned.

Compilation of References 289

About the Contributors 315

Index 323

Preface

Despite certain challenges, the positive impact that Information and Communication Technologies (ICTs) have on worldwide public administration systems is now well accepted by all the stakeholders of e-governance. It is also recognized that, in the paradigm shift "from government to e-governance," ICTs play the major role, among others, in transforming public administration systems from hierarchic, command-and-control-oriented to a non-hierarchic, citizen- and-stakeholder-oriented one. These revolutionary changes also enhance the interests of researchers culminating in increased numbers of research works and consequent publications on this subject area. This volume is also a product of that trend. With fast-paced developments in the field of e-governance, it is always essential to disseminate new knowledge to the wider audience for expanding the horizon of the field. With that objective in mind, this project was initiated quite some time ago. At the end of the day, with the efforts of everyone involved, this book largely deals with the current practices, potentials, and challenges of ICT-enabled public administration systems in understanding what e-governance broadly entails along with its effects on all the stakeholders. In doing so, this book not only deals with aspects of public administration systems but also covers the business sector, academia, and many other relevant factors (e.g., citizens and community engagement) that come into play in the smooth functioning of e-governance, be it in the developed or developing world.

Drawn from different backgrounds, disciplines, and experiences from different parts of the world, the authors in this volume offer a wide range of issues relevant to the digital transformation of governance and public administration. All the contributors in this book, however, agree and subtly conclude that e-governance indeed brings a paradigm shift in public administration. Given that the application of ICTs and the pace of e-governance maturity are not the same across nations, the experiences and challenges to adapting to the new governance regimes are also essentially different in these countries in their efforts of digital transformation. With the mix of cases from the developed and developing countries, this book is thus suitable for a wider audience aspiring to bring citizen-centred changes through e-governance. Though the book is not divided into different sub-sections, it covers areas that are visibly separated and thus can be distinctly grouped. Forming such a group, the first five chapters shed light on the theoretical issues, the transformative journey from e-government to e-governance, the challenges of e-governance in a politically driven environment, as well as ICTs' applications in public administration systems. The focuses of the next three chapters are citizens, their engagement in the non-hierarchic e-governance regime, and the broader area of community value, as well as the virtual community under which the prospect of Web 2.0 and other social media applications are discussed.

The subject matter of the third group of chapters includes the business-IT alignment in cross-governmental partnerships, as well as the ever-increasing significance of the impact of e-governance and its measurement and evaluation. Starting with the dialogue on the transition from e-government to e-governance in the context of Europe, the fourth segment covers the selective country cases with focuses on core values, anti-corruption, ICT policy, and e-governance's prospects and challenges in the context of developing countries. Specific countries that are covered include Bangladesh, Brunei, India, and Ghana. Examples from other countries, such as the USA, are also made in the earlier chapters. The second to last chapter brings in the issue of e-governance savvy global knowledge enterprise and academics' role in it. The book concludes with excellent arguments made in favor of how to successfully introduce and sustain e-governance in any country context.

From Government to E-Governance: Public Administration in the Digital Age offers an up-to-date and comprehensive account of the concepts, developments, challenges, and prospects of the field of digital or e-governance. It does not only have contributors from around the world, but also covers different aspects of e-governance from different areas of the world as they differ in their status quo, challenges, and prospects. In this way, this book contributes to the better understanding of e-governance and helps the readers in understanding the issue from diverse perspectives. Though the application of ICTs in governance is not without pitfalls and challenges, it is our hope that this publication, with the latest information on those challenges and new knowledge on innovation, would assist students, researchers, academics, experts, and civil society groups around the world in their understanding of this fascinating area of study.

Muhammad Muinul Islam
Jahangirnagar University, Bangladesh

Mohammad Ehsan
University of Dhaka, Bangladesh

Acknowledgment

From Government to E-Governance: Public Administration in the Digital Age is the outcome of close collaboration and the patience of everyone involved in the publication project. For several reasons it took more than the usual time to finish the project, and we thank everyone—the authors, reviewers, publisher, and advisory board members—for being with us over the last several months. We would like to acknowledge our gratitude to all the authors, thirty-six in total, from all over the world, for their time, energy, and productive contributions to this volume. We express our sincere appreciation to the authors who have also reviewed a few of the chapters in the volume. Without your contribution, this project would not have been completed even at this time. At this point, we would also like to remember the assistance and encouragement we received from Matti Malkia of the Police College of Finland and Jiri Vanicek of the Czech University of Life Science, who passed away last year. Both of them were very enthusiastic about this publication project.

The editorial advisory board members, from time to time, have advised us on the project. They have always been available when we needed their advice, and we are indeed grateful for that. Their association with this book project has been an encouraging incentive and learning experience for us. We are grateful to Professor Paul Brown, as well, for his advice on various aspects of the publication process. The staff at IGI Global have always been very helpful with their professional suggestions and assistance in completing this project. Especially, we would like to mention the hard work and patience of Christine E. Smith, who has been with us in the final phase of the publication. Without you, Christine, and your colleagues' contributions, the publication would not have been possible. Finally, we would like to appreciate the direct and indirect assistance that we received from our host institutions, i.e., the University of Dhaka and Jahangirnagar University, both situated in Bangladesh.

Last but not the least, we would like to thank our family members for giving us the time and space needed for completing such a publication project. We are thankful, respectively, to our wives, Farzana and Naveen. Our kids, Abrar, Raafi, and Wasil, provided us with the much needed fun break from this and other academic responsibilities. We thank you all for that!

Finally, for innumerable reasons, we dedicate this work, respectively, to the memories of Peter Aucoin, who was a wonderful human being and a great scholar and renowned teacher of political science and public administration at Dalhousie University, as well as to Professor Gregory B. Lewis of Georgia State University's Andrew Young School of Policy Studies for being a great mentor.

Muhammad Muinul Islam
Jahangirnagar University, Bangladesh

Mohammad Ehsan
University of Dhaka, Bangladesh

Chapter 1
From e-Government to e-Governance:
A Holistic Perspective on the Role of ICTs

Wolter Lemstra
Delft University of Technology, The Netherlands

ABSTRACT

In this chapter, the author elaborates on the concepts of e-Government and e-Governance and place these concepts in the broader context of the introduction and diffusion of information and communication technologies. The question that is being addressed is "what should an effective and pro-active e-Government do?" For that purpose, the ICTs are not only considered general purpose technologies, but recognized as the driving force of an unfolding technological revolution. As such, a recurrent pattern in techno-economic and socio-economic development can be discerned that goes back to the First Industrial Revolution in Britain around the late 1700s and early 1800s. It is argued that appreciating the fifth instance of this pattern and recognizing the challenges each transition invokes can inform the policy formation process and make policy action more effective. In the broader context of roles that governments may assume, a stepwise approach is introduced to address the many challenges the diffusion of ICTs is bringing about in the economic and social realm, aimed at reaping the benefits implied in the new techno-economic paradigm.

INTRODUCTION

The term e-Government is generally associated with the use of ICTs by governments. The OECD defines e-Government as the use of information and communication technologies, and in particular the Internet, as a tool to achieve better government (OECD, 2004). This notion of 'better government' is typically translated, first of all, into improving the administrative service delivery. In a next step, the improvement of the policy-making process through ICTs is being envisioned. In introducing information technologies the principle objective is improvements of process efficiency, the use of communications technologies makes the government more accessible, it may extend the reach and

DOI: 10.4018/978-1-4666-1909-8.ch001

improve the involvement of the citizenry, thereby further improving efficiency, as well as the quality of the governing process.

Enabling Technologies

By introducing information and communication technologies, the peculiarities of these technologies are being introduced. For instance in the business environment, the early investments in IT hardware and software did not lead to productivity improvements being measured at the national aggregate level. This has led to the hypothesis of a so-called 'productivity paradox.' A variety of possible explanations have been brought forward by economists for this lack of improvement in total factor productivity, all inconclusive. However, more recently detailed analysis by Brynjolfson has shown a significantly positive relationship between IT investments and productivity in a series of firm level studies. Their findings suggest that productivity improvements are resulting *if and only if* investments in hardware and software are complemented with investments in skills development *and* the related processes are being re-engineered to take the full benefit of the capabilities that the information technologies are offering (Brynjolfson, 1992; Brynjolfson & Hitt, 1998)[1]. For the application of IT in governments, this conclusion poses additional challenges as the user is often the public at large, and investments in skills development and process re-engineering can only be realised in an indirect way.

Nonetheless, the role of ICTs for economic growth and social development are broadly recognized, as illustrated by the various e-Action Plans formulated by governments around the globe. For instance, eEurope as part of the Lisbon Agenda in 2000, the Information Super Highway initiative in the USA in 2000, the e!Japan initiative in 2001, and the eKorea Vision of 2002.

To appreciate the challenges governments are facing, one should recognise that the impact of 'informatisation' on the economy and society through information technologies is comparable to the impact of the earlier principle of 'mechanisation.' In this respect we can refer to what Lipsey et al. call 'General Purpose Principles' (GPPs) or concepts that are employed in many different technologies that are widely used across the economy for many purposes and that have many spill-over effects (2005). Information technologies are the embodiment of such a principle and the computer, in its many manifestations is a recent representation of what is called a 'General Purpose Technology' (GPT), being defined as: "a single generic technology, recognizable as such over its whole lifetime, that initially has much scope for improvement and eventually comes to be widely used, to many uses, and to have many spill-over effects" (Lipsey, et al., 2005). This also applies to the Internet. Both are part of the general GPT category of information and communication technologies, which includes writing and printing. Both GPTs started out as a product, but have become transformational in terms of process and organisational change.

From e-Government to e-Governance

The recognition of IT and CT as GPTs is crucial and moves the issue beyond the domain of e-Government into the realm of e-Governance.

To govern is to steer and in the context of governments, this is typically understood as the exercise of authority, to control and direct the making and administration of policy. Government is hence the act or process of governing. Governance is a much broader notion and can be understood as the political and economic processes that coordinate activity among economic actors (Campbell, Hollingworth, & Lindberg, 1991). Groenewegen defines governance as the coordination of transactions, which includes information, negotiation, and decision-making. The governance can be both private, i.e. between private actors in private governance structures, and public, i.e. between public actors in public governance structures. Part of the governance is spontaneous, it takes place automatically between actors without an authority that steers. Another part of governance is a matter of steering, of authority. Governing is about the steering part in the process of coordination. Private governing is concerned with private governance structures, such as firms, and public governing, which equates to government at various levels, is concerned with public governance (Groenewegen, 2008).

The introduction of information and communication technologies to facilitate existing forms of public governing or enable new forms has led to the popular acronym of e-Government. In first instance, this related to the automation of internal tasks and processes. With the emergence of the Internet, the notion extended to include the electronic delivery of governmental service, also denoted as e-Administration.

Through ICTs the participation of citizens and businesses in government can be improved, hence, e-Government has obtained a developmental dimension associated with 'good' government. The "e" has become associated with: efficiency, effectiveness, empowerment, economic and social development (Misuraca, 2007)[2].

For e-Government to become broadly accepted and part of the daily routines, the ubiquity of information and communication technologies forms a necessary condition, albeit it will not be sufficient as the adoption will depend on the public at large. In this respect, the e-Readiness index has been introduced. Each year, the Economist Intelligence Unit produces in cooperation with the IBM Institute for Business Value a ranking of e-readiness across countries, based on six dimensions: (1) connectivity and technology infrastructure; (2) business environment; (3) social and cultural environment; (4) legal environment; (5) government policy and vision; and (6) consumer and business adoption[3].

There is a direct link to the role governments have assumed in the roll-out of communication networks, i.e. the telephone network, either through telecom service providers becoming governmental entities (mainly in Europe) or through tight regulatory supervision of private firms providing communication services (predominantly in the USA). The provision of interconnection between cities and the provision of services in rural areas have been stated as the main reason for government intervention, see also Mueller (1993). The Universal Service Obligation (USO), mostly enforced upon the incumbent operator, is an early example of the results of political governance of the telecommunications sector. Other examples of sectors in the economy that have attracted a high degree of government involvement are health and education. Improvement of service delivery through ICTs has been placed high on the various political agendas, hence, the notion of e-Health and e-Education. The network effects and spillovers associated with ICTs drives the strong involvement of governments with these sectors.

The involvement of government with the development of the Internet, considered critical to the realisation of e-Government objectives, has been quite different. The trigger of what would lead to the Internet is a response to a perceived threat to the integrity of the telecommunications infrastructure. In 1957—the Cold War period—the former Soviet Union had launched Sputnik, the first satellite, leading in the United States to an awareness of a 'science gap' which prompted a surge in government investments in science and technology (Abbate, 1999). The threat of a nuclear attack that could destroy essential communications facilities and affecting the ability of the military to maintain 'command and control' led to a recognized need of 'survivable communications' in the early 1960s. At Rand, a non-profit corporation dedicated to research on military strategy and technology, explorations into 'survivable communications' started in 1959 through the efforts of Baran. This would lead in 1960 to the invention of packet switching by Baran and independently by Davies at the UK National Physics Laboratory. In 1959, the Advanced Research Projects Agency (ARPA) was initiated, funded by the US government. ARPA's efforts to link computers through packet switching, whereby the first connection through the ARPANET was made in 1969, is generally considered as the start of what is now called the Internet, this term to be introduced in 1984 (Slater, 2002)[4]. From 1972, the ARPANET and later the NSFNET formed the backbone network and had been run on behalf of the US government to connect regional networks and supercomputer sites for research and education purposes. In 1989, Performance Systems International (later PSINet) was established, as a 'spin-off' of one of the regional research networks, and started to provide TCP/IP network services to business customers. This was soon followed by other 'regional operators' and long-distance operators, such as AT&T, MCI, and Sprint (Abbate, 1999). In 1991, a non-profit organisation Commercial Internet Exchange (CIX) was created, to connect the networks through gateways. RIPE, established in 1989, provided a similar function in Europe (1999)[5]. Many networks subsequently joined the CIX and RIPE, and thus a commercial alternative backbone network was formed, on a worldwide basis. In 1995, the NSFNET backbone that connected by that time to 22,000 international networks, was retired and the Internet 'transitioned' from the public to the private sector.

Hence, the role of governments in the operation of the current Internet is minimal, except for the role the US government has assumed in coordinating the domain name system through IANA and ICANN (Mueller, 2002, 2010). Albeit, the use of the Internet increasingly invokes the role of governments in protecting public interests and public values, e.g., in fighting spam, in the protection of privacy, the protection of minors, and addressing national security. As the involvement of governments is expanding, at the same time the nature of the problems shows the limitations of the instruments available to governments. These instruments tend to be linked to the national sovereignty and national jurisdiction while the issues are increasingly of a global nature (Mathiason, 2009).

Thus, in following Kooiman, e-Governance can be perceived as: "the totality of theoretical conceptions on governing that involves ICTs" (2003). This includes the ICTs as enabling (public) governing, the policy making and the execution, as well as the governance of ICTs by the (public) government.

Role Perception of Governments

To distinguish the public from the private domain Parsons provides the following guidance: "The idea of public policy presupposes that there is a sphere or domain of life which is not private or purely individual, but held in common. The public comprises that dimension of human activity which is regarded as requiring governmental or social regulation or intervention, or at least common action" (Parsons, 1995). Parsons emphasizes the tensions that have always existed (and will remain to exist) between what is held to be 'public' and what 'private.' The way this tension is being resolved depends on the societal model and the role perception of governments. On one end of the spectrum, we find the (pure) market economies, based on the idea that through market forces the maximization of individual interest could best promote the 'public interest.' On the other end of the spectrum, we find the centrally coordinated or planned economies, where the private interests are considered best served through the public domain.

Today, with a prevalence of capitalistic market economies, the distinction is more gradual, whereby a distinction can be made between the regulatory state and the developmental state. In both the market is centre point, in the regulatory model the government is focused on the proper functioning of the market process, while in the developmental model the outcome of the process is important. In the developmental model a government also develops explicit targets and plans for economic development in general, and articulates the desirable developments at sector level (Groenewegen, 1989). Common to both models is for policy makers to assure a proper functioning of the markets. Therefore, one of the more fundamental tasks of governments is the establishment and maintenance of a formal institutional environment that is conducive to the proper functioning of the markets, e.g. in terms of the operation of a system of property rights, the legal system, and the monetary system. Moreover, governments may wish to intervene if markets are considered to be failing[6]. Market failure is said to occur "...when markets fail to allocate scarce resources efficiently in and through the pursuit of monetized private interest" (Jessop, 2005). Wolf argues that this criterion should be applicable to static or allocative, efficiency as well as to dynamic efficiency (Wolf, 1990)[7]. Furthermore, Wolf adds 'distributional inequity' of income or wealth as a possible market failure.

The conscious awareness of choice between two main alternatives for steering societies' is what Dror calls policy-making (Reference made to Dror in Parsons, 1995). Policy is thereby understood to have become a term expressing political rationality: "To have a policy is to have rational reasons or arguments which contain both a claim to understanding of a problem and a solution. It puts forward what is and what ought to be done. A policy offers a kind of theory upon which a claim for legitimacy is made."

THEORETICAL FRAMING: REASONED HISTORY

To appreciate the broader nature of ICTs and the implications for governing and governance, we have to assess and understand the impact of a new general-purpose principle and a general-purpose technology on the economy and society at large. In this respect, we can learn from historical developments. For this purpose Freeman and Louçã in their theory of reasoned history distinguish five interacting subsystems in society: science, technology, economy, politics, and culture (Freeman & Louçã, 2001)[8]. Perez in this context refers to 'three spheres of change in constant reciprocal action': technological, economic, and institutional (Perez, 2002).

Based on the notion of Schumpeter that 'any satisfactory explanation of the evolution of capitalist economies must place innovations, their profitability, and their diffusion at the center of analysis,' Freeman and Louçã identify a recurring pattern of technological revolutions, whereby to date five successive technology-driven revolutions can be distinguished, see also Table 1 (Freeman & Louçã, 2001). The most recent revolution being ICT-driven.

Freeman and Louçã identify Perez as the person first suggesting that some technologies are so pervasive that they dominate the behaviour of the whole economy for several decades and influence major social and political changes. She also suggested the following characterization for each revolution (Freeman & Louçã, 2001; Perez, 1983):

- Key factors (core inputs) becoming so cheap and universally available leading to a wide range of new factor combinations;
- Key factors plus complimentary inputs would lead to new products, which would give rise to the emergence of fast growing new industries (carrier branches), that would give a major impetus to the growth of the entire economy;
- A new type of infrastructures would serve the needs of the new industries and in turn stimulate the development of the new industries;
- The structural transformation invoked by the new technologies, industries, products and services would lead to organizational innovations needed to design, use, produce, and distribute them.
- The transformation process from the application of old to new technologies, pro-

Table 1. Condensed summary of the Kondratieff waves

Constellation of technical and organizational innovations	Examples of highly visible, technically successful, and profitable innovations	"Carrier' branch and other leading branches of the economy	Core input and other key inputs	Transport and communications infrastructures	Managerial and organizational changes	Approx. timing of the 'upswing' (boom) 'down swing' (crisis of adjustment)
(1)	(2)	(3)	(4)	(5)	(6)	(7)
1. Water-powered mechanization of industry	Arkwright's Cromford mill (1771) Henry Cortt's 'puddling' process (1784)	Cotton spinning Iron products Water wheels Bleach	Iron Raw cotton	Canals Turnpike roads Sailing ships	Factory systems Entrepreneurs Partnerships	1780s-1815 1815-1848
2. Steam-powered mechanization of industry and transport	Liverpool-Manchester Railway (1831) Brunel's 'Great Western'Atlantic steamship (1838)	Railways and railway equipment Steam engines Machine tools Alkali industry	Iron Coal	Railways Telegraph Steam ships	Joint stock companies Subcontracting to responsible craft workers	1848-1873 1873-1895
3. Electrification of industry, transport and the home	Carnegie's Bessemer steel rail plant (1875) Edison's Pearl St. New York Electric Power Station (1882)	Electrical equipment Heavy engineering Heavy chemicals Steel products	Steel Copper Metal alloys	Steel railways Steel ships Telephone	Specialized professional management systems 'Taylorism' giant firms	1895-1918 1918-1940
4. Motorization of transport, civil economy, and war	Ford's Highland Park assembly line (1913) Button process for cracking heavy oil (1913)	Automobiles Trucks Tractors, tanks Diesel engines Aircraft Refineries	Oil Gas Synthetic materials	Radio Motorways Airports Airlines	Mass production and consumption 'Fordism' Hierarchies	1941-1973 1973-
5. Computerization of entire economy	IBM 1401 and 360 series (1960s) Intel microprocessor (1971)	Computers Software Telecommunication equipment Biotechnology	'Chips' (integrated circuits)	'Information Highways' (Internet)	Networks; internal local, and global	??

Reprinted from Freeman and Louçã (2001) with permission from the publisher Oxford University Press.

cesses, and organizational forms will not be smooth; there will be a transitional period of structural adjustment.

The new approach to management and organization has been described by Perez: "...variously as a 'new technological style' and a 'new techno-economic paradigm'" (Freeman & Louçã, 2001). An insight that evolved from the role of innovation (Schumpeter), through the 'clustering of innovation' or notion of 'constellation' (Keirstead), and through 'technological trajectories' (Nelson and Winter) and technological paradigms (Dosi) to 'technology systems' (Freeman, Clarke, and Soete) (in Freeman, 1998).

The succession of a recurring pattern is consistent with the fact that sustained economic growth can be observed since the First Industrial Revolution, starting in England around 1770, being characterized by the water-powered mechanisation of industry (Lipsey, et al., 2005). Moreover, this framing of economic development links to the observation of long waves in economic development, also known as the Kondratieff cycle. Different explanations of the occurrence of the long wave have been made, none has been satisfactory until Schumpeter in 1939 provided a link to the clustering or discontinuous nature of technical innovation as the driving force (Freeman, 1998; Kleinknecht, 1987). Kleinknecht also points to other, complementary forces that are driving the long wave, identified by Van Gelderen and De Wolff: "In each upswing of the long wave, the production of investment goods will expand more rapidly than the production of consumer goods." And the hypothesis by Van Gelderen on the availability of cheap loan capital together with a low price level at the end of a long wave depression (1987)[9].

The concept of the long waves is not without its critics, in particular where empirical evidence could not be provided for the claims being made. Kleinknecht refers to the critique by Kuznetz in his review of Schumpeter's 'Business cycles' (1939) and cites two basic questions posed by Kuznetz (1987):

- Is there any evidence of Kondratieff long waves in important indicators of general economic activity?
- Is there any evidence of a bunching of Schumpeter's heroic innovations (and if yes, what is the theoretical explanation)?

From his research, Kleinknecht concludes that Schumpeter's hypothesis about long waves in economic life and an uneven distribution over time of radical innovations can be defended, not only in time but also in certain sectors. The theoretical explanation is to be found in the "…reallocation of R&D and other investments towards new technological paradigms in response to the *rien ne va plus* during the long wave depression" combined with "...an endogenously caused over-expansion and depreciation of capital stock… [that] is caused by an expansionary self-ordering feedback loop: to satisfy demand for investment goods from the consumer goods sector, the capital goods producing sector itself has to expand its capacity, ordering capital goods for the production of capital goods." Furthermore, Kleinknecht argues, "…the hypothesis seems plausible that prolonged depressions not only trigger a reallocation of innovative resources but also create strong pressure towards social, political and institutional change" (Kleinknecht, 1987). Also Freeman and Soete argue for a broader perspective that "… clusters of radical technical innovations do also lead to major disruptions not just in the production sphere but also in the broad social, institutional and organizational sphere" (Freeman & Soete, 1997).

Freeman and Louçã also caution the reader: "we should re-emphasize here our belief that this recurrence is limited in scope and content. Each technological revolution and each phase of economic growth has its own unique features. This does not mean, however, that we cannot learn a great deal from even this limited recurrence as well as from unique events…The work of Carlota Perez (1983, 1985, 1988) on long waves has shown that, even if identical behaviour is ruled out, as it must be, there may still be striking similarities or dissimilarities and some hidden ones too, which are helpful in understanding the phenomena and even in making probabilistic forecasts and indications for policy" (Freeman & Louçã, 2001).

Stylized Model of Great Surges

In her 2002 contribution 'Technological Revolutions and Financial Capital: The Dynamics of Bubbles and Golden Ages,' Perez expanded on her suggestions captured by Freeman (Perez, 2002)[10]. In observing the 'boundless rise of two forces: the information revolution and financial markets' in the last quarter of the twentieth century, she argues that: "productivity explosions and bursts of financial excitement leading to economic euphoria and subsequent collapse of confidence have occurred together before. They are interrelated and interdependent phenomena; they share the same root cause and are in the nature of the system and its workings. They originate in the way technologies evolve by revolutions, in the peculiar manner in which these great upsurges of wealth creating potential are assimilated by the economic and social system and in the functional separation of financial and production capital."

Figure 1. Great surge model

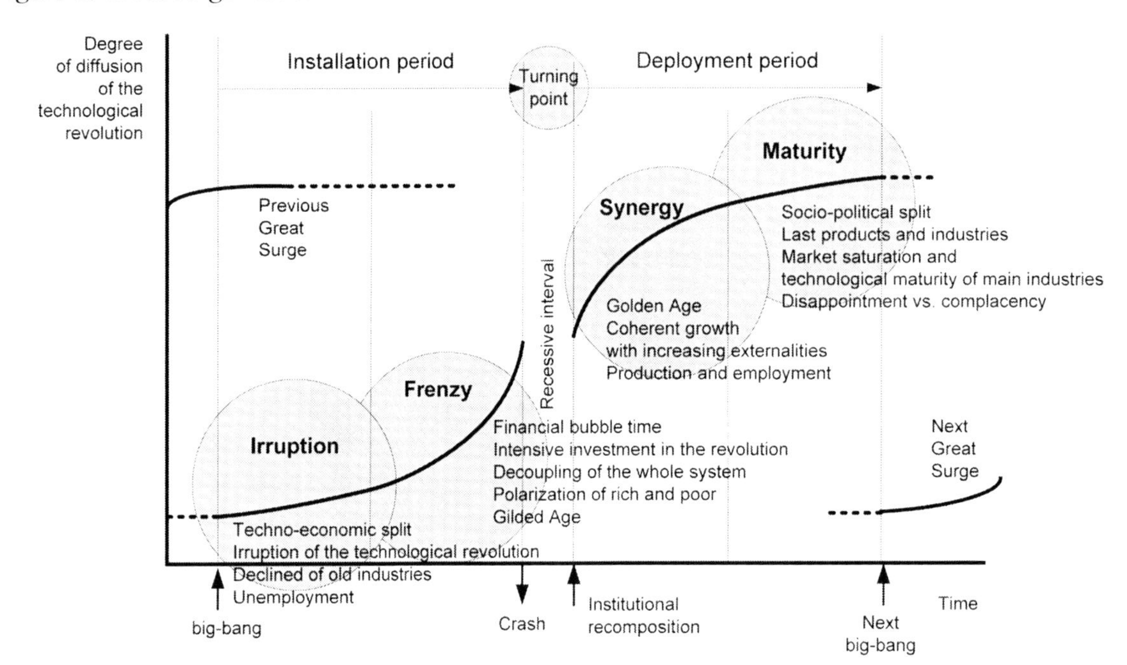

Reprinted from Perez (2002) with permission from the publisher Edward Elgar Publishing and the author Carlota Perez.

Based on historical analysis she shows that the sequence of 'technological revolution – financial bubble – collapse – golden age – political unrest,' is recurring about every half century. This recurrence is considered to be based on "causal mechanisms that are the nature of capitalism, which stem from the features of the system, which interact with and influence one another:

- The fact that technological change occurs by clusters of radical innovations forming successive and distinct revolutions that modernize the whole productive structure;
- The functional separation between financial and production capital, each pursuing profits by different means; and
- The much greater inertia and resistance to change of the socio-institutional framework in comparison with techno-economic sphere, which is spurred by competitive pressures."

The ensuing life cycle of a technological revolution is shown in Figure1. In the early phases, there is the battle of the new paradigm with the power of the old paradigm, which is "ingrained in the established production structure and embedded in the socio-cultural environment and in the institutional framework." When this battle is won, the new paradigm diffuses across the whole of the economy and society. Hence, the diffusion of the new paradigm can be seen as two distinct periods, the 'installation period' and the 'deployment period,' both typically lasting 20-30 years. The 'turning point' from the installation to the deployment is "usually a period of serious recession, involving a re-composition of the whole system, in particular of the regulatory context that enables the resumption of growth and the full fructification of the technological revolution."

General Principles

A technological revolution can be defined as "...a powerful and highly visible cluster of new and dynamic technologies, products and industries, capable of bringing about an upheaval in the whole fabric of the economy and of propelling a long-term upsurge of development." Moreover, "[e]ach techno-logical revolution results from the synergistic interdependence of a group of industries with one or more infrastructural networks." A techno-economic paradigm is "...a best-practice model made up of a set of all-pervasive generic technological and organizational principles, which represent the most effective way of applying a particular technological revolution and of using it for modernizing and rejuvenating the whole economy. When generally adopted, these principles become the common-sense basis for organizing any activity and for structuring any institution." The emergence of a new techno-economic paradigm "...affects behaviours related to innovation and investment in a way that could be compared to a gold rush or the discovery of a vast new territory. It is the opening of a wide design, product, and profit space that rapidly fires the imagination of engineers, entrepreneurs, and investors, who, in their trial and error experiments applying the new wealth-creating potential, generate the successful practices and behaviours that gradually define the new best-practice frontier. The action of these pioneering agents blazes the trail, giving rise to increasing externalities and conditionings—including production experience and the training of consumers—that make it easier and easier for others to follow suit. Their success becomes a powerful signal in the direction of the most profitable windows of opportunity. That is how the new paradigm eventually becomes the new generalized 'common sense,' which gradually finds itself embedded in social practice, legislation and other components of the institutional framework, facilitating compatible innovations and hindering incompatible ones."

It should be noted that Perez points out that the model that is being constructed, is "...a heuristic device. Not a straitjacket to force upon history. In spite of regularities and the isomorphism the model claims to be identifying, there is full awareness that the subject matter rebels and refuses. It is full of exceptions and of huge independent events that constantly twist and break the proposed regularity. Wars, droughts and gold discoveries, are not included in the 'clean' model, nor are many other significant social and political occurrences. The sequence has been stripped of all those events not causally related to the absorption of technologies, which leads inevitably to streamlined simplifications that hardly ever occur as such. Nevertheless, this risky attempt at gleaming the strains of causal order underlying chaos, at structuring the unwieldy mass of historical events into a meaningful sequence, is still worthwhile. After the job is done—if it ever can be—the infinite enrichment of real life can be brought back in, but this time with the benefit of an organizing background, which highlights even more all the unique unexplained events"[11].

The Periods and Phases of a Technological Revolution

With reference to Figure 1, the different periods and phases in the life cycle of a technological revolution can be described as follows[8]:

Gestation Period

Preceding the 'big bang' or the public landmark that signifies the start of a new technological revolution, there is a period of gestation in which the ideas and inventions happen that will become central to the new revolution.

Installation Period

Phase One: Irruption

This is the period in which the new paradigm is configured. It is characterized by explosive growth and fast innovations in the new industries, introducing new products. The exhaustion of the old paradigm brings with it both the need for radical entrepreneurship and the idle capital to take the high risks of trial and error. The idle money in the hands of non-producers looking for a profit starts a 'love affair' with the new technological revolution. A fast learning process takes place among engineers, managers, sales and service people and obviously consumers, about the production and use of the new products. Learning involves acquiring the new organizational notions embodied in the new paradigm. At the same time, a mismatch between the old socio-institutional framework and the requirements of the new paradigm is becoming apparent. The production infrastructure is becoming focused on the realities of the new paradigm.

Phase 2: Frenzy

This is the period of fast diffusion of the new technologies, the introduction of successive new products, industries, and technology systems, plus the modernization of existing ones. In this period, a full constellation of the new industries, technology systems, and infrastructure is in place. Financial capital takes over; its immediate interests overrule the operation of the whole system. The paper economy decouples from the real economy, finance decouples from production while there is a growing rift between the forces in the economy and the regulatory framework, turned impotent. The financial frenzy is a powerful force in propagating the technological revolution, in particular its infrastructure. A time of speculation, corruption, and unashamed (even widely celebrated) love of wealth. Diverging and explosive growth in the new industries in stark contrast with the decline in the industries tight to the old paradigm. The mismatch of the old socio-institution framework with the new paradigm becomes apparent.

Turning Point

This is a conceptual device, denoting the transition between the installation and deployment period of a new technological revolution. With the collapse of the bubble, which ends the period of frenzy, comes recession and sometimes depression, which brings financial capital back to reality. Together with mounting social pressure, this creates the conditions for institutional restructuring and for re-routing growth onto a sustainable path. A swing of the pendulum from the extreme individualism typical for the frenzy period to giving greater attention to collective well being.

Deployment Period

Phase 3: Synergy

In this period, we see the full expansion of innovation and market potential offered by the new technologies, yielding fast growth. There is the introduction of successive new products, industries, and technology systems, plus the modernization of existing ones. Production rules, financials are linked again to production realities. The socio-institutional framework is being adapted to and shaping the new paradigm. Converging growth in most of the industries, aligned with the new paradigm.

Phase 4: Maturity

In this period, we see a diminished potential offered by the new, now old, paradigm. The last introduction of new products and industries occurs. Earlier ones are approaching maturity and market saturation sets in. Idle finance is looking for new opportunities. A mature production infrastructure is looking for market opportunities. While the signs of prosperity and success are still around social dissatisfaction sets in, the qualities of the system are being questioned.

THE ICT-DRIVEN REVOLUTION: THE NEW TECHNO-ECONOMIC PARADIGM

For the fifth successive technological revolution the 'cluster of new and dynamic technologies' is a combination of computing and communication technologies, which have in common the miniaturization of semiconductor devices, following 'Moore's Law (Moore, 1965).' The 'core input' of this revolution are the semiconductors. What sets it aside from the semiconductor developments that already started in 1948 with the invention of the transistor in Bell labs, was the invention of the microprocessor by Intel in 1971. This is the landmark or 'big bang' of the Fifth Great Surge. The microprocessor will become 'the heart' of the personal computer or PC, which marks the shift from computers as 'business tools' to 'personal tools,' changing the diffusion by orders of magnitude. Moreover, the PC will become the device that will provide the access to the Internet. These microprocessors will also become the 'controllers' of the 'switches' and 'routers' that, together with the transmission systems, are part of the worldwide telecommunications network, that will provide the connectivity between the PCs, to facilitate the operation of the World-Wide-Web. Information technologies, computers and software, have already had a major impact on the 'fabric of the economy,' and combined with 'networked computing' facilitated through the Internet, one may speak of an 'upheaval.'

Timing of the Fifth Surge

In Figure 2 the timing of the Fifth Surge is reflected vis-à-vis the performance of the NASDAQ.

The Techno-Economic Paradigm of the Fifth Surge

Perez defined a techno-economic paradigm as "…a best-practice model made up of a set of all-

Figure 2. Timing of the fifth surge

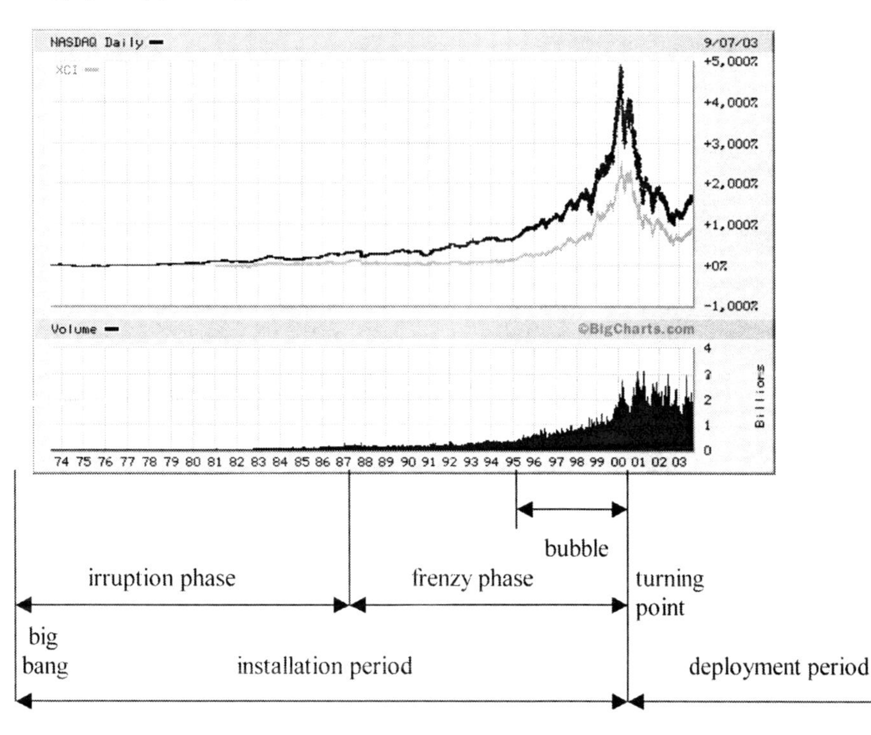

pervasive generic technological and organizational principles, which represent the most effective way of applying a particular technological revolution and of using it for modernizing and rejuvenating the whole economy. When generally adopted, these principles become the common-sense basis for organizing any activity and for structuring any institution." This would apply to both private and public actors, as well as private and public governance. The understanding of a new techno-economic paradigm becomes clearer if it is contrasted with the old paradigm. Using the main dimensions identified by Perez, the transitions from the old to the new paradigm are reflected in Table 2 (based on Perez, 2002, 2004a).

With reference to the stylized model described above, the different periods and phases in the life cycle of the Fifth Surge can be described:

The Gestation Period: 1948-1971

Earlier we noted that preceding the 'big bang' or public landmark that signifies the start of a new technological revolution, there is a period of gestation when the ideas and inventions emerge that will become central to the new revolution. In the context of the Fifth Surge, this is the period leading up to the invention and introduction of the microprocessor in 1971. One could argue that this period starts with the application of semiconductors, i.e. the invention of the transistor in 1948. Important events that took place during the gestation period are: UNIVAC, the first commercial computer introduced by Remington Rand in 1951, the co-invention of the integrated circuit by Texas Instruments and Fairchild in 1957, the invention of packet switching at Rand Corporation and National Physics Labs in the UK in 1960, invention of the minicomputer by Cray in 1961, the first stored program controlled switch by AT&T in 1965, the first digital switch Plato by the CNET of France in 1970[12].

Table 2. The three levels in the 4th and 5th techno-economic paradigm

	Techno-Economic Paradigm	
	4th Fordist	**5th ICT**
Level 1	**Technology & Infrastructure**	
A constellation of technologies, products and industries with wide generic applicability, and a supporting infrastructure	Internal combustion engine (for autos, tractors, electricity generation, aeroplanes, etc.)	Micro-processor (as information processing engine)
	Oil and gas as fuels	Data as fuel
	Petrochemical industry (refinery, synthetic materials and chemicals)	Communications and Information Technology industry (hardware, software, services)
	Motorways, airports, airlines	Internet, broadband access
Level 2	**Organization**	
An organizational model for best practice in all productive activities	Dedicated mass production	Adaptable production systems
	Compartmented hierarchical pyramids	Flexible networks, flat and broad ranging
	Materials and energy intensive	Information intensive
Level 3	**'Common sense' principles**	
A general set of 'common sense' principles for guiding organisational and institutional innovation	Centralization	Decentralization
	Separation of work and organizations by function	Re-integration of functions
	Massification	Diversification
	Negotiation of conflicts	Consensus building
	Regulation and supervisory control	Guidelines, trust and monitored control

The Installation Period: 1971-2000

Marked by the launch of the microprocessor, as the major visible event of this technological revolution, the installation period starts in 1971 and runs until the collapse of the Internet bubble in 2000. This is also the time of the famous Carterfone decision (1968), marking the start of the telecom reform process. As well as the emergence of the ARPANET, the first five university nodes being connected—the precursor of the Internet (1970), and the invention of email (1971).

Phase One: Irruption – 1971-1987

The exhaustion of the old Fordistic paradigm can be illustrated by a declining growth rate in labor productivity from the 1950-60s into the 1970-80s (Dekker & Kleinknecht, 2003; Freeman & Louçã, 2001). A number of events occurring in the period illustrate the point: (1) the concerns that arose at the end of the 1960s with respect to the use of natural resources and the growth of world population and consumption, which culminated in the Club of Rome Project and the publication of the report The Limits to Growth in 1972 (Meadows, 1972); (2) the oil crises of 1973 and of 1979.

The early increase in IT investments have, for a variety of reasons, not directly resulted in measurable productivity improvement at the aggregate level, and as a consequence have led to the suggestion of an IT productivity paradox by Roach of Morgan Stanley in 1987 (Brynjolfson, 1992; Brynjolfson & Hitt, 1998)[13]. However, firm level data that became available in the early 1990s showed that a dollar of IT capital is associated with a substantial increase in revenue each year. The returns suggest that they represent more than a return on technology, as there are large associated expenditures on training, process redesign and other organizational changes involved (Brynjolfson & Hitt, 1998). The firm level information also showed large variation. Moreover, long-term benefits were substantially larger than short-term benefits, by 2 to 8 times. These long-term benefits are "…not just the returns from IT but from a system of technology and organizational changes. In other words, for every dollar of IT there are several dollars of organizational investments that, when combined, generate the large rise in measured firm productivity and value"[14]. In this respect the authors refer to Drucker's 1988 article on "The Coming of the New Organization," predicting that "…technology rich firms will increasingly shift to flatter, less hierarchical organizations where highly skilled workers take on increasing levels of decision-making responsibility" and the notion of business process redesign, the shift from 'mass production' to flexible manufacturing[15]. Hence, a production infrastructure that exploits the low cost communications and information processing capabilities of the new IT based paradigm[16].

The 'delay' in productivity improvement to show through should not come as a surprise as in the Third Wave the productivity from electric motors took almost 40 years to emerge. Initially the steam engine was replaced by a large electro motor transmitting power via the existing shafts, pulleys, and belts. The big productivity gains came when factories were re-engineered and electric motors were being 'distributed across the factory floor,' allowing machines to be arranged in accordance with the logic of the work flow instead of proximity to the once central power unit (David, 1990; Schurr, Burwell, Devine, & Sonenblum, 1990).

The learning process around the new paradigm is twofold, in the information technology environment, and in the telecom environment. In IT, the shift from specialists using computers to broad based learning and familiarization with IT applications comes with the introduction of the PC (Apple in 1977, IBM in 1981). It also marks the shift from mainframe based computing to client-server architectures.

The learning around the new communications paradigm starts with the early Internet development in the open environment of university research labs. This is in sharp contrast with the closed environment of the corporate research labs that are focused on the circuit mode paradigm, used in voice communication over the telephone network. The new learning mode becomes even more pronounced when Cisco, provider of routers, founded in 1984, starts to engage users in the product development process and starts to publish their product and training manuals, allowing many to participate in the new industry[17].

The mismatch in the socio-institutional framework becomes apparent with the "coming of the new organization," with a shift from mass production to more flexible production, from hierarchical organizations to flatter organizations increasingly as part of broader and flexible networks of suppliers. The 'Fordist' mass production paradigm gives way to networked organizations with flexible manufacturing processes, providing users with a much wider range of choice in products and services. This shift requires adjustments to create more flexible labor and product markets.

Phase Two: Frenzy – 1987-2000

In 1984 the Internet is named and the then existing 1000 hosts were converted *en masse* to TCP/IP. In 1990, the WWW application is introduced, and in 1993, the Mosaic browser is launched. One could argue that thereby the 'basic constellation' is being in place for the take off of the Internet. In 1984, Cisco is founded to become the leading provider of routers for the Internet. In 1995, the landmark IPO of Netscape occurs. In the same year, Microsoft 'converts' to the Internet and Amazon.com starts business and eBay is founded. By this time, one could argue that the 'full constellation' is in place.

The year 1995 can be defined as the beginning of the Internet bubble and 1997 as the year in which speculation starts for real. In the terminology of Perez, in 1995 'financial capital' takes over, while in 1997 the paper economy takes over from the real economy. During the unraveling of the bubble a range of inappropriate behaviour has come to the surface, the major cases being the failure of WorldCom and Enron (for details see Lemstra, 2006).

Compared to the diffusion of other technologies that affect the consumer, the diffusion of the Internet has indeed been fast: to reach a market in which 50 million people participate it took radio 38 years, television 13 years, and the Internet, once it was open to the general public, just 4 years (Slater, 2002).

The Turning Point

Perez characterizes the turning point as a conceptual device, denoting the transition between the installation and deployment period, more a notion of transition than an interval or time period *per se*. She characterizes the turning point as follows: With the collapse of the bubble, which ends the period of frenzy, comes recession and sometimes depression, which brings financial capital back to reality. Together with mounting social pressure, this creates the conditions for institutional restructuring and for re-routing growth onto a sustainable path. A swing of the pendulum from the extreme individualism typical for the frenzy period to giving greater attention to collective well-being. Following the crash in 2000, economic growth stagnated.

The Deployment Period: 2000-

Perez points to the fundamental difference in the role capital plays in the 'installation period' versus the 'deployment period': financial capital versus production capital. In that respect, she points to the different types of prosperities in the two periods, in particular the 'frenzy phase' and the 'synergy phase.' The 'frenzy phase' is typified by change, instability, and short term gains, while the 'synergy phase' is typified by increasing stability, emerging successful business models, and long-term gains.

The current ICT-driven paradigm replaces the previous "Fordist" paradigm characterized by mass production. As a consequence, the new paradigm raises tensions with the institutional environment that has been optimised to accommodate the previous "Fordist" paradigm. The new techno-economic paradigm has a major impact on forms of coordination and hence impacts governance, for instance:

- New modes of communication (increasingly computer mediated, the Internet),
- New modes of exchange (e.g. electronically mediated),
- New perspectives on (information) property (e.g. business *and* private interests),
- New modes of contracting (e.g. authentication, authorization),
- New forms of organising (e.g. the networked organisation), and
- New modes of cultural engagement (e.g. virtual identity, virtual communities).

Some of these new modes can be accommodated within the existing institutional environment; others require adaptation of the existing institutions or the introduction of new institutions to allow the full benefits of the new ICT-driven techno-economic paradigm to be realized. This represents the broader e-Governance agenda.

THE E-GOVERNANCE CHALLENGE

The installation period of the new techno-economic paradigm has given us the opportunity to observe the changes and the tensions that the paradigm shift is bringing about. These changes have been summarized and clustered using an adapted version of the four layer framework for institutional analysis of complex technological systems as defined by Groenewegen and Koppenjan, based on Williamson, see Figure 3 (adapted from Koppenjan & Groenewegen, 2005; Williamson, 1998). As this framework links economic activity to governance, the institutional environment, and the cultural context, it provides a useful framework to capture the changes a techno-economic paradigm is bringing about.

The framework facilitates the process of identifying the tensions that emerge between the old institutional environment and the characteristics of the new paradigm. The placement of items is sometimes a forced choice. The compilation of the information has been based on the work by Freeman, Louçã, and Perez (Freeman & Louçã, 2001; Perez, 2002, 2004b), as well as Shapiro and Varian (1999), Tidd et al. (1997, 2001), Castells (2000, 2001), Kuhn (1996), Mansell (Mansell, 2002; Mansell & Steinmueller, 2002), Fransman (2002), and was inspired by the philosophical contributions by Wentink (2002, 2004). It builds upon the research executed in relation to the project: "Rethinking the European ICT agenda" for the Dutch Ministry of Economic Affairs (PriceWaterhouseCoopers, 2004) and is expanded through the MSc thesis work performed by Boelsma, which included, in addition, the review of work by Florida (2002), Graham (Graham & Marvin, 2001), (Jonk & Van Velzen, 2002), (De Mul, Müller, & Nusselder, 2001), (Russell, 1983), (Schaap, 2002), (Trommel, 1999), (Zuurmond, 1994), and others.

The attributes being listed are considered to be linked directly or indirectly to the new techno-economic paradigm. While the new paradigm is an important driver of change, it is not necessarily the only source of change. For the purpose of this contribution, we will not elaborate on the causal links, inclusion is considered more important than origin. For certain attributes, the distinction between the paradigms seems more gradual, and increasing over time, as the old paradigm fades to the background.

The paradigm shift from the 4th to the 5th Technological Surge as reflected in Table 3 and Table 4, points to the mismatch between the old socio-institutional framework and the require-

Figure 3. Five layer model: levels of institutional analysis, technology enabled

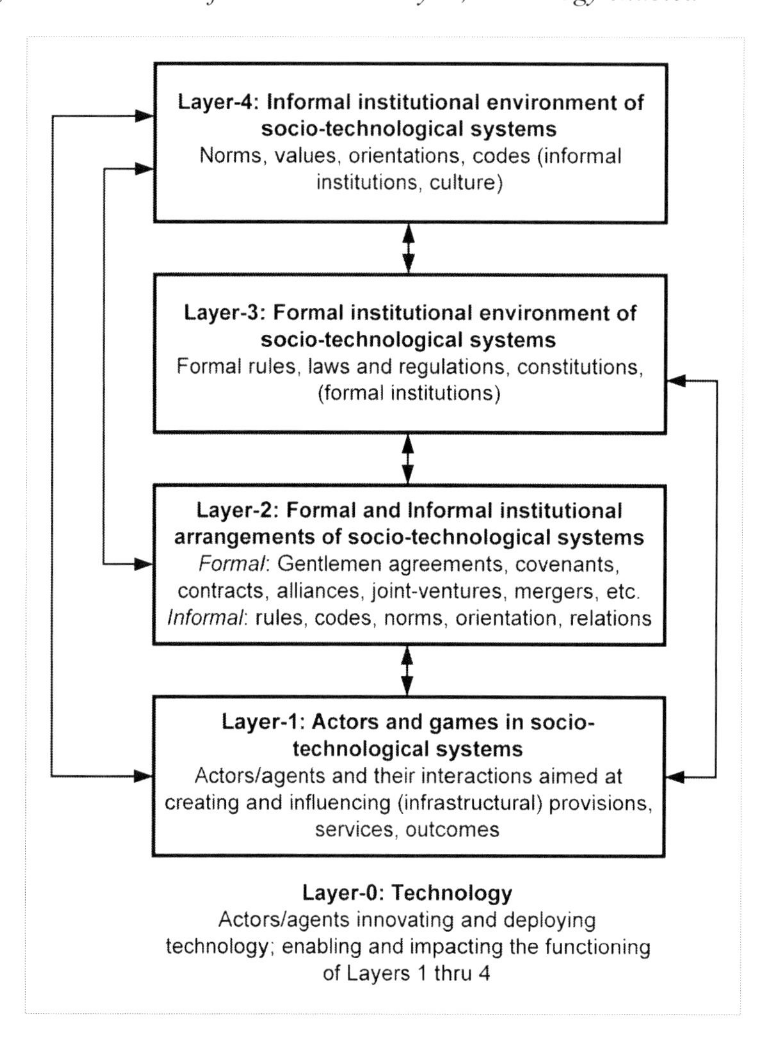

ments of the new paradigm. It should be noted that the technological changes are not captured in these tables, see for an extensive discussion of this dimension Lemstra (2006).

SUMMARY AND IMPLICATIONS FOR GOVERNING AND GOVERNANCE

The productivity improvements related to the current techno-economic paradigm are resulting from two related phases in the current revolution: the information technology revolution phase and the telecommunications/Internet revolution phase.

As discussed, the IT related productivity improvements are a result of investments in hardware and software, complemented by investments in human resources and in organisational change. The telecommunications related productivity improvements are related to the lowering of production and transaction costs, facilitated in particular by the Internet. The relationship between investments in ICT-assets, the complementary investment in human resources and the linkage to organizational change, in order to realise the productivity improvements that are embedded in the new technologies, can be considered a 'best practice' of the current paradigm. The changing

Table 3. Paradigm attributes in the layered framework: level 1-2

	Topical area	4th TEP Fordist Mass production	5th TEP Information and Telecommunication
Layer 1			
Social embeddedness Informal institutions, customs, traditions, norms, religion		Identity is a given	Identity more conscientiously addressed (physical and virtual)
		Traditional role patterns	Individualized role patterns
		Importance of strong ties (primary and secondary relations)	Importance of weak ties (tertiary relations); Networked individualism
		Less segregation	More segregation along various lines (enclaves; creative class; elites)
		Social status and esteem derived from position and wealth ('conspicuous consumption')	Additional status in virtual world: derived from contribution to the network ; peer and social reciprocity ('conspicuous contribution')
		Pluriformity and reliability of information linked to stratification of society	Abundance but diffuse supply of information; open 'can be surfed by all'
		Communication primarily based on necessity	Communication also as part of social awareness and consensus building
		Time delays in and between contact and action	Instant contact and action; instant global communication
			Increasingly computer mediated communication; increasingly multi-mode (voice, text, image)
		No environmental concern	Environment as guide to innovation
Layer 2			
Institutional environment Formal rules of the game, esp. property (policy, judiciary, bureaucracy)	Bureaucracy	Strong bureaucracy	Crumbling bureaucracy; emerging infocracy
		Government control and sometimes ownership	Government information, coordination and regulation
	Policy	Welfare state	Well being and individualized responsibilities
		Keynesian demand management	Minimal government idea
		Universal service	Service differentiation by location/geography (poly-nucleated city)
	Property	Financial divide	In addition Information (digital) divide
		Intellectual property rights an issue for firms (few transactions)	Intellectual property rights an issue for individuals (many transactions; digital rights management)
			Awareness of public versus private ownership of information
	Judiciary	Physical authentication, authorization	Electronic authentication, certification, authorization
		Enforcement within national boundaries	Enforcements requires cross border action (cybercrime; spam)
		Burglary	Cybercrime (privacy, security, identity theft)
	Monetary	Physical and electronic funds transfer	Electronic payment, multiple forms of payment, including micro-payments

role of information in organizations is another element of 'best practice.'

The regularity of technological revolutions suggests that the productivity improvements related to the new techno-economic paradigm are not limited to productivity improvements that can be obtained in relation to firms alone, as demonstrated by e.g. Brynjolfson and Hitt (1998, 2003), but that they can be obtained across the economy and society at large. The ICT-driven revolution thereby affects our routines and habits, not only in terms of work routines, but also in the social realm, e.g. in the way we communicate and transact.

Table 4. Paradigm attributes in the layered framework: level 3-4

	Topical area	4th TEP Fordist Mass production	5th TEP Information and Telecommunication
Level 3			
Governance Play of the game, esp. contracts (aligning governance structure with transactions)	Industrial organisation	Large firm dominated (vertical integration)	Networked firms, local & global
		Economies of scale	Economies of scope and specialization
	Firm organisation	Hierarchies; pyramids	Flat networks; internal & external
		Departmental	Integrated; project orientation
		Automation of separate activities	Systemation, linking activities along the value chain
	Production level	Mass production, standardized goods	Rapid changes in product variety and mix
		Dedicated plant and equipment	Flexible production systems
		Stable routines	Continuous improvement
		Separation of mind and hand	Integration of mind and hand
		Specialized skills	Multi-skilling
		Sequential design	Concurrent engineering
	Market level	Three tier stable market	Highly segmented market: dis-intermediation and re-mediation
		Traditional marketplaces	Increasingly electronic (Internet enabled) marketplaces
		Product competition	Systems competition
		High search costs (access)	Lower search costs (abundance of information) off set by uncertainty in quality (authentication)
		Customer interaction is remote	Customer engagement in product development; in the after-sales process
		Multi-national markets	Global markets
	Business level	Fixed plans	Flexible strategies
		Economies of scale	Economies of scope and specialization
		Mainly competition in the market	Increasingly competition for the market (creating critical mass; network effects)
		Centralized control, vertical information flow	Distributed control, horizontal information flows
		Centralize intelligence	Localized intelligence
		Closed, localized innovation systems (fragmented knowledge base; few innovators)	Open dispersed innovation systems (common knowledge base; many innovators)
		Collectively wage bargaining	More tailored employment contracts
Level 4			
Resource allocation and employment Prices and quantities; incentive alignment	Transaction level	Trust embedded by institutions and familiarity with transaction partners	Use of the Internet requires new arrangements to establish and maintain trust
		Physical/tangible transactions	Increasingly computer mediated transactions
	Factors of production	Capital, labour, natural endowments	Knowledge, creativity, relations
	Product level	Standardized (limited choice)	Customized (wide choice)
		Highly Tangible	Highly intangible (information goods) with tangible complementarities; Increasingly experience goods
		Product with service	Service with product
		Low degree of network effects; low degree of lock- in	High degree of network effects; high degree of lock-in
	Pricing	Cost based	Value driven (incl. quality and convenience)
	Incentives	Monetary and fixed; social security;	Monetary and variable; high powered; increasing attention to social aspects in job motivation
		Identification with the firm (job-for-life)	Identification with the job; mobile jobs, job migration
		Human resources	Human capital
		Skill specialization	Multi-skilling
		Spatial division of labor	Flexible workplace

ICT-technologies have changed also the nature of products, from tangible to increasingly intangible. In this respect Shapiro and Varian refer to information goods and experience goods (Shapiro & Varian, 1999). The change in the nature of products also invokes changes in the way these products are produced and transacted. Replication cost of information goods are extremely low, and communications technologies allow for the instantaneous distribution of these products. This affects the way we tend to deal with products in terms of institutional arrangements and their enforcement, e.g. with respect to ownership, taxation and illegal reproduction. It also affects the 'economic order' as the relatively low entry costs of the ICT-revolution has allowed new regions to participate, as demonstrated by the uptake of IT-related economic growth in e.g. India and China (as illustrated in PriceWaterhouseCoopers, 2004). Hence, the impact of the ICT-driven revolution affects economies and societies at large. It has become apparent that the institutional structures and governance arrangements that have been created to facilitate the production, distribution and consumption or use of the very tangible products of the previous 'Fordist'-revolution are not fit for the intangible products of the ICT-driven revolution. Tensions have emerged and are emerging as a result of the 'installation' of the ICT-driven techno-economic paradigm. These have been described and summarized as tensions between 'best practice' in the Fourth Wave and perceived 'best practice' for the current and Fifth Wave. They also reflect the tensions that have arisen between the vested interests linked to the previous Wave and those emerging in the current Wave.

Historical regularity has shown that adaptation has taken place, both spontaneously and directed, and has shown that the benefits of the new techno-economic paradigm have been accrued. However, there is no guarantee that this will happen again, as the post 1929 depression period has shown. Hence, there is a role for the actors involved, private as well as public, to understand and appreciate the tensions that have emerged and reflect on the question whether we may expect these tensions to be resolved spontaneously or whether their resolution is important enough to warrant government action, to allow the benefits of the new paradigm to be obtained in a timely manner.

A first essential step is to interpret the tensions summarized in Tables 3 and 4. From the tables it becomes apparent that the implication of the paradigm shift does not remain restricted to the ICT sector, but is affecting essentially all dimensions of economic and social activity. The differences in the dimensions between the Fourth and the Fifth techno-economic paradigm provides an indication of the need for adaptation of the institutional environment. As historical analysis suggests, this is a task in which governments will have to play an important role. It is thereby feasible to consider "replacing structures that are unwanted, unneeded and restrictive" by those structures that are "wanted, needed, and empowering" (Lawson, 1997).

The differences between the Fourth and the Fifth techno-economic paradigm are considered important sources to inform the policy formation process. It should be acknowledged that many topics identified are already featuring on the various political agendas, in one form or the other. However, very few topics have been resolved, which should not come as a surprise when considered in the context of a shift in techno-economic paradigms.

Freeman and Soete have argued that the economic and social potential of new clusters of technologies will only be realized over fairly long historical periods, and that through the process of 'learning' the necessary changes in management strategies and institutional environments will occur (Freeman & Soete, 1997). Recent research related to the ICT productivity has shown that the benefits of the new paradigm do not emerge through investments in computer hardware and software alone. Essential is linking these investments in physical assets to investments in human capital and to organisational change. Perez observed that "...each technological revolution brings with it, not only a full revamping of the productive structure but eventually also a transformation of the institutions of governance, of society and even of ideologies and culture, so deep that one can speak about the construction of successive and different modes of growth in the history of capitalism" (Perez, 2002).

Initial analysis of the changes involved by the new techno-economic paradigm show that the changes are profound, affecting Layers 1 through 4 of the layered framework for institutional analysis, that is including culture. Hence, irrespective the differences in role perception of governments, it would not be realistic to expect, if at all desired, that all of the issues that have emerged can be addressed and resolved through government action alone. However, what may improve the possibility for the benefits of the ICTs to be captured is government policy informed and inspired by the lessons than can be learned from historical regularities.

Assuming that this recommendation will be adopted, the need to understand the techno-economic paradigm shift will follow, as "... each techno-logical revolution is different, each paradigm is unique, each set of solutions needs to be coherent with the problems to overcome and with the logic of the paradigm, its opportunities, and its best practices" (Perez, 2002). Many authors have made contributions towards a better understanding of the new paradigm as reflected in Table 3 and 4. This could serve as a starting point for sharing the insights obtained across multiple disciplines involved.

Recommended Next Steps

In a diagnosis of the of the current process of adaptation, we may conclude that many of the changes the paradigm shift has evoked are still very much emergent, and hence have certainly not led to a full alignment between the technological, economic and social domains. We are in the process of adaptation. It also appears that many of the issues are being addressed in isolation, i.e. not linked to the broader phenomenon of diffusion of a technological revolution. When considered in isolation, a resolution may be problematic, as the broader goal that should be pursued is not being perceived. Hence, the results may be suboptimal. Therefore, it is being recommended to revisit the current policy formation process against the backdrop of the diffusion process of the 'ICT-driven' technological revolution.

To facilitate this process the following stepwise approach is recommended for public and private actors alike:

1. To understand and appreciate the attributes of the paradigm shift, using the notion of a new Techno-Economic Paradigm,
2. To understand and appreciate the relationships to economic activity, using the four layer framework for institutional analysis,
3. To identify the tensions between the current institutional framework that has been optimized for the previous 'Fordist' paradigm and the (emerging) needs of the 'ICT-driven' paradigm,
4. To assess the scope and impact of these tensions, to determine whether spontaneous or directed coordination is appropriate,
5. To identify the solution space (local, regional, global),
6. To identify the stakeholders involved (governments, industry, and citizens),
7. To engage and resolve the issues, as they relate to policy and strategy formation and implementation.

It should be noted that we are at a unique juncture, the transition from the 'installation' phase to the 'deployment' phase of the new techno-economic paradigm. This is a period in which the state, economic and social actors can adjust the rules and regulations, i.e. to implement new e-Governing and e-Governance schemes, to facilitate the solid expansion of production capital, with the prospect of a 'golden age' to develop.

REFERENCES

Abbate, J. (1999). *Inventing the internet*. Cambridge, MA: MIT Press.

Brynjolfson, E. (1992). *The productivity paradox of information technology: Review and assessment*. Retrieved 2004-02-09, from http://ccs.mit.edu/papers/CCSWP130/ccswp130.html

Brynjolfson, E., & Hitt, L. M. (1998). Beyond the productivity paradox - Computers are the catalyst for bigger changes. *Communications of the ACM.* Retrieved from http://ebusiness.mit.edu/erik/bpp.pdf

Brynjolfson, E., & Hitt, L. M. (2003). *Computing productivity: Firm-level evidence*. Cambridge, MA: MIT.

Brynjolfson, E., & Kahin, B. (Eds.). (2000). *Understanding the digital economy*. Cambridge, MA: MIT Press.

Campbell, J. L., Hollingworth, J. R., & Lindberg, L. N. (Eds.). (1991). *Governance of the American economy*. Cambridge, UK: Cambridge University Press. doi:10.1017/CBO9780511664083

Cassidy, J. (2002). *Dot.con - The real story of why the internet bubble burst*. London, UK: Penguin.

Castells, M. (2000). *The rise of the network society* (2nd ed.). Oxford, UK: Blackwell Publishers.

Castells, M. (2001). *The internet galaxy - Reflections on the internet, business, and society*. Oxford, UK: Oxford University Press.

David, P. A. (1990). The dynamo and the computer: A historical perspective on the modern productivity paradox. *The American Economic Review*, *1*(2), 355–361.

De Mul, J., Müller, E., & Nusselder, A. (2001). *ICT de baas? Informatietechnologie en de menselijke autonomie*. Utrecht, The Netherlands: Center for Public Innovation.

Dekker, A., & Kleinknecht, A. H. (2003). *Flexibiliteit, technologische vernieuwing en de groei van de arbeidsproductiviteit* (No. A203). Delft, The Netherlands: TUDelft, Faculteit Techniek, Bestuur en Management.

E-Readiness. (2012). *Wikipedia.* Retrieved from http://en.wikipedia.org/wiki/E-readiness

Florida, R. (2002). *The rise of the creative class*. New York, NY: Basic Books.

Fransman, M. (2002). *Telecoms in the internet age - From boom to bust to.?* Oxford, UK: Oxford University Press.

Freeman, C. (1998). Lange wellen und arbeitslosigkeit. [Long waves and unemployment] In Thomas, H., & Nefiodow, L. A. (Eds.), *Kondratieffs Zyklen der Wirtschaft - An der Schwelle neue Vollbeschäftigung?* Herford, Germany: BusseSeewald.

Freeman, C., & Louçã, F. (2001). *As time goes by - From the industrial revolutions to the information revolution*. Oxford, UK: Oxford University Press.

Freeman, C., & Soete, L. (1997). *The economics of industrial innovation* (3rd ed.). Cambridge, MA: The MIT Press.

Graham, S., & Marvin, S. (2001). *Splintering urbanism - Networked infrastructures, technological mobilities and the urban condition*. London, UK: Routledge. doi:10.4324/9780203452202

Groenewegen, J. P. M. (1989). *Planning in een markteconomie*. Delft, The Netherlands: Eburon.

Groenewegen, J. P. M. (2008). *Personal communication*. Delft, The Netherlands: Academic Press.

Jessop, B. (2005). Capitalism, steering and the state. In *Formen und Felder politischer Intervention: Zur Relevanz von Staat und Steuerung* (pp. 30–49). Munster, Germany: WestfÃlisches Dampfboot.

Jonk, A., & Van Velzen, G. (2002). *De politieke partij in de netwerksamenleving*. Den Haag, The Netherlands: Academic Press.

Kleinknecht, A. H. (1987). *Innovation patterns in crisis and prosperity - Schumpeter's long cycle reconsidered.* New York, NY: St. Martin's Press.

Kooiman, J. (2003). *Governing as governance*. London, UK: Sage.

Koppenjan, J. F. M., & Groenewegen, J. P. M. (2005). Institutional design for complex technological systems. *International Journal of Technology. Policy and Management, 5*(3), 240–257.

Kuhn, T. S. (1996). *The structure of scientific revolutions* (3rd ed.). Chicago, IL: The University of Chicago Press.

Lawson, T. (1997). *Economics and reality*. London, UK: Routledge.

Lemstra, W. (2006). *The internet bubble and the impact on the development path of the telecommunication sector.* Delft, The Netherlands: TUDelft.

Lipsey, R. G., Carlaw, K. I., & Bekar, C. T. (2005). *Economic transformations: General purpose technologies and long term economic growth*. Oxford, UK: Oxford University Press.

Mansell, R. (Ed.). (2002). *Inside the communications revolution - Evolving patterns of social and technical interactions*. Oxford, UK: Oxford University Press.

Mansell, R., & Steinmueller, W. E. (2002). *Mobilizing the information society - Strategies for growth and opportunity*. Oxford, UK: Oxford University Press.

Mathiason, J. (2009). *Internet governance - The new frontier of global institutions*. London, UK: Routledge.

Meadows, D. L. (1972). *Rapport van de club van Rome*. Utrecht, The Netherlands: Spectrum.

Misuraca, G. C. (2007). *e-Governance in Africa - From theory to action*. Trenton, NJ: Africa World Press.

Moore, G. E. (1965). Cramming more components onto integrated circuits. *Electronics, 38*(8).

Mueller, M. L. (1993, July). Universal service in telephone history - A reconstruction. *Telecommunications Policy*, 352–369. doi:10.1016/0308-5961(93)90050-D

Mueller, M. L. (2002). *Ruling the root: Internet governance and the taming of cyberspace*. Cambridge, MA: The MIT Press.

Mueller, M. L. (2010). *Networks and states: The global politics of internet governance*. Cambridge, MA: MIT Press.

O'Mahony, M., & Van Ark, B. (2003). *EU productivity and competitiveness: An industry perspective*. Luxembourg, Luxembourg: European Communities.

OECD. (2004). *The e-government imperative*. Paris, France: OECD.

Parsons, W. (1995). *Public policy - An introduction to the theory and practice of policy analysis*. Cheltenham, UK: Edward Elgar.

Perez, C. (1983). Structural change and the assimilation of new technologies in the economic and social system. *Futures, 15*, 357–375. doi:10.1016/0016-3287(83)90050-2

Perez, C. (2002). *Technological revolutions and financial capital: The dynamics of bubbles and golden ages*. Cheltenham, UK: Edward Elgar.

Perez, C. (2004a). *The new techno-economic paradigm.* Amsterdam, The Netherlands: Ministry of Economic Affairs DG Telecom & Post.

Perez, C. (2004b). *Various TEP*. Cambridge, UK: Cambridge University Press.

PriceWaterhouseCoopers. (2004). *Rethinking the European ICT agenda - Ten ICT-breakthroughs for reaching Lisbon goals*. The Hague, The Netherlands: Ministry of Economic Affairs.

RIPE. (2005). *About RIPE*. Retrieved 2005-09-26, from www.ripe.net/ripe/about.html

Russell, P. (1983). *The global brain - Speculation on the evelutionary leap to planetary consciousness*. Los Angeles, CA: J.P. Tarcher.

Schaap, F. (2002). *The words that took us there*. Amsterdam, The Netherlands: Aksant Academic Publishers.

Schurr, S. H., Burwell, C. C., Devine, W. D. Jr, & Sonenblum, S. (1990). *Electricity in the American economy - Agent of technical progress*. New York, NY: Greenwood Press.

Shapiro, C., & Varian, H. R. (1999). *Information rules - A strategic guide to the network economy*. Boston, MA: Harvard Business School Press.

Slater, W. F. (2002). *Internet history and growth*. Retrieved 2003-01-08, 2003, from www.isoc-chicago.org

Tidd, J., Bessant, J., & Pavitt, K. (2001). *Managing innovation - Integrating technological, market and organizational change* (2nd ed.). Chichester, UK: John Wiley & Sons.

Trommel, W. (1999). *ICT en nieuwe arbeidspatronen*. Den Haag, The Netherlands: Government Press.

Van Ark, B., & Piatkowski. (2004). *Productivity, innovation and ICT in old and new Europe*. Groningen, The Netherlands: Rijksuniversiteit Groningen.

Van Ark, B., Inklaar, R., & McGuckin, R. H. (2003). *The contribution of ICT-producing and ICT-using industries to productivity growth: A comparison of Canada, Europe and the United States. The Haag*. The Netherlands: Centre for the Study of Living Standards.

Wentink, V. (2002). *Digital alterego and the new economy*. Zeist, The Netherlands: Comparc.

Wentink, V. (2004). *Een nieuwe horizontalisering*. Zeist, The Netherlands: Comparc.

Williamson, O. E. (1998). Transaction cost economics: How it works, where it is headed. *The Economist, 146*(1), 23-58.

Wolf, C. Jr. (1990). *Markets or governments - Choosing between imperfect alternatives*. Cambridge, MA: The MIT Press.

Zuurmond, A. (1994). *De infocratie*. Den Haag, The Netherlands: Phaedrus.

ENDNOTES

1. For more recent contributions see *Understanding the Digital Economy*, edited by Brynjolfsson and Kahin (2000), and also publications by Van Ark (O'Mahony & Van Ark, 2003; Van Ark, Inklaar, & McGuckin, 2003; Van Ark & Piatkowski, 2004)
2. Misuraca provides a comprehensive overview of e-Government and e-Governance in relation to socio-political developments in Africa.
3. See E-Readiness (2012).
4. Cassidy traces the origins of the Internet back to concept of the Memex proposed by Bush in 1945, as a precursor of the hypertext principle (Cassidy, 2002, p. 9-12).
5. RIPE (Réseaux IP Européens) is a collaborative forum open to all parties interested in wide area IP networks. The objective of RIPE is to ensure the administrative and technical co-ordination necessary to enable the operation of the Internet within the RIPE region (RIPE, 2005).
6. The implied suggestion is that governments can remedy market failure. However, there is also 'non-market' failure. For government intervention to be legitimate from an economic point of view, the cost of intervention should be lower than the cost of the market failure. See for a more extensive review e.g. Markets or Governments by Wolf (1990).

[7] Static or allocative efficiency: resources are being allocated to the production of the goods and services most valued by society. Dynamic efficiency: relates to the time, effort, and resources applied to innovation or the renewal of products and production processes.

[8] Freeman and Louçã claim three central innovative features in their approach: (1) "It is a description based on the overlapping of subsystems, since their relationship is more adequate to explain reality than the artificially isolated description of each of the subsystems; (2) it analyses the crises and phase transitions from the viewpoint of the lack of synchronicity and maladjustment between subsystems, which defines the time band of major fluctuations; (3) the social conflicts of all types are generated and articulated by the coordination process, that is by power under all its forms, from the production of legitimacy to strict coercion. This coordination process proceeds at several simultaneous levels. The first level is that of the actions embodied in the social working of the economic system, the tension to integrate the conflicts, the conventions, and the institutions, and the second level is that of power, strategy, and domination" (Freeman & Louçã, 2001).

[9] According to Kleinknecht, Van Gelderen anticipated almost everything that is now being rediscovered and rewritten with respect to the phenomenon of the long wave, predating the classic study by Kondratieff of 1926 ("Springvloed: Beschouwingen over Industriële Ontwikkeling en Prijsbeweging," articles published in 1913). Van Gelderen and De Wolff worked at the University of Amsterdam.

[10] The summary that follows is derived from Perez (2002) as are the citations.

[11] Perez also points to the uneven development and time lags in the diffusion of technological revolutions across different countries and regions.

[12] To this list could be added the invention of the cellular concept by Bell Labs in 1947, albeit the first commercial application is in 1981, and the launch of Sputnik by the USSR in 1957.

[13] Including intangible benefits such as quality, variety, and convenience.

[14] The analysis is based on three inputs that influence the firm value-added: ordinary capital stock, computer capital stock, and labor. The resulting regression estimates for the computer coefficient (the contribution of computerization to three-factor production growth) range from 0.01 to 0.02 in the first year to 0.04 and 0.05 in the seventh year. The data represents 527 firms and covers the period 1987-1994 (Brynjolfson & Hitt, 2003).

[15] The research showed that firms that couple IT investments with the decentralized work practices are about 5% more productive than firms that do neither. Firms can be worse off if they invest in computers without the new work systems. The more extensive treatment can be found in Brynjolfson and Hitt (2003).

[16] A comparative study on productivity between the USA and the EU shows that the post 1995 acceleration in productivity in the USA is dominated by the ICT producing sectors, wholesale and retail trade and banking and auxiliary financial services (O'Mahony & Van Ark, 2003).

[17] This is of particular importance in avoiding a resource bottleneck in a high growth industry. A book search on Amazon in December 2005 yields over 1100 books related to Cisco products and training, many for self-study. The title of Cisco Certified Engineer and more recent Cisco Certified Network Associate have become valuable tickets in the industry.

Chapter 2
E-Government: Some Factors for a Conceptual Model

Mehdi Sagheb-Tehrani
Bemidji State University, USA

ABSTRACT

Some state, national, and local governments around the world have long played active roles in the use of Information Technologies (IT) to stimulate economic development. Electronic government utilizes information technology to provide the citizenry access to a wide range of public services. Governments in many countries around the world indeed perceive IT as a way to improve the quality of life of their citizens. Today, governments at all levels respond to millions of citizen demands electronically. Many public organizations are implementing Electronic Government (e-Government) projects. There is a need to put forward a conceptual model focusing on steps towards implementing more successful e-Government projects. This chapter argues that several key success factors are appropriate for e-Government project implementation. About twelve e-Government websites are examined upon those key success factors. This chapter puts forward a conceptual model for a better implementation of electronic government.

INTRODUCTION

In recent years, nearly all countries have integrated Information Technology (IT) into their national economic development strategies. Governments see IT as a way to improve the quality of life of their citizens. The scale of activity on the part of public sectors in leveraging IT has increased in volume (Smith, 2008). E-government is enabling government companies to provide better services to their customers. The ability to improve citizens' access to services online has made e-government a desirable application for government organizations (Gorla, 2008; Donna & Yen, 2006). Governments around the world are implementing e-government. In every part of the world—from industrialized countries to developing ones, governments are putting information online to provide better services for citizens (Working Group, 2002; Chircu & Lee, 2005; Palmer, 2006). Transactions

DOI: 10.4018/978-1-4666-1909-8.ch002

Table 1. e-Government increase by region of the world

	2001	2002	2003	2004	2005	2006	2007	2008
North America	51.0%	60.4%	40.2%	39.2%	47.3%	43.1%	45.3%	53.1%
Western Europe	34.1	47.6	33.1	30.0	29.6	35.2	36.8	37.2
Eastern Europe	--	43.5	32.0	28.0	27.1	29.2	31.7	30.1
Asia	34.0	48.7	34.3	31.6	37.3	35.9	39.5	39.7
Middle East	31.1	43.2	32.1	28.1	27.4	29.4	33.5	32.3
Russia/Central Asia	30.9	37.2	29.7	25.3	25.0	30.6	27.8	31.2
South America	30.7	42.0	29.5	24.3	25.9	28.0	32.1	33.3
Pacific Ocean Islands	30.6	39.5	32.1	29.9	27.9	32.4	33.8	39.0
Central America	27.7	41.4	28.6	24.1	24.1	25.0	29.2	31.2
Africa	23.5	36.8	27.6	22.0	22.0	24.3	26.0	26.3

Source: West (2008a, p. 3)

such as renewing driver's licenses, applying for jobs, and filing tax forms can now be conducted online, quickly and efficiently (West, 2008b). To be able do these services, e-government uses Information Technology (IT). The increase in e-government operation throughout the world, although significant, is due mostly to small number of countries, including Taiwan, Singapore, United State, Hong Kong, and Canada. Table 1 shows some differences in e-government by region of the world (please see appendix A for various e-government website URLs).

Developing countries are behind in this race to provide e-government services to their citizens. This can be due to many reasons such as lack of a good communication infrastructure, low computer literacy, and limited access to the Internet and so on (Akther, Onishi, & Kidokoro, 2007; Kottemann, & Boyer-Wright, 2010). These issues have to be addressed before developing e-government applications. Officials should be aware of the obstacles before starting an e-government project because; they are long and costly project (Working Group, 2002). In the following sections of this chapter, the author makes an effort to disclose the concept of e-government in a way that leads to more successful e-government project development.

CONCEPTS DEVELOPMENT

E-Government

One frequently asked question regarding electronic government (e-government) is "What is e-government?" E-Government is more than just providing some public information and specific citizen services available to people via a website (Lee, Wu, Lin, & Wang, 2008; Curtin, et al., 2003). E-government serves as a portal focused mainly on access to the public sector; these portals are aimed at citizens (G2C), businesses (G2B), other governments (G2G), and anyone else who are interested in the government and its services. Over 160 countries worldwide have already begun some kind of e-government project, creating a major market for IT vendors and service providers that are competent of helping public agencies in their technology initiatives (Greiner, 2005). E-government is an emerging concept and recent researches focus on applying the new concept of e-commerce and management in e-government such as knowledge management enterprise resource planning (Raymond & Bergeron, 2006). E-government is being considered as one of the tools that can be used to meet the many challenges faced by governments (Jupp, 2003). Governments are facing increased service expectations by their citizens. Some of the services that can be offered by e-governments are as follows (Evolution of e-Government, 2002).

Government to Citizens (G2C)

1. Income taxes: notification of assessment
2. Job search services by labor offices
3. Social security contributions
4. Personal documents (passport and driver's license)
5. Car registration (new, used, and imported cars)
6. Application for building permission
7. Declaration to the police
8. Public libraries (availability of catalogues, search tools)
9. Certificates (birth, marriage): request and delivery
10. Enrollment in higher education / university
11. Announcement of moving (change of address)
12. Health related services (e.g., interactive advice on the availability of services in different hospitals; appointments for hospitals)

Government to Business (G2B)

1. Social contribution for employees
2. Corporation tax: declaration, notification
3. Value Added Tax (VAT): declaration, notification
4. Registration of a new company
5. Submission of data to statistical offices
6. Customs declarations

According to a study made by Cap Gemini and Ernst and Young (2001), in Europe, the most used services are the job search, income taxes, VAT. and corporate tax services; the least used are the health related services, building, and environment-related permits (See Figure 1).

Further, the study emphasizes that in Europe the biggest customer of e-government services is business (G2B, 53%) whereas services for citizen (G2C, 40%) scores significantly lower.

Though US companies lead the e-commerce initiatives among businesses, e-government portal efforts in the US are not ahead of the world. FirstGov or USA.gov (see appendix A for URL) is the US federal government's portal, providing access to both state and federal government agency websites. US portal now offers Americans a complete source of information, and the options to apply for student loans and even Social Security benefits online. FirstGov has about 186 million pages across 22 different sites and receive 6 million visitors per month (Greiner, 2005). The US e-government initiative is divided in three main groups as follows (Murra, 2003; Hipp & Warner, 2008).

Government to Citizens (G2C)

1. Free online tax filing;
2. Job search;
3. Social security;
4. Personal documents (birth and marriage certificates, passport applications, driver license);
5. Immigration services;
6. Health and related services;
7. Government benefits;
8. Student loans;
9. Disaster help;
10. Other useful information (for sales, weather forecast, recreation).

Government to Business (G2B)

1. Comment on federal regulation;
2. Corporation tax;
3. Business opportunities;
4. Registration of a new company;
5. Business laws and regulations;
6. Central contractor registration;
7. Government auctions and sales;
8. Employer ID number;
9. Wage reporting;
10. Subcontracting opportunities;

Figure 1. Ranking of e-government public services in Europe

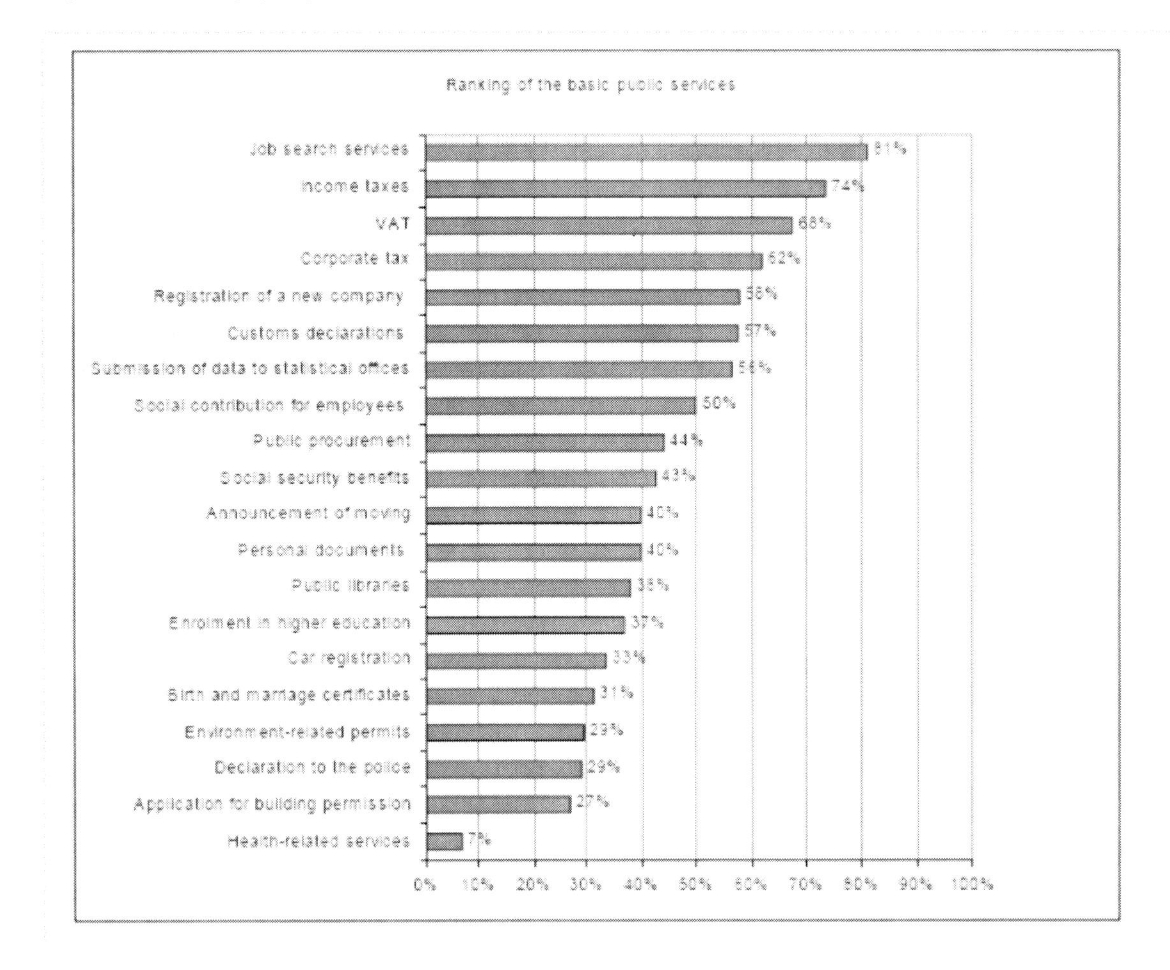

Source: Cap Gemini and Ernest andYoung survey

11. Patents and trademarks filing;
12. Export portal.

Government to Government (G2G)

1. 2003 Federal Pay Tables;
2. Grants;
3. Background Investigation Application;
4. E-Training Initiative for Federal Workers;
5. For Sale to Government Buyers;
6. FirstGov Search for Federal Agencies;
7. Per Diem Rates;
8. Employee Directory;
9. Federal Personnel-Payroll Changes.

The US e-government strategy is to improve the quality of the services to the citizens and businesses. According to one study by (West, 2008a), 11% of the websites examined have no services, 12% provide one service, 10% have two services, and 67% have three or more services. Obviously, both federal and state governments are making important strides in providing services online (see Table 3).

The United States has fallen behind many countries in Internet access and broadband usage. America has fallen behind Sweden, Denmark, Switzerland, Australia, Germany, and The Netherlands in Internet subscribers per 100 inhabitants.

Whereas 36 percent of Swiss residents have access to Internet subscription services, 31 percent of Americans have access to the Internet (West, 2008a).

In this complex world in which we live, everyone must learn more about e-government. Even our politicians may not fully understand the concept and application of e-government. Surveys carried out by the United Nations Conference for Trade and Development (UNCTAD, 2002) on the development of e-commerce in various parts of the world identifying the need in developing countries for transparency within government operations (Mitra, 2005). Electronic Commerce (EC) has revolutionized the way the business and individuals interact. In the United States and Europe, the use of the Internet in the public sector has initiated a discussion about new forms of democracy. The e-government will change the course of democracy by providing all citizens access to government operations.

Vision, Principles, and Priorities

A fully implemented e-government can break down bureaucratic barriers and move to a better service level, connection, and protection that a government may want and need in every aspect of government's activity. This provides an opportunity not merely to manage business but also to get wide access to what government is doing or intending to do, and how, and why. This will allow citizens more than ever before to take part in government decisions and become more knowledgeable of the performance of their elected representatives. Citizens will have the chance to become stable players in the process of determining and making government task (McGinnis, 2003; Sagheb-Tehrani, 2007). A broad vision of e-government should be shared by all citizens, i.e., encouraging stakeholders (citizens, officials, businesses, civil society groups, and others) to participate in determining the vision. A shared vision can lead to a more successful implementation of e-government, i.e., supporting e-government project from beginning to end (please see Table 2). Broad categories of goals that are commonly shared by citizens are as follows (Working Group, 2002):

- Improving the productivity of government,
- Improving services to citizens,
- Improving the quality of life for disadvantaged communities, and
- Improving the legal system and law enforcement.

Putting it differently, e-government would make government become closer to this vision: an institution of the citizens, run by citizens, owned by citizens and for the citizens. Fighting corruption should be included in the vision. This may be announced to the public as "anti-corruption" goal of e-government. More, a challenge for public sectors is to recognize today's trends and apply effective tools for creating and implementing policies that optimize the role of IT in their societies. These strategies are often novel, a result of the spread of relatively new technologies, such as the Internet, mobile devices, opens sources (e.g., Linux, games, information) viewing movies via Internet, and e-mailing. This means that government agencies will need to choose wisely the most appropriate strategies. Naturally, each government's vision should also be accompanied by a short list of priority areas for the e-government project. Improvements in the following areas are recommended:

- Employee productivity,
- Service delivery,
- Information security,
- IT infrastructure,
- Data management,
- IT management,
- Human resource management,
- Disaster recovery/management, and

Table 2. Comparison of various e-government websites based on key e-government concepts

Website URL	G2C	G2B	G2G	VISION	PUBLICATION	INTERACTION	TRANSPARENCY	ACCESSIBILTY	TRANSACTION
www.direct.gov.uk	A	A	NA	A	A	A	A	A	A
www.agimo.gov.au	NA	NA	NA	A	A	A	A	NA	NA
www.e-gov.com	A	A	NA	A	A	A	A	NA	A
w.developmentgateway.org	NA	NA	A	A	A	A	NA	NA	NA
www.usa.gov	A	A	NA	A	A	A	A	A	A
www.australia.gov.au	A	A	NA	A	A	A	A	A	A
www.canada.gc.ca	A	A	NA	A	A	A	A	A	A
www.ecitizen.gov.sg	A	A	NA	A	A	A	A	A	A
english.www.gov.tw/e-gov/index.jsp	NA	A	NA	NA	A	A	A	NA	NA
www.info.gov.hk	A	A	NA	A	A	A	A	A	A
www.unpan.org	NA	NA	A	A	A	A	A	NA	NA
www.gateway.gov.uk	A	A	NA	A	NA	A	A	NA	A

A = Available
NA = Not available

- Others.

Up to 2007, research not only confirms the historic role governments played in affecting the employment of IT, but more significantly that IT is considered to be a main component of a national economic development policy. Today, no advanced nation can ignore the role of IT strategies in its economy (Ghapanchi, Albadvi, & Zarei, 2008; Haigh & Griffiths, 2008; Cortada, Gupta, & Le Noir, 2007; Vintar, & Nograˇsek, 2010).

One strategy is the effective use of IT by government agencies themselves to improve their internal productivity and increase their ability to serve citizens (as in providing 24 x 7 services). By providing citizens convenient access to information around the clock over the Internet, a government encourages citizens to access public services, data, and application forms using PCs and the Internet. This action encourages citizens to utilize that technology in other ways in their private and professional lives. Another popular tactic is requiring suppliers to provide do their services using online procurement systems. In addition to helping lower the costs of acquiring these by a government, it makes suppliers start using the Internet and related technologies in an "e-business" environment (Abramson & Harris, 2003). In recent years, connectivity has been all about making it possible for individuals, firms, and other institutions to access mobile (wireless) networks and the Internet. Wireless Internet service should be available all over cities in any country. This should be one of the main strategies for any government. Costs of these services to individuals and organizations should be affordable for all citizens. The expansion of the Internet plays a vital role in the economic development activities of the public sector. This is a development that is attracting renewed attention to the topic of IT by economists and public policy experts (Breznitz, 2007; Baumol, Litan, & Schramm, 2007; Weer-

Table 3. Percentage of government sites offering online services

	2000	2001	2002	2003	2004	2005	2006	2007	2008
No Services	78%	75%	77%	56%	44%	27%	23%	14%	11%
One Service	16	15	12	15	18	11	16	15	12
Two Services	3	4	4	8	11	8	12	13	10
Three or More Services	2	6	7	21	27	54	49	58	67

Source: West (2008b, p. 4)

akkody, Dwlvedi, & Kurunananda, 2009). This reality is made more passionate by the fact that more people and firms dependent on IT to go about their job and private lives has enlarged over the past quarter century.

Publications

The ability to search for a specific website is a basic tool needed by citizens. In this regard, one significant new development has been the formation of online service portals. This service is an important advantage for ordinary citizens because it reduces the need to log on to various websites to order services or find information (please see Table 2). Citizens can connect in "one-stop" shopping, and locate what they require through a single site that integrates a range of government websites. One of the main dissatisfaction for citizens is going through enormous amounts of information to locate useful material (Haigh & Griffiths, 2008). Admission to updated publications, contact information, and databases are vital to citizen access to information and improve democratic responsibility. Another way that e-government websites can provide the available information to citizens is by personalizing the website or letting citizens register to receive update publications (Sagheb-Tehrani, 2007). This is known as "push technology" (Murru, 2003). All these services utilize IT to expand access to government information, so that citizens do not need to go to the government offices in person and wait in long lines. This is the leading frame of e-government. Naturally, knowledge is required on how to manage publications, how to present information clearly online and how users likely to use the information.

Interaction, Transparency, and Accessibility

A state should aim to have broadband connections for all public administrations. Broadband services can be offered on various technological platforms. Public Internet Access Points (PIAP), preferably with broadband connections, should be provided for all citizens in their communities. Internet is a perfect tool for obtaining public access to government information. Accessible and clear information can improve citizens' understanding and knowledge and may lead them to take part in the decision-making process, developing democracy. With the increase of the Internet, the value of well-designed e-government website will become even more obvious (please see Table 2). Making it easier for citizens to access public information will improve participation and democracy. Knowledge is required on how citizens or government officials look for information and like to receive it.

E-government sites should also consider disability access. World Wide Web Consortium (W3C) has introduced some standards regarding disability access. There has been some progress in this area on US government websites (see Figure 2). Further, e-government sites should provide foreign language accessibility as well. Public outreach is one of the most important character-

Figure 2. Percentage of state and federal sites meeting W3C disability accessibility

	2003	2004	2005	2006	2007	2008
Federal	47%	42%	44%	54%	54%	25%
State	33	37	40	43	46	19

Source: West (2008b, p. 5)

istics of any e-government. Put another way, one of the most promising benefits of e-government is its ability to draw citizens closer to their governments. In my examination of US Citizenship and Immigration (USCIS) websites, a visitor to the USCIS website cannot email or phone a person in any particular department.

Transaction

The goal of transaction is to provide government services online (please see Table 2). Government agencies can computerize particular processes and procedures, such as fine collection, tax collection, and credit card purchases. By providing these services online, government can attempt to restrict corruption and improve citizens' trust in government. Further, this can lead to increased productivity in both private and public sectors. Knowledge of efficiency and security is required for designing such a computerized system. In the study by West (2005a), it is mentioned that there are several novel services available on US state portals, such as live online help desk and state tourism sites featuring online planning for travelers. At the same time, the study mentioned some aspects of e-government privacy and security issues (see Table 4).

Research Method, Questions, Process, and Limitation

This study attempts to explain the concept of e-government by defining various vital perceptions and their relationships involved in embracing e-government. The research introduced here draws upon social system theory in the functionalist sociology defined by Burrell and Morgan (1979). The focus of social system theory is on the "holistic view," i.e., all parts of a system are related to each other. This chapter approaches its subject matter from an objectivist perspective. Objectivist is one of several doctrines holding that all reality is objective and external to the mind and that knowledge is reliably based on observed objects and events. Put differently, objectivism holds that reality exists independent of consciousness; that individual persons are in contact with this reality

Table 4. Assessment of e-government privacy and security statements

	2001	2002	2003	2004	2005	2006	2007	2008
Prohibit Commercial Marketing	12%	39%	32%	40%	64%	58%	64%	53%
Prohibit Cookies	10	6	10	16	21	16	32	40
Prohibit Sharing Personal Information	13	36	31	36	65	54	37	51
Share Information with Law Enforcement	--	35	35	39	62	49	50	49
Use Computer Software to Monitor Traffic	8	37	24	28	46	60	65	57

Source: West (2008b, p. 4)

through sensory insight; that human beings can gain objective knowledge from perception through the process of concept creation and deductive and inductive logic The conceptual model presented here is based on the "holistic view" school (Social System Theory). The methodology is based on a literature review and personal experiences as an IT consultant in numerous organizations. This study attempts to answer the following main research problem:

- What concepts are involved in implementing e-government?
- What are the steps towards implementing e-government?

Many public organizations are implementing Electronic Government (e-government) projects. Therefore, there is a need to put forward a conceptual model focusing on steps towards successful planning e-government projects. From the author's point of view, good research requires a sequence of well-defined steps planned in advance. The steps in this study include:

- Generate research idea
- Review literature
- Develop concepts
- Collect data via literature studies
- Develop conceptual model
- Publish result

Most of the research on e-government has utilized the Internet to examine government websites. The use of the Internet has been suggested by e-government scholars as a method to assess e-government development (Mofleh & Wanous, 2009). As with any research, this study has limitations. The data were presented in various tables extracted from the literature review and about twelve websites (appendix A) were examined (see Table 2). Therefore, one limitation would be the number of the articles and websites that were reviewed in this study. Furthermore, the questions in this study are based upon the author's understanding of the literature review.

CONCEPTUAL MODEL

Designing an efficient and successful e-government is very challenging and demanding process (Sagheb-Tehrani, 2007). Theory is important for researchers. Researchers who proceed without theory rarely conduct top-quality research. Concepts are the main building blocks of theory. A concept can be an idea expressed as a symbol or in words (Neuman, 2003). Thus, the conceptual model presented here may contribute to the theory of correlated fields. Figure 3 shows an e-government conceptual model with its relationships.

The conceptual model may suggest a number of propositions regarding the impacts of some concepts related to e-government. In this section, seven particular research proposals are stated in general terms. The aim is to suggest important issues that need to be investigated further. Deeper discussion of the research propositions may also reveal that potential efforts are often complex with both positive and negative connotations.

- **Proposition 1:** Knowledge management is required to shape the concepts of IT, vision, publication, interaction, and transaction.
- **Proposition 2:** The concept of IT is based upon other concepts such as: IT blue print, communication, IT priorities, and IT strategies.
- **Proposition 3:** The concept of vision is derived by concepts of citizen centered and vision priorities.
- **Proposition 4:** The concept of publication is formed by the concepts of update, personalizing, and push technology.
- **Proposition 5:** The concept of interaction is created by the concepts of PIAP, transparency, and accessibility.

Figure 3. E-government conceptual model

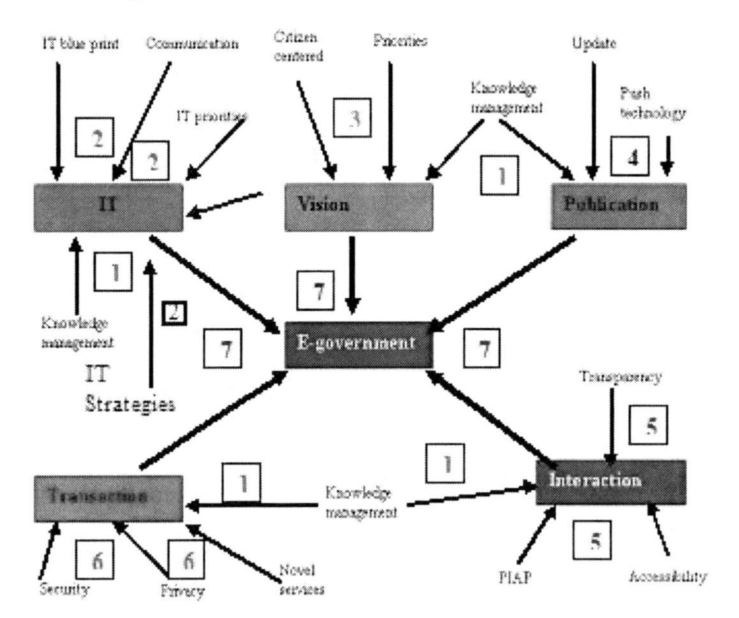

- **Proposition 6:** The concept of transaction is based upon the concepts of security, privacy, and novel services.
- **Proposition 7:** The concept of e-government is generated by the concepts of IT, vision, publication, interaction, and transaction.

CONCLUSION

All government or business operations require an effective management, as it is for e-government. One may say that there is no "one size fits all" IT strategy that works for all societies. To be able to deliver a project within a budget and on time and to coordinate effectively between all partners requires skillful management (Working Group, 2002). This study has introduced a conceptual model of e-government. The chapter has argued that several key success factors are appropriate for e-government implementation. About twelve e-government websites were examined upon those key success factors (please see Table 2). The conceptual model allows one to comprehend very broadly the concept of e-government. This helps to design more successful e-government projects. Further, this work may supply a basis for future research in the associated disciplines. One direction would be to use the conceptual model presented here in a case study. Moreover, the propositions launched here are meant to provide a starting point for supplementary research on this subject. Research in this theme should make a contribution to the knowledge of e-government development so that these projects can be implemented more effectively.

REFERENCES

Abramson, M. A., & Harris, R. S. (2003). *The procurement revolution*. Sterling, CT: Rowman and Littlefield.

Akther, M. S., Onishi, T., & Kidokoro, T. (2007). E-government in a developing country: Citizen-centric approach for success. *International Journal of Electronic Governance*, *1*(1), 38–51. doi:10.1504/IJEG.2007.014342

Baumol, W. J., Litan, R. E., & Schramm, C. J. (2007). *Good capitalism, bad capitalism and the economics of growth and prosperity*. New Haven, CT: Yale.

Breznitz, D. (2007). *Innovation and the state: Political choice and strategies for growth in Israel, Taiwan, and Ireland*. New Haven, CT: Yale.

Burrell, G., & Morgan, G. (1979). *Social paradigms and organization analysis*. New York, NY: Heinemann.

Cap Gemini & Ernest and Young. (2001). *Web-based survey on electronic public services*. Retrieved from http://europa.eu.int/information_society/eeurope/news_library/documents/bench_online_services.doc

Chircu, A. M., & Lee, D., & Hae-Dong. (2005). E-government: Key success factors for value discovery and realization. *Electronic Government, 2*(1), 11–24. doi:10.1504/EG.2005.006645

Cortada, J. W., Gupta, A. M., & Le Noir, M. (2007). *How the most advanced nations can remain competitive in the information age*. Palo Alto, CA: IBM.

Council of the European Union. (2002). *Evolution of e-government in the European Union*. Retrieved from http://www.map.es/csi/pdf/egovEngl_definitivo.pdf

Curtin, G. G., Sommer, M. H., & Vis-Sommer, V. (2003). *The world of e-government*. New York, NY: The Haworth Press.

Donna, E., & Yen, D. C. (2006). E-government: Evolving relationship of citizens and government domestic, and international development. *Government Information Quarterly, 23*(2), 207–235. doi:10.1016/j.giq.2005.11.004

Ghapanci, A., Albadavi, A., & Zarei, B. (2008). A framework for e-government planning and implementation. *Electronic Government: An International Journal, 5*(1), 71–90. doi:10.1504/EG.2008.016129

Gorla, N. (2008). Hurdles in rural e-government projects in India: Lessons for developing countries. *Electronic Government: An International Journal, 5*(1), 91–102. doi:10.1504/EG.2008.016130

Greiner, L. (2005). *State of the marketplace: E-government gateways*. Retrieved from http://www.faulkner.com/products/faulknerlibrary/00018297.htm

Haigh, N., & Griffiths, A. (2008). E-government and environmental sustainability: Results from three Australian cases. *Electronic Government: An International Journal, 5*(1), 45–62. doi:10.1504/EG.2008.016127

Hipp, L., & Warner, M. E. (2008). Market forces for the unemployed? Training vouchers in Germany and the USA. *Social Policy and Administration, 42*(1), 77–101.

Jupp, V. (2003). Realizing the vision of egovernment . In Curtin, C. G., Sommer, M. H., & Vis-Sommer, V. (Eds.), *The World of E-Government* (pp. 129–147). London, UK: Routledge.

Kottemann, J. E., & Boyer-Wright, K. M. (2010). Socioeconomic foundations enabling e-business and e-government. *Information Technology for Development, 16*(1), 4–15. doi:10.1002/itdj.20131

Lee, T.-R., Wu, H.-C., Lin, C.-J., & Wang, H.-T. (2008). Agricultural e-government in China, Korea, Taiwan, and the USA. *Electronic Government: An International Journal, 5*(1), 63–70. doi:10.1504/EG.2008.016128

McGinnis, P. (2003). Creating a blueprint for e-government . In Curtin, C. G., Sommer, M. H., & Vis-Sommer, V. (Eds.), *The World of E-Government* (pp. 51–63). London, UK: Routledge.

Mitra, A. (2005). Direction of electronic governance initiative within two worlds: Case for a shift in emphasis. *Electronic Government, 2*(1).

Mofleh, S. I., & Wanous, M. (2009). Reviewing existing methods for evaluating e-government websites. *Electronic Government: An International Journal*, *6*(2), 129–142. doi:10.1504/EG.2009.024438

Murra, M. E. (2003). *E-government: From real to virtual democracy*. Retrieved from http://unpan1.un.org/intradoc/groups/public/documents/other/unpan011094.pdf#search='egovernment%3Afrom%20Real%20to%20virtual%20democracy

Neuman, W. L. (2003). *Social research methods: Qualitative and quantitative approaches* (5th ed.). Boston, MA: Allyn & Bacon.

Palmer, I. (2003). *State of the world: E-government implementation*. Retrieved from http://www.faulkner.com/products/faulknerlibrary/00018297.htm

Raymond, L., Uwizeyemungu, S., & Bergeron, F. (2006). Motivations to implement ERP in e-government: An analysis from success stories. *Electronic Government: An International Journal*, *3*(3), 225–240. doi:10.1504/EG.2006.009597

Sagheb-Tehrani, M. (2007). Some steps towards implementing e-government. *Journal of ACM . Computers & Society*, *37*(1), 22–29. doi:10.1145/1273353.1273356

Smith, A. D. (2008). Business and e-government intelligence for strategically leveraging information retrieval. *Electronic Government: An International Journal*, *5*(1), 31–44. doi:10.1504/EG.2008.016126

UNCTAD. (2002). *Reports on e-commerce and development of the united nations conference on trade and development*. New York, NY: United Nations.

Vietor, R. H. K. (2007). *How countries compete: Strategy, structure, and government in the global economy*. Boston, MA: Harvard Business School.

Vintar, M., & Nograˇsek, J. (2010). How much can we trust different e-government surveys? The case of Slovenia. *Information Polity*, *15*, 199–213.

Weerakkody, V., Dwlvedi, Y., & Kurunananda, A. (2009). Implementing e-government in Sri Lanka: Lessons from the UK. *Information Technology for Development*, *15*(3), 171–192. doi:10.1002/itdj.20122

West, D. (2008a). Improving technology utilization in electronic government around the world. Retrieved from http://www.brookings.edu/reports/2008/0817_egovernment_west.aspx

West, D. (2008b). *State and federal e-government in the United States*. Retrieved from http://www.brookings.edu/reports/2008/0826_egovernment_west.aspx

Working Group. (2002). *Roadmap for e-government in the developing world*. Los Angeles, CA: Pacific Council on International Policy.

World Bank. (2007). Country brief: Russian federation, economy. Retrieved from http://web.worldbank.org

APPENDIX A

List of some related links to electronic governments

- Australian Government Information Management Office: http://www.agimo.gov.au/
- Directgov: http://www.direct.gov.uk/
- E-Gov.com: http://www.e-gov.com/
- E-Government Development Gateway: http://www.developmentgateway.org/
- FirstGov (USA): http://www.usa.gov, http://www.firstgov.gov/
- Government of Australia Portal: http://www.australia.gov.au/
- Government of Canada Portal: http://www.canada.gc.ca/
- Singapore eCitizen Portal: http://www.ecitizen.gov.sg/
- Taiwan e-Government: http://english.www.gov.tw/e-Gov/index.jsp
- Hong Kong e-Government: http://www.info.gov.hk
- United Nations Online Network in Public Administration and Finance: http://www.unpan.org
- UK Government Gateway: http://www.gateway.gov.uk/

Chapter 3
Understanding E-Governance: A Theoretical Approach

Muhammad Muinul Islam
Jahangirnagar University, Bangladesh

Mohammad Ehsan
University of Dhaka, Bangladesh & Dalhousie University, Canada

ABSTRACT

Another new paradigm shift is in the offing and slowly becoming distinct from the amorphous sphere of public administration. It is the ICT-blessed governance, or e-Governance. The adoption of ICTs and the new approach to management in symbiosis are e-Governance. E-governance speaks of a new way and style in every beat and pulse of the system of public administration. It brings about changes in the structure and functions of public services, ushering transformation through effectively engaging the government, businesses, and citizens—all stakeholders. It not only ensures efficiency in public service delivery but also offers unlimited potential to combat corruption and many other bureau-pathologies in the public administration system. Based on secondary sources, this chapter offers brief theoretical discussions of e-governance, including its emergence, types of service delivery, transformation stages, and relevant other issues.

INTRODUCTION

A system of public administration is all-pervasive and has been ubiquitous since times immemorial. Today, what we understand as the public administration existed even before the birth of modern states. The nature, functions, and mode of public service delivery, however, have gone through radical changes from those earlier times. This chapter offers an extension of public administration paradigms proposed and postulated by Henry (1975) and Golembiewski (1977). It also deals with the basic theoretical backgrounds of e-governance, its types of operation and transformation phases.

The birth of public administration as a separate field of study is marked by Woodrow Wilson's seminal publication from 1887. Since then to the late 1970s, Henry identified five paradigms of public administration. These are: paradigm one: the politics/administration Dichotomy 1900-1926;

DOI: 10.4018/978-1-4666-1909-8.ch003

paradigm two: the principles of administration 1927-1937; paradigm three: public administration as political science 1950-1970; paradigm four: public administration as management 1956-1970; and paradigm five: public administration as public administration, 1970-? Henry did not mention the exact end year of the fifth paradigm, as it was still dominant in the intellectual discourse of the discipline of Public Administration. However, we can cautiously suggest the finishing line of the fifth paradigm to be the early 90s. Since 2001, a new idea[1] slowly permeated in the theories and practice of public administration forcing a paradigm shift in the discipline. The sixth paradigm, as an extended version of Henry's paradigm, can thus be called as "*Public Administration as New Public Management (NPM), 1991-?*" However, as others suggested, we can also think of another paradigm shift in the discourses of public administration that concurrently exists with the sixth one and is very likely to be a dominant one for years to come. This seventh paradigm can be called "*Public Administration as E-governance*[2] *1995-?*" By now, there is widespread agreement among the academicians and practitioners of the role that Information and Communication Technology (ICT) plays in the day to day operations of public administration. ICTs dramatically revolutionized the structure, processes, and radically transformed the way public administration systems work around us (Roy, 2011). E-governance has been ubiquitously adopted and adapted, to various degrees, by the governments of developed, transitional, and developing countries[3]. The glaring transformation that public administration so far has gone through with e-governance has been amazing.

In fact, over the past three decades, information and communication technology has brought changes in the operations of government organizations both in developed and developing countries. This is due to the process of converting information from analog to digital forms. The lifeblood of government is information and the digital revolution has allowed government organizations to store, analyze, and retrieve information more effectively and efficiently. This process has also been strongly affected by changes in telecommunications technology and the convergence of computer and communication technologies. The most recent manifestation of this process of technological change is the advent of Internet or World Wide Web (Bretschneider, 2003).

This reform of government administration and the provision of improved services to citizens have long been acknowledged as a major criterion for development and today's drive towards e-governance in many parts of the world can be considered as a part of this wider developmental goal (Madon, 2004). Throughout the world, governments, businesses and NGOs are working together to adopt e-governance—from Singapore to South Africa, Andhra Pradesh to Washington, or Bangladesh to Malaysia. These are not just experiments in new modes of service delivery. E-governance inevitably also embraces— and is driven by—new models of policy formulation, new forms of citizenship, new patterns of relationship and power, new options for economic development, and the search for new ways to connect people with the political process. As indicated, the rapid adoption of e-governance is facilitated by dynamic technological and telecommunication innovations. In many countries, ICTs are seen as a catalyst for e-governance. "After e-commerce and e-business," as *The Economist* (2000) predicted long ago, the next Internet revolution would be e-government." It is naturally expected that e-governance will ensure transparency, speedy information dissemination, and improved service in public administration. In the era of informed citizens, e-governance is also seen as a vehicle for cost-effective and efficient way of public service delivery (Agarwal, et al., 2003). Furthermore, e-governance is expected to empower the citizens, increase the profit margin of the businesses and enterprises, enhance flexibility in government service delivery, force data digitization, and also strengthen the anti-corruption movement (Bhogle, 2008).

Governments around the world with a view to transform their governance structure and processes from traditional to electronic form are putting critical information online, automating once cumbersome processes, and interacting electronically with their citizens. This enthusiasm comes in part from a belief that technology can transform government's often-cited negative image. In many places, citizens view their governments as bloated, wasteful, and unresponsive to their most pressing needs. Mistrust of government is rife among the public and businesses. Civil servants are often seen as profiteers. The spread of information and communication technology brings hope that governments can transform. Indeed, forward-looking public officials everywhere are using technology to improve their governments (PCIP, 2002).

Governments, over the years, have taken multifarious decisions on how best to accommodate ICTs to transform public administration for enhancing the wellbeing of citizens. However, e-government and e-governance mean more than ICTs. Traditional pathways to better governance and government have been transformed through integrating innovation in ICTs, innovation in organizational and business practices, and changes in people's skills and expectations. E-governance is often embraced as facilitating changes in citizen involvement, transforming traditional hierarchical approaches to coordinated network approaches, and breaking down barriers—among the government departments, among layers of government (national, regional, local, etc.), and between the public and private sectors—to enhance a citizen/customer-focused approach to service delivery. This can bring about improvements in the accountability, transparency, and openness of public institutions. However, e-governance also presents new challenges to citizens' trust in governments. Security of digital transactions and communications also needs to be assured; privacy needs to be protected, and citizens need to be able to control their personal data and must be ensured that only selective government officials would have access to it (Dugdale, 2004).

This chapter, which is primarily based on secondary sources, endeavors to shed some lights on the theoretical issues of e-governance. Our effort here remains brief, as many authors, by now, have done excellent theoretical discussions on the issue (Annttiroiko, 2008). In so doing, the chapter at the outset presents a brief overview of the notion of e-governance in the intellectual discourse, and how it is conceptualized by academician, practitioners, and also by different international agencies. Later on, it focuses on the ways e-governance is transforming the public administration system.

UNDERSTANDING E-GOVERNANCE: SOME THEORETICAL ISSUES

From Governance to e-Governance

The word 'e-governance' can be viewed at the crossroad of two major shifts—governance and information revolution. The issue of 'governance' has been around for decades. According to concise Oxford Dictionary, the word 'governance' has been developed from a Greek word *'kuberna,'* which means to steer. The first classic political science essay on the subject discuss about the concept of 'governability,' which made *rule of law* as the core to development (Johnson, 1997).

Clearly, 'governance' gets into development discourse around the period of late 1980s. Human Development Report (1991) accepts the fact that freedom and democracy, though not a necessary condition, are entirely consistent with growth and development. 'International development' therefore shifted its focus from *'economic growth'* of the 1950s (UN Development decades) to *'sustainable human development'* that includes concerns for people and nature to be widely accepted by state, market and civil society. The environmental movement considered 'governance' as an urgent issue to deal the development agendas in a holistic manner: to include not only the sector at hand and the obvious stakeholders, but also others affected by them in other areas. It has forced a redefinition of the public interest with nature itself as a recognized stakeholder (Carino).

Intense globalization of market and trade, after the collapse of Soviet Union in 1989, left most countries but to join the World Trade Organization (WTO), which gave concern about the markets being opened to cheaper products and labor forces. This resulted in creation of safety nets and bureaucracies to be involved with regulatory governance mechanism. However, at the same time because of trade globalization and networked economy, the countries cannot avoid creating a level playing field with transparent governance mode (Carino). This openness and predictability of government functioning is further echoed at e-governance.

Globalization with the emergence of new information and communication technologies had a profound impact in the development of the notion of e-governance. ICT replaced two basic elements of productions—'labor' and 'capital' by 'information' and 'knowledge' for the first time in the last two centuries. Internet created the same break-through as the printing press did in the 15th century. It shapes the ability to communicate, share, distribute, exchange, formalize, use and network information at a speed that is not experienced before. Moore's law[4] pointed out that, the processing power of microchips is doubling every 18 months with a trend of 20-30% decline in quality adjustment prices for computers. This means computers are getting cheaper, powerful, and ubiquitous, making the network and automation of services viable to government. Political activism on the other hand, is also using the space with increased number of public interest groups, community, or voluntary organizations propagating their demands and activities in the electronic network.

Over the years, the Weberian[5] principles of bureaucratic governance are being replaced with the trends of horizontal, leaner, dynamic, and networked governance. Administrative reform and development have experienced TQM[6] in 1980s, and 're-engineering and re-invention' in the 1990s. E-governance reflects this process of re-invention and re-engineering in governance and *'is aimed at adapting administration to the further increasing flow of information: accelerating the process of decision making by optimizing resources, and making the mechanism for decision making self-regulating'* (Baev, 2003). This led 'Governance' be defined independently from the 'the act of government' to the practice of getting the consent and cooperation of the governed. The concrete objective of e-governance is to support and simplify governance for all parties—government, citizens, and businesses through online services and other electronic means. In other words, e-governance uses electronic means to support and stimulate good governance (Backus, 2001).

However, the core ideas and techniques associated with "putting government online" first emerged in the most technologically advanced Western countries, especially those whose populations were pioneers in the adoption of the Internet in the 1990s. In the United States, Bill Clinton's administration's objective of "reinventing government" closely followed the managerial path, and the Bush administration had remained on the same track, with an even greater emphasis on cost reduction through efficiency gains. The current administration is no exception to this. In the British case, the managerial use of ICTs emerged as a strong theme in the labour administration's obsession with "joined-up government"—a phrase that has recently crossed the Atlantic to the U.S. At the level of the European Union, despite greater recognition of the democratic potential of new ICTs, most discussion has centered on issues of efficiency and "service delivery." It was the United States and Britain (along with other countries, notably Canada and Australia) that led the way, both in establishing a basic informational form of Web presence in the mid-1990s and in developing what became known as "e-government" in the late 1990s (Chadwick & May, 2003).

As e-governance supports and facilitates good governance for all stakeholders, we need to understand that e-governance is not just about a website or not merely a digitization of service delivery. It certainly stands on a greater definition of engagement and depth of relationship that surrounds both the citizens and the government (Fang, 2002).

From among the popular definitions of e-governance, the US 2002 E-Government Act defines it as "the use by the Government of Web-based Internet applications and other information technologies, combined with processes that implement these technologies, to a) enhance the access to and delivery of Government information and services to the public, other agencies, and other Government entities, or b) bring about improvements in Government operations that may include effectiveness, efficiency, service quality, or transformation" (US Congress, 2002). OECD defines e-governance as "the use of ICTs, and particularly the Internet, as a tool to achieve better government" (OECD, 2003, p. 23). The e-governance efforts by the European Union are based on the following definition: "e-government is the use of Information and Communication Technologies in public administrations combined with organizational change and new skills in order to improve public services and democratic processes" (EU, 2004).

There is a difference of meaning between e-governance and e-government. Governance is the manner or the process to guide a society to best achieve its goals and interests, while government is the institution or the apparatus to perform that job. This means government is one (of the many) institutions of governance. Interestingly, international bodies define e-governance as per their focuses to frame governance in general. For example, World Bank's concern on governance is exclusively related to the contribution they make to social and economic development by economic and structural liberalization. Therefore to them, e-governance implies the use of ICT channels to change the way citizens and business interact with government to enable citizen's involvement in decision making, increased access to information, more transparency and civil society strengthening (Deane, 2003). Grönlund provides rather a vivid distinction of e-government and e-governance. To him, e-government refers to what is happening within government organizations while e-governance, on the other hand, refers to the whole system involved in managing a society. The system includes activities not only by government organizations but also companies and voluntary organizations, as well as citizens. Moreover, it features the processes and flows of governance, dimensions that are critical to understanding the context of information systems deployment and use (Atkinson, 2003).

Most North American and European nations reach interactive stage of e-governance maturity and it poses increased reflexivity in relationship between administration and citizenry. Being e-government is not the major concern now for these developed nations; rather an effective and timely service to citizen through e-governance is probably one of major goals of these countries. In the words of Marche and McNiven, "it is not just a question of e-government, it is also a question of e-governance." According to them, "e-governance is the provision of routine government information and transactions using electronic means, most notably those using Internet technologies, whether delivered at home, at work, or through public kiosks. E-governance, on the other hand, is a technology-mediated relationship between citizens and their governments from the perspective of potential electronic deliberation over civic communication, over policy evolution, and in democratic expressions of citizen will" (Marche & McNiven, 2003, p. 75).

Bhatnagar observes that the World Bank refers to e-governance as the use of information technologies (such as Wide Area Networks, the Internet, and mobile computing) by the government agencies. In other words, e-governance may be defined as the delivery of government services and information to the public by using electronic means. These technologies enable the government to transform its relations with its other wings, citizens, businesses. Such an exercise leads to better delivery of government services to citizens, improved interactions with business and industry, citizen empowerment through access to information and a more efficient government management. The resulting benefits can be lesser corruption, increased transparency, greater convenience, revenue growth, and/or cost reductions (Bhatnagar, 2001).

UNDP (2003) relates the concept of governance to that of sustainable human development. It views e-governance as a process of "creating public value with the use of modern ICT" where public value is defined as a notion "rooted in people's preferences." Therefore, e-government is justified if it enhances the capacity of public administration to increase the supply of public value—the outcome of a high quality of life. Focusing more on the 'governance' possibilities, UNDP is of the view that e-governance can "equip people for genuine participation in an inclusive political process that can produce well-informed public-consent, the ever more prevalent basis for the legitimacy of governments." The UN's five guiding principles on e-government objectives are: (1) building services around citizens choices; (2) making government and its services more accessible; (3) social inclusion; (4) providing information responsibly; and (5) using IT and human resources effectively and efficiently (UN Survey, 2002). The Public Administration (PUMA) Group of the Organization for Economic Cooperation and Development (OECD) focuses on three main components of online and participatory e-governance: "information, active participation, and consultation" (Riley, 2003).

The Government of India took the basis of SMART for its vision statement on e-governance. This relates to "application of IT to the process of government functioning to bring out *Simple, Moral, Accountable, Responsive,* and *Transparent governance* (SMART)." This vision helped India outlining further objectives and strategic initiatives on e-governance. Rogers W'O Okot-Uma of Commonwealth Secretariat in London thinks that e-governance seeks to realize processes and structures for harnessing the potentialities of information and communication technologies at various levels of government and the public sector and beyond for the purpose of enhancing good governance.

E-government encompasses a broad spectrum of activities involving improving government operations and services as well as enabling a more cooperative and meaningful relationship with citizens and other non-state actors. E-government initiatives, however, have predominantly focused on changing government operations, structures, and services rather than redefining a new role and set of responsibilities for citizens." Heeks observes that e-governance should be seen to encompass all ICTs, but the key innovation is computer networks—from intranets to the Internet—creating a wide range of new digital connections:

- **Connections within government:** permitting 'joined-up thinking;'
- **Connections between government and NGOs/citizens:** strengthening accountability;
- **Connections between government and business/citizens:** transforming service delivery;
- **Connections within and between NGOs:** supporting learning and concerted action;
- **Connections within and between communities:** building social and economic development.

The novelty of using new technologies in governance is that it expands beyond internal government operations to include electronic service delivery to the public and the subsequent interaction between the citizen and the government. This potential for interactivity can be identified as one of the most important elements in the way e-governance will change the nature of government (Heeks, 2001).

Brown (2005) proposes a comprehensive definition of e-governance. To him, "a broader view of e-government is that it relates to the entire range of government roles and activities, shaped by and making use of information and communications technologies. A high-level statement of this view is 'knowledge-based government in the knowledge-based economy and society.' More concretely, e-government brings together two elements that have not been naturally joined in the past. One is the environment, within government and in the society at large, created by the use of electronic technologies such as computing, e-mail, the World Wide Web, wireless and other ICTs, combined with management models such as client/citizen centricity and single-window convergence. The other is the basic model of the state and of public administration within that, linking the dynamics of democracy, governance, and public management."

Among the four different dimensions of e-government that are based on the functions of government, e-governance is one of the prominent, the others being e-administration, e-services, and e-democracy (Annttiroiko, 2008). E-governance in the context of public sector is about managing and steering multi-sectoral stakeholder relations on a non-hierarchical basis with the help of ICTs for the purpose of taking care of the policy, service, and development functions of government. In practical terms, it is about cooperation, networking, and partnership relations between public organizations, corporations, NGOs, civic groups, and active citizens, utilized by public organizations to gather and coordinate effectively both local and external resources to achieve public policy goals (Gronlund, 2007; Finger & Langenberg, 2007; cited in Annttiroiko, 2008). Despite variations in definition, it is true that e-governance has the subversive potential to ensure that citizens are no longer passive consumers of services offered to them as it also allows them to play a more proactive role in deciding the kind of services they want and the structure through which this service can best be provided.

Types of Service Delivery through e-Governance

The quest to improve service delivery through the use of ICTs in governments typically focuses on four main dimensions. These are:

1. **G2C (Government-to-Citizens):** This focuses primarily on developing user-friendly one-stop centers of service for easy access to high quality government services and information. Citizens-to-Government (C2G) has also emerged as an important stakeholder relationship.
2. **G2B (Government-to-Business):** This aims to facilitate and enhance the capability of business transactions between the government and the private sector by improving communications and connectivity between the two parties. Business-to-Government (B2G) has also emerged as an important stakeholder relationship.
3. **G2G (Government-to-Government):** This is an inter-governmental effort that aims to improve communication and effectiveness of services between federal, state, and local governments in the running of day-to-day administration.
4. **Intra-Government:** This aims to leverage ICT to reduce costs and improve the quality of administration and management within government organization (Karim, 2003, p. 192).

The Stages of e-Government Transformation

Gartner (2000) suggests that e-governance matures following four stages in their E-governance Maturity Model (Baum & Maio, 2000). As an addition, we propose 'institutionalization' as the fifth stage of e-governance maturity:

1. Information → Presence
2. Interaction → Intake processes
3. Transaction → Complete transactions
4. Transformation → Integration & change
5. Institutionalization → Interactive democracy and public outreach

In the *first stage,* e-governance means being present on the Web, providing the public (G2C and G2B) with relevant information. The format of the early government websites is similar to that of a brochure or leaflet. The value to the public is that government information is publicly accessible; processes are described and is more transparent, which improves democracy and service. Internally (G2G), the government can also disseminate static information with electronic means, such as the Internet.

In the *second stage,* the interaction between government and the public (G2C and G2B) is stimulated with various applications. People can ask questions via e-mail, use search engines, and download forms and documents. These ultimately save time. In fact, the complete intake of (simple) applications can be done online 24 hours per day. Normally this would only have been possible at a counter during opening hours. Internally (G2G), government organizations use LANs, intranets and e-mail to communicate and exchange data.

With stage *three,* the complexity of the technology is increasing, but customer (G2C and G2B) value is also higher. Complete transactions can be done without going to an office. Examples of online services are filing income tax, filing property tax, extending/renewal of licenses, visa, and passports, and online voting. State three is made complex because of security and personalization issues. For example, digital (electronic) signatures will be necessary to enable legal transfer of services. On the business side, the government is starting with e-procurement applications. In this stage, internal (G2G) processes have to be redesigned to provide good service. Government needs new laws and legislation to enable paperless transactions.

The *fourth stage* is to reach a stage where all information systems are integrated and the public can get G2C and G2B services at one virtual counter. One single point of contact for all services is the ultimate goal. The complex aspect in reaching this goal is mainly on the internal side, e.g., the necessity to drastically change culture, processes, and responsibilities within the government institution (G2G). Government employees in different departments have to work together in a smooth and seamless way. In this stage, cost savings, efficiency, and customer satisfaction reach highest possible levels.

The fifth stage is interactive democracy with public outreach and a range of accountability measures. Here, government websites move beyond a service-delivery model to system-wide political transformation. In addition to having integrated and fully executable online services, government sites offer option for website personalization. Through these and other kinds of advanced features, visitors can personalize websites, provide feedback, make comments, and avail themselves of a host of sophisticated features designed to boost democratic responsiveness and leadership accountability (West, 2004).

BENEFITS AND POTENTIALS: E-GOVERNANCE AS A CATALYST FOR DEVELOPMENT

The major task of public sector reform today is based on introducing and implementing e-governance projects. Innovative and sustainable application of Internet and ICTs bring good governance among the government, private sector, and citizens.

Application of Internet technologies and ICTs in public management helps achieve development goals of the nations. There are great deals of empirical evidences that suggest that e-governance contributed enormously to development goals. They can do so at both micro and national levels by increasing the effectiveness and reach of development interventions, enhancing good governance, and lowering the costs of service delivery. A project study in Italy comments that "e" means efficiency, effectiveness, empowerment, and economic and social development (UNDESA, 2002).

Among different benefits of e-governance are reduced transaction costs to better capacity to target groups, increased coverage, and quality of service delivery, enhanced response capacity to address issues of poverty and increase in revenue. Besides, e-governance provides increased accountability and transparency, which may greatly reduce the risk of corruption and raise the perception of good government among citizens. Citizens' trust in their government may have an impact on their willingness to invest and to pay taxes and levies for services.

Other benefits include employment creation in the third sector, improvements in the education and health system, providing better government services, and developing increased capacity for the provision of safety and security.

E-governance can offer to reduce functional insularity in public administration often referred to as "silos" or "stove-piping." This means the tendency on the part of bureaucracy not to integrate service provisioning across government departments when responding to citizen's needs. E-governance in this regard provides easy access and opportunities that permit cross-organizational services through internet and other communication technologies. It not only improves services and offer convenience to the users, but is also cost-effective (Marche & McNiven, 2003, p. 75).

RISKS AND CHALLENGES: NEED FOR A CAUTIOUS STEP

One of the major challenges of e-governance is digital divide within the country as well as among different countries of the world. A study of Chatfield and Alhujran (2009, p. 151) confirm a wide digital divide that remains between the Arab countries and the leading developed countries.

Heeks (2003) made an extensive research on e-governance projects. In his analysis, most e-government projects, both in industrialized and developing countries, fail either totally or partially. There are very little data about rates of success and failure of e-government projects, but some baseline estimates indicate that behind the high-tech glamour of these projects lies a dirty reality—the majority of projects are failures. The reasons for such failures are, as Heeks observed, lack of "e-readiness," and the oversize gaps between project design and on-the-ground reality (known as "design-reality gaps"), meaning the lack of assessment of needs prior to the implementation of a project.

CONCLUSION

As evident, e-governance facilitates development and offers many benefits to the citizens. It has the potential, which made governments around the world to initiate innovative changes in the delivery of public services. The issues of poverty reduction, economic underdevelopment, illiteracy, pervasive corruption can be minimized, if not completely eliminated, through the skillful application of e-governance initiatives. Despite its enormous potential, it is also true that the benefits of e-governance are not duly reaped by the governments, both in developed and developing countries. The main stumbling blocks in the way are basically the political leadership and bureaucratic inertia. Another major concern for global equitable access of e-governance is the 'digital divide'[7] often called as "information black hole." As indicated, e-governance can very positively direct a paradigm shift, from traditional bureaucratic administration to a more responsive, accountable and effective public administration, that many governments around the world are aspiring for a long time.

REFERENCES

Accenture. (2002). *E-government leadership – Realizing the vision*. Retrieved from http://www.accenture.com/xd/xd.asp?it=enWeb&xd=industries%5Cgovernment%5Cgove_welcome.xml

Agarwal, V., Mittal, M., & Rastogi, L. (2003). *Enabling e-governance – Integrated citizen relationship management framework – The Indian perspective*. Retrieved from http://www.e11online.com/pdf/e11_whitepaper2.pdf

Anderson, K. (1999). Reengineering public sector organizations using information technology . In Heeks, R. (Ed.), *Reinventing Government in the Information Age*. London, UK: Routledge.

Annttiroika, A. (2008), Introductory Chapter: A Brief Introduction to the Field of E-Government (ed.) *Electronic Government: Concepts, Methodologies, Tools, and Applications*, Hershey: PA, IGI Global

Atkinson, R. (2003). *Network government for the digital age*. Washington, DC: Progressive Policy Institute.

Backus, M. (2001). *E-governance and developing countries: Introduction and examples*. Research Report, No. 3. The Hague, The Netherlands: International Institute for Communication and Development. Retrieved from editor.iicd.org/files/report3.doc

Baev, V. (2005). Social and philosophical aspects of e-governance paradigm formation for public administration. In *Razon Y Palabra, 42*.

Baum, C., & Maio, D. (2000). *Gartner's four phases of e-government model*. Washington, DC: Gartner's Group.

Bhatnagar, S. C. (1999). *E-government: Opportunities and challenges*. Ahmadabad, India: Indian Institute of Management.

Bhatnagar, S. C. (2001). *Philippine customs reform*. Washington, DC: World Bank.

Bhogle, S. (2008). E-governance . In Anttiroiko, A.-V. (Ed.), *Electronic Government: Concepts, Methodologies, Tools and Applications*. Hershey, PA: IGI Global.

Bretschneider, S. (2003). Information technology, e-government, and institutional change. *Public Administration Review, 63*(6). doi:10.1111/1540-6210.00337

Brown, D. (1999). Information systems for improved performance management: Development approaches in U.S. public agencies . In Heeks, R. (Ed.), *Reinventing Government in the Information Age*. London, UK: Routledge.

Brown, D. (2005). Electronic government and public administration. *International Review of Administrative Sciences, 71*(2), 241–254. doi:10.1177/0020852305053883

Cariño, L. V. (2006). From traditional public administration to governance: Research in NCPAG, 1952-2002. *Philippine Journal of Public Administration, 50*(1-4), 1–24.

Chadwick, A., & May, C. (2003). Interaction between states and citizens in the age of internet: E-government' in the United States, Britain and European Union. *Governance: An International Journal of Policy, Administration and Institutions, 16*(2). doi:10.1111/1468-0491.00216

Deane, A. (2003). *Increasing voice and transparency using ICT tools: E-government, e-governance*. Washington, DC: World Bank.

Dugdale, A. (2004). *E-governance: Democracy in transition*. Paper presented at the Annual Meeting of the International Institute of Administrative Sciences. Washington, DC.

Economist. (2000, June 22). A survey of government and the internet: The next revolution. *The Economist.* Retrieved from http://www.economist.com/node/80746

Economist. (2000, June 24). The next revolution – A survey of government and the internet. *The Economist,* 3.

Ehsan, M. (2004). Origin, ideas and practice of new public management: Lessons for developing countries. *Administrative Change, 31*(2), 69–82.

EU. (2004). *eGovernment research in Europe*. Retrieved from http://europa.eu.int/information_society/programmes/egov_rd/text_en.htm

Fang, Z. (2002). *E-government in digital era: Concept, practice, and development.* Bangkok, Thailand: School of Public Administration, National Institute of Development Administration (NIDA).

Finger, M., & Langenberg, T. (2007). Electronic governance . In Anttiroiko, A. V., & Malkia, M. (Eds.), *Encyclopedia of Digital Government* (*Vol. 2*). Hershey, PA: IGI Global.

Fountain, J. (2001). *Building the virtual state: Information technology and institutional change*. Washington, DC: Brookings Institution.

Gotembiewsky, R. T. (1977). *Public administration as a developing discipline, part 1: Perspectives on past and present*. New York, NY: Marcel Dekker.

Gronlund, A. (2007). Electronic government . In Anttiroiko, A. V., & Malkia, M. (Eds.), *Encyclopedia of Digital Government* (*Vol. 2*). Hershey, PA: IGI Global.

Heeks, R. (Ed.). (2001). *Reinventing government in the information age*. London, UK: Routledge.

Heeks, R. (2003). *Most egovernment-for-development projects fail: How can risks be reduced?* iGovernment Working Paper Series, No 14. Manchester, UK: IDPM, University of Manchester.

Henry, N. (1995). *Public administration and public affairs*. New York, NY: Prentice-Hall, Inc.

Johnson, I. (1997). *Redefining the concept of governance*. Gatineau, Canada: Political and Social Policies Division, Policy Branch, Canadian International Development Agency (CIDA).

Karim, M. R. A. (2003). Technology and improved service delivery: Learning points from the Malyasian experience. *International Review of Administrative Sciences, 69*.

Madon, S. (2004). Evaluating the developmental impact of e-governance initiatives: An exploratory framework. *The Electronic Journal on Information Systems in Developing Countries, 20*(5).

Marche, S., & McNiven, J. D. (2003). E-government and e-governance: The future isn't what it used to be. *Canadian Journal of Administrative Sciences, 20*(1), 74. doi:10.1111/j.1936-4490.2003.tb00306.x

Moon, M. J., & Bretschneider, S. (1997). Can state government actions affect innovation and its diffusion? An extended communication model and empirical test. *Technological Forecasting and Social Change, 54*(1), 57–77. doi:10.1016/S0040-1625(96)00121-7

Nye, J. Jr. (1999). Information technology and democratic governance . In Karmarck, E. C., & Nye, J. Jr., (Eds.), *Democracy.com? Governance in Networked World*. Hollis, NH: Hollis Publishing Company.

OECD. (2001). *Understanding the digital divide*. Paris, France: OECD Publications. Retrieved from http://www.oecd.org/dataoecd/38/57/1888451.pdf

OECD. (2003). *The e-government imperative*. Paris, France: OECD e-Government Studies.

PCIP. (2002). *Roadmap for e-government in the developing world*. Retrieved from http://www.pacificcouncil.org/pdfs/e-gov.paper.f.pdf

Riley, T. B., & Riley, C. G. (2003). E-governance to e-democracy - Examining the evolution . In *International Tracking Survey Report 2003*. Ottawa, Canada: Riley Information Services.

Roy, J. (2011). The promise (and pitfalls) of digital transformation . In Leone, R. P., & Ohemeng, F. L. K. (Eds.), *Approaching Public Administration: Core Debates and Emerging Issues*. Toronto, Canada: Edmond Montgomery Publications.

UNDESA. (2002). *Plan of action - E-government for development*. Rome, Italy: Government of Italy, Ministry for Innovation and Technologies. Retrieved from http://www.palermoconference2002.org

US Congress. (2002). *US 2002 e-government act*. Washington, DC: US Congress Printing Office.

West, D. M. (2004). E-government and the transformation of service delivery and citizen attitudes. *Public Administration Review*, *64*(1). doi:10.1111/j.1540-6210.2004.00343.x

ENDNOTES

1. The concept of 'New Public Management (NPM)' is coined by Christopher Hood in 2001 in his famous article "A public management for all seasons?" Apart from him, other scholars like Pollitt, Lan, and Rosenbloom and Osborne and Gaebler have also worked on the same idea but with different names in the early period of the 1990s.
2. E-governance as a paradigm shift in the discipline of public administration can be benchmarked in the mid-1990s with the introduction of Internet and Web-presence by the government.
3. The dramatic changes occurring due to the technological revolution are in the domains of politics (Nye, 1999), government institutions (Fountain, 2001), performance management (Brown, 1999), red tape reduction (Moon & Bretschneider, 2002), and re-engineering (Anderson, 1999) in the administration.
4. A founder of Intel Corporation, Gordon Moore, made his famous observation in 1965, just four years after the first planar integrated circuit was discovered. The press called it "Moore's Law." More information can be found at: http://www.intel.com/research/silicon/mooreslaw.htm
5. Max Weber has given the ideal typical model of bureaucracy. The Weberian model categorically focuses on two dimensions: (1) the structural, relating to the hierarchical arrangement of positions, legal rational basis of authority, with system of compensation, and (2) the behavioral, relating to the merit-based selections of officials with the emphasis on training.
6. For more detail on TQM, please visit http://www.google.com/search?hl=en&lr=&oi=defmore&q=define:TQM
7. Digital divide refers to the gap between individuals, households, businesses, and geographic areas at different socio-economic levels with regard to both the opportunities to access information and communication technologies their use of the Internet for a wide variety of activities (OECD).

Chapter 4
E-Government, M-Government, L-Government:
Exploring Future ICT Applications in Public Administration

Alberto Asquer
University of Cagliari, Italy

ABSTRACT

The development of Information and Communication Technologies (ICT) brings about considerable changes in the ways public administration provides information and delivers services to citizens, businesses, and other public administration systems. This chapter reviews the application of ICT in the provision of public administration services. e-Government tools have been introduced in various countries in the world and enabled the strengthening of existing public administration services and the activation of innovative ones. m-Government tools, which are related to the emergence and diffusion of Internet mobile technology and devices, allow both overcoming infrastructure deficits and providing innovative services, which are particularly sensitive to users' context conditions. Finally, l-Government tools—i.e., ubiquitous, seamless, user-centric, and automated application of Internet technology to public administration services—have the potential to further redefine the terms of access of users to public administration services and to enhance the ties among citizens, businesses, and the government.

INTRODUCTION

The aim of this chapter is to outline and discuss possible future scenarios concerning the use of Information and Communication Technologies (ICT) in public administration. This aim is accomplished by discussing, first, the ways in which e-government allows to enrich the interaction between public administration, citizens and businesses, as well as between different public administrations. Then, we will discuss how the currently emerging 'm-government' trend (i.e., 'mobile government') may reconfigure the structural and functional features of the interaction with

DOI: 10.4018/978-1-4666-1909-8.ch004

public administration. Lastly, we will outline a possible future scenario in which governments tap the full potential of ICT in conjunction with complementary technological innovations (i.e., artificial intelligence). We call this possible scenario, which is based on the disentanglement of the interaction between citizens and public administration from both the physical space and time, as 'l-government' or 'liquid government.' In short, liquid government could be one in which citizens interact with an 'augmented public administration,' namely one which seamlessly combines 'brick and mortar' service delivery with digital ones, provides ubiquitous (i.e., everywhere and anytime) information and service delivery, is centered on the individual user, and makes use of enhanced automation or Artificial Intelligence (AI) technologies.

The use of ICT in public administration has significantly increased in the last about twenty years. As highlighted by Islam and Ahmed (2007), the intensification of ICT in public administration has revolutionized the structure, processes, and operations of governmental entities, or at least introduced the potential for doing so. Various authors related this trend to the emergence of a new paradigm for public administration, namely *e-governance*, which has been adopted or is in the process of being adopted by both industrialized and developing countries alike (Heeks, 2001; Finger & Pécoud, 2003; Chadwick, 2003; Saxena, 2005; Islam & Ahmed, 2007; Marche & McNiven, 2009). Although definitions abound and partially differ, e-governance can be understood as the process through which society is directed to achieve the most desirable long-term goals by means of improved interaction between public administration, citizens, and business through the opportunities offered by state-of-the-art ICT.

Within the frame of reference of e-governance, e-government is a tool (Hood, 2006; Solomon, 2002) which broadly consists of the use of ICT for the provision of information and the delivery of public services. The World Bank conceives it particularly suited to "improve the efficiency, effectiveness, transparency, and accountability of government" (World Bank, 2007). The United Nations highlight its potential for transforming the internal and external relationships of public administration (UNDESA, 2003). The Organization for Economic Cooperation and Development (OECD) generally considers it as a tool to achieve better government (OECD, 2003). e-Government can be considered the centerpiece of e-governance, in the sense that the delivery of government service through ICT is an essential requisite for directing the efforts of public administration, citizens, and businesses towards desirable long-term goals. E-governance, hence, generally takes a broader meaning than e-government, in the sense that—apart from the ICT-based instruments by itself—attention is also placed on the mechanisms of ICT-based interaction between public administration, citizens, and businesses, i.e., what enables and stimulates actors to participate in the formulation and implementation of public policies.

Despite considerable interest and investments on the part of governments, the effects of e-government have been controversial so far. Welch et al. (2005) found that government website use is positively associated with e-government satisfaction and website satisfaction, and that e-government satisfaction is positively associated with trust in government. Wong and Welch (2004) also found, however, that the use of e-government does not reduce fundamental differences in the level of accountability between different national bureaucracies. West (2004) held, instead, the view that e-government fell short of its potential to transform service delivery and public trust in government, although it can enhance democratic responsiveness and boost beliefs that government is effective. Moon (2002) highlighted that at the municipal level the implementation of e-government did not deliver what its rhetoric promised, possibly because of the lack of financial, technical and personnel capacities and legal (privacy) issues. Others (Dada, 2006; Ciborra & Navarra,

2005) also pointed out that e-government may be inadequate in developing countries because of the lack of bureaucratic and governance prerequisites for effective e-government use.

Where is current trend of e-governance heading to? Will public administration be able to implement more and more ICT-based solutions, which improve service quality, cost-effectiveness, accountability, and democratic participation? Past experience suggests that future scenarios may take the shape of a somehow checkered landscape of 'forerunners' and 'laggards' in the use of ICT in public administration. Plausibly some national bureaucracies, or some specific organizations within public administration, will make use of state-of-the-art ICT while others will find it harder to upgrade extant routines to novel technological standards. Possibly, part of the public administration will genuinely re-engineer its organizational processes, while others will partially implement ICT-based solutions as mere 'window dressing' operations of services, which will be still delivered in more traditional ways.

Within this scenario, particular attention should be placed on how ICT can be used at its maximum potential in public administration. If we believe that e-governance does bring the possibility to revolutionize the way governments deliver services to citizens, and the ways in which it is held accountable to citizens, then we should be concerned with exploiting ICT-based solutions at best. For a couple of decades, e-government offered several tools for improving public administration's service delivery. Currently, e-governance has been characterized by an enhanced focus on the potential offered by ICT-based solutions related to Internet mobile technology—generally referred to as 'm-government.' Further development of ICT, however, may result in novel modes of interaction between citizens and public administration, where service delivery is less and less dependent on specific time and space conditions. Some authors have already labelled this set of tools as 'ubiquitous government' (u-government). For the reasons which are articulated below, 'liquid government' (l-government) could be a better characterization of the features that future ICT-based solutions might offer for improving public administration services.

The rest of this chapter is organized into four sections. Next section will outline the general characteristics of e-government tools, both in general terms and with respect to various instances of application of Internet-based technologies to the delivery of public services. The current 'Action Plan for e-government' issued by the government of Italy (2008-2012) is used to provide instances of the variety of e-government tools employed in developed countries. Section three will discuss the potential for public administration's service delivery offered by Internet mobile technologies. Selected instances of m-government in Estonia will provide an illustration of the implementation of this trend. Section four will outline the ways in which future ICT-based solutions might take the shape of diffuse and automated public administration service delivery. Finally, section five will draw the conclusions.

E-GOVERNMENT: SOME INSTANCES OF ICT-BASED TOOLS FOR PUBLIC ADMINISTRATION

In broad terms, e-government is a set of tools, which allows public administration to make use of ICT-based solutions for the delivery of services to citizens, businesses, and other public administrations. In the early stages of diffusion of ICT in public administration, e-government mostly consisted of simple provision of information or services via the Internet. Nowadays, e-government includes various instruments, which aim to reshape the ways in which citizens and businesses interact with public administration, and public administrations interact with each other (Jackson & Curthoys, 2001). The original set of ICT-based solutions in public administration (e.g., automating information access and delivery, self-retrieval of information and documents, etc.) has been progressively enlarged to include a wide range of innovative services. The growth of capacity of ancillary technologies (e.g., database management) also allowed to track and target the features of public administration's users, diversify service according to different segments of citizens and businesses, and reinvent or transform the way in which governments operate (Heeks, 2000; Al-Kibsi, et al., 2001; Prins, 2001; Silcock, 2001; Fountain, 2001; Ho, 2002; Gant & Gant, 2002; Mellor & Parr, 2002; Moon, 2002).

The move from the traditional modes of service delivery of government to e-government took place through various initiatives around the world. Within Europe, for instance, in 2001 the EU Commission launched the *eEurope* plan in order to stimulate Member States to undertake programs for developing ICT-based solutions in their respective public administrations. The plain aimed to "bring everyone in Europe—every citizen, every school, every company—online as quickly as possible" (EU Commission, 2000, p. 5). In 2005, the revision of the *eEurope* plan led to further specify the objectives of Member States' efforts as aiming to create a favorable context for public and private investments in new technologies, stimulate new jobs, increase productivity, modernize public services, and guarantee to all citizens greater participation to the global information society ('e-inclusion').

The 'Action Plan for e-Government' of Italy provides an example of e-government policy adopted and implemented among EU Member States. As a country context, Italy provides an interesting case because of contrasting features between the country's economic development and its ICT infrastructure. On the one hand, Italy ranks 10th among world countries in terms of GDP per capita (Purchasing Power Parity) according to the International Monetary Fund (2009) and the World Bank (2008). On the other one, the country's ranking is just 38th in the UN e-government development index (UN, 2010) and wide differences in the diffusion of x-DSL technology exist between northern and southern regions. Italy, therefore, is a case of county context for the diffusion of e-government in which actual use of ICT-based solutions in public administration still lags relatively behind full potential.

Launched in 2008, the 'Action Plan for e-Government' of Italy aims to fill the gap in the use of e-government tools in Italian public administration with respect to average EU and OECD countries. The program is intended to bring about innovation in several policy areas: in school and university, it aims to innovate teaching methods, strengthen interactions between families and teachers, and improve administrative services; in health, to allow general practitioners to better share information about patients; in justice, to reduce paperwork and offer online services to citizens; in registry services, to unify and integrate public records; in taxation, to spread the use of e-invoices; in general public administration services, to implement the use of certified email for communications between public administration, citizens, and businesses, and to integrate public administration information systems.

The case of the 'Action Plan for e-Government' of Italy provides several instances of e-government tools. For the sake of simplicity, these tools (or applications) can be classified according to various criteria. First, e-government tools differ in terms of the actors, which are involved in the interaction, e.g., public administration vs. citizens or businesses and public administration vs. other public administration. Second, e-government tools differ in terms of the mode of interaction, e.g., unidirectional (one actor is 'sender' and the other is 'receiver' of information or one is the 'provider' and the other the 'user' of services) or bidirectional (both actors interact by exchanging information and/or co-producing information or services). Third, e-government tools differ in terms of whether they strengthen existing services (e.g., providing online access to information or

Table 1. Instances of e-government tools of the 'action plan for e-government' of Italy, which strengthen existing services

	Unidirectional	Bidirectional
PA vs. citizens/businesses	Providing more Internet connectivity in schools and universities Providing a PC to primary school pupils Providing online access to jurisdictional documents Providing health certificates online Receiving citizens' reports to police online	Providing online information and services in schools (e.g., students' marks, tutoring, parents' meetings) Electronic identity card Providing online services in consular offices Providing services for setting up businesses
PA vs. PA	Providing electronic transmission of documents related to jurisdictional cases between public administrations	Provide a system for online communication between health care operators

services already provided offline) or expand the range of services offered (e.g., by providing innovative services).

We can consider, first, e-government tools, which aim to strengthen existing services. Table 1 shows some examples of such kind of applications, which are included in the 'Action Plan for e-Government' of Italy. Providing health certificates online and receiving citizens' reports to police online are examples of unidirectional e-government applications between public administration and citizens. Providing online services for setting up businesses is an instance of bidirectional e-government applications between public administration and citizens. The electronic transmission of documents related to jurisdictional cases exemplifies unidirectional relationship between different public administrations. Finally, the possibility to establish online communication between health care operators is an illustration of bidirectional e-government application between public administrations.

We can move, then, to consider e-government tools, which aim to expand the range of services offered. Table 2 illustrates some examples of this kind of applications out of the 'Action Plan for e-Government' of Italy. An instance of innovative service (which consists of unidirectional information between public administration and citizens) is the provision of online information to parents

Table 2. Instances of e-government tools of the 'action plan for e-government' of Italy, which expand the range of services offered

	Unidirectional	Bidirectional
PA vs. citizens	Providing online information in schools (e.g., notification of student's absence from school) Providing the issue of jurisdictional certificates also by other public administrations Provide online access to geographical maps and environmental information	Introducing innovative teaching methods with online platforms, in both schools and universities Providing online attendance of schools to pupils hospitalized Creating the national tourism portal Creating the national cultural heritage portal Creating the national non-profit portal
PA vs. PA	Providing automated transmission of variations of registry from local governments to other administrations	Creation of a national registry of school registries and performance records. Providing unified electronic health registry of citizens Creation of a national registry of roadways

concerning notification of student's absence from school. Another instance (which refers to bidirectional information between public administration and citizens) is the creation of comprehensive portals in the areas of tourism, cultural heritage, and non-profit. Innovative services between public administration include the automated transmission of variations of registry (e.g., residential address) from local governments to other administrations, and the formation of a national registry of school registries and performance records, which should allow policy-makers and school managers (and, possibly, also citizens) to draw comparisons between different schools.

The examples drawn from the 'Action Plan for e-Government' of Italy illustrate the kind of ICT-based solutions currently applied in the public administration of an industrialized Western country. E-government tools which strengthen existing services generally aim to improve efficiency, cost-effectiveness, and quality of service delivery. To some extent, e-government applications are generally related to the use of the Internet as timesaving device, e.g., allowing citizens remote access to information and services. To some extent, however, some of these applications are also related to the (partial) re-engineering of workflows, e.g., providing automated transmission of information between public administrations. E-government tools, which expand the range of services offered, instead, can significantly affect behavioral patterns of citizens, businesses, and public administrations alike. For example, providing centralized storage of information which is made available for retrieval by other public administrations can bring about significant changes in public officers' understanding of integration between public administrations' activities and collaborative ties.

M-GOVERNMENT: ICT-BASED TOOLS IN CONTEXT

During about the last decade, the diffusion of mobile Internet devices (e.g., smart phones) led to a growing interest towards the possibility to provide information and deliver services from public administration to users 'wherever and whenever' they wish. M-government is currently considered the 'new frontier' of public administration services, which are based on countries' mobile network infrastructure. Setting the supporting technological infrastructure aside, m-government applications do not significantly differ from e-government ones insofar as both consist of the use of the Internet for mediating the relationship between public administration and citizens, businesses, and other public administrations. If we consider the physical features of the mobile network infrastructure, however, then it becomes evident that m-government considerably enlarges the kinds of information and services that public administration can deliver to users.

The emergence of m-government is plainly related to the rapid diffusion of mobile ICT, especially laptop computers, GSM telephones, Personal Digital Assistants (PDAs), microwave local area networks (e.g., Bluetooth), and Wireless Application Protocol (WAP) (Kakihara & Sørensen, 2002a), and related 'killer' applications (e.g., email, instant messaging, social networks, global positioning systems) (Sørensen, 2003; Kakihara, 2003). Technologies enabled the development of innovative ways of structuring social relationships and behave in inter-personal settings. With respect to the space dimension, mobile ICT set individuals free from the rigid constraints of fixed line point of access to the Internet. With respect to the time dimensions, mobility access to the Internet leads individuals to redefine the time of social interactions (e.g., sporadic for linear clock time but conceived as continuous). The overall effect is to establish 'ties' between the contextual place and time where individual live and physically remote space and time situations, which are nevertheless perceived and conceived as proximate to individuals' experience.

What are the effects of mobile ICT on the delivery of public administration services? The diffusion of Internet mobile is mostly understood to bring about significant benefits in developing countries and in regions, which are negatively affected by the digital divide (Roggenkamp, 2004). It seems possible, indeed, that mobile ICT can allow countries and regions, which have not been deeply endowed of fixed line infrastructure to 'leapfrog' the stage of implementation of x-DSL technology and enjoy the benefits of Internet access through mobile devices. The diffusion of Internet mobile, however, is also understood to trigger significant changes in the way the government works and public administration provides information and service delivery. Features of the spatial, temporal, and contextual aspects of mobility open up the possibility of radically innovative interactions between public administration, citizens, and businesses which respect to traditional 'brick and mortar' technologies and more conventional e-government applications (Kakihara & Sørensen, 2002a; Song, 2005).

Table 3 illustrates how the way to m-government (passing through e-government) allows to extend the conditions to access information and service delivery. In the old days of traditional government, users could access information and services by showing up in a specific place (e.g., public bureau) at a specific time (e.g., opening hours). Technology of the time allowed to serve citizens and businesses provided that they approached public officers ('human tellers') where resources for delivering public services where physically stored. To some extent, users could access a limited range of services (mostly, access to information, like the one provided in a bulletin board) at undefined time (e.g., outside opening hours) provided that they went to the specific place (bureau) where services where delivered. Some technological tools (e.g., telephone) also allowed to provide some information and services at undefined space (e.g., calling the public administration from home), but generally within well-specified time frames (e.g., working hours of public administration's call centers).

The advent of e-government marked an important development of the conditions to access information and service delivery. Within the framework illustrated in Table 3, e-government basically opened up the possibility for public administration to provide information and service delivery irrespective of the place and time of the user. Because of their very technological features, e-government applications are not generally affected by the physical location of the user's access to the Internet (possibly apart from legal restrictions to access copyrighted material outside a country's jurisdiction, or from matter of security). Automation of response to users' queries, moreover, allows the provision information and service delivery even in the absence of any human operator located at public administration's side. In other words, e-government allows the disentanglement, to some extent, of the provision of information and service delivery from any specific place and time constraints. However, if access to the Internet is constrained by fixed-lines (e.g., home x-DSL or Wi-Fi connectivity), then no provision

Table 3. From confined place-time to contextualized place-time

	Confined time	Undefined time	Contextualized time
Confined place	Human tellers	Bulletin boards	X
Undefined place	Call centers	e-government	e-government
Contextualized place	X	m-government	m-government

of information and delivery of service can take place in proximity of the specific events, which make users approach public administration.

Enter m-government. Quite obviously, the possibility to access the Internet through mobile devices amplifies the places and time at which any user can interact with public administration through 'conventional' e-government applications. The potential of m-government, however, seems to rest quite beyond the possibility to surf Internet-based public administration's services anywhere and anytime. The portability of mobile Internet devices rather suggests that access to Internet-based services is made available in any *context* in which users may find themselves embedded. The point, then, is to make citizens and businesses able to access information provision and service delivery *at the place* and *at the time* in which they most need them. Differently from e-government, then, m-government should take into account the global positioning of the user (a function which is typically embedded in last generation mobile devices) and should allow the user to rapidly access place and time-relevant information and services.

Needless to say, the possibility to access information and service delivery at a certain place (e.g., while traveling far from the city of residency and in the need for directions urgently) may be hampered by the limited time frame within which 'traditional' public administration operators work (e.g., working hours of call centers). Similarly, the request for information and service delivery at a certain time (e.g., during night hours and in the need for health check urgently) may be subject to condition that the user approaches public administration at specific locations (e.g., hospital's emergency unit). Access in specific time and from undefined place may be provided by 'conventional' e-government tools, however. In addition, m-government also provides the possibility to access information and service delivery at anytime, from whatever location the users are present—hence, according to particular needs that related to the context conditions in which users are embedded.

Instances of m-government tools abound in the world. For the sake of brevity, a limited range of examples can be drawn from the experiences conducted in Estonia. As a country context, Estonia is interesting because it is an instance of former developing country, which has been recently 'upgraded' to the rank of developed ones in 2010. The country, which ranks 47th in the world in terms of GDP per capita (PPP; International Monetary Fund, 2009; 37th, according to the World Bank ranking, 2008), has undertaken various initiatives which fall within the m-government kind (Rannu, et al., 2010). They include 'm-teacher' services (e.g., online communication between school teachers and families), 'm-police' (e.g., online notification from the police to taxi drivers, bus drivers, security companies, and citizens about a missing person, stolen car, or other incidents that require cooperation of citizens), m-parking (e.g., using mobile phones for paying parking and public transport), and notifications to city public administration (e.g., concerning the need for maintenance of city infrastructure).

L-GOVERNMENT: EXPLORING FUTURE SCENARIOS FOR PUBLIC ADMINISTRATION

A new trend in the use of ICT in public administration may be already underway. Some authors have explored the vision of a 'digital inclusive society' where citizens and businesses can access and share information online as an essential part of their civic and democratic status (Lam & Lee, 2006). Others pointed out that changes of technology and users' behavior lead towards the progressive erosion of place and time constraints and could ultimately result in total (i.e., 24x7) availability of online public administration services through multiple channels of communication and delivery (Gouscos, et al., 2001). Others have conceived the emergence of a new topology of the structure of relationships between citizens, businesses, and public administration as a 'fluid' one, which is characterized by the variability and dissolution of space and time boundaries (Kakihara & Sørensen, 2002a; Pica & Kakihara, 2003).

A common threat of these works is that the development of ICT can potentially result in a range of novel applications if combined with ancillary technologies and new service models. Next generation Internet-based tools for public administration can include the provision of information and the delivery of services in a way that (1) makes constant and pervasive use of computing, (2) builds on deep integration between different public administrations so that users experience fragmented services as wholes, (3) focuses on the specific identity and issues that citizens and businesses face, and (4) processes information, problems, and solutions through the use of automation or, whenever made possible, Artificial Intelligence (AI). For the sake of brevity and contiguity with other 'buzzwords' of this field, we can label this trend as the coming of 'l-government' or 'liquid government,' i.e., one which can potentially interact continuously with citizens and businesses in any aspect of their daily matters.

The first feature of l-government is understood as the constant and pervasive use of computing. Since the early nineties, the development of computing capacity led to outline the possibility that computers could—one day—permeate any aspect of daily life. According to Weiser (1991), computers are expected to become more and more integrated in the physical environment, up to the point that users do not even notice their presence (Ridgway, 2006). Within the field of e-government, the diffusion of computing is generally associated to the emergence of 'ubiquitous government' (or 'u-government'), that is, the constant availability, context-awareness, and integration of computers in the provision of information and delivery of services of public administration to citizens and businesses (Kim, 2005). The idea of u-government, in essence, restates the elimination of time and place constraints and the possibility to provide access to online services through multiple channels and around the clock (Anttiroiko, 2005).

The second feature of l-government is the deep integration between different public administrations, which are perceived by users as a seamless, coherent whole. Various e-government tools have facilitated the development of coordination and of collaborative ties between different public administration bodies. Generally, however, citizens and businesses access online services through specialized venues, e.g., organizational websites or portals. Over time, we can expect public administration to strengthen the degree of integration between different bodies and departments in both 'back office' and 'front office' areas. Rather than facing separated 'silos' providing administrative functions (e.g., registry, schooling, health, etc.), citizens, and businesses could move across governmental services without noticing any boundaries between different public administration bodies.

The third feature of l-government is user-centric perspective. Various e-government tools have been implemented by targeting individual needs, which are generally associated to electronic ID attributed to citizens and businesses. Generally, however, electronic ID has been used as a way to segment public administration users and to reengineer administrative processes around user-specific characteristics and needs. In a user-centric perspective, the provision of information and the delivery of services should be able to *anticipate* citizens and businesses' needs, depending on their characteristics and the context in which they are embedded. In l-government, public administration can take a proactive stance and fulfill the emergence of users' needs on the basis of personalized profiles which store the social, professional, economic and cultural attributes of users.

Finally, the fourth feature of l-government is the enhanced use of automation or, if possible,

AI. The pervasive presence of l-government services in time and space necessarily calls for the use of alternative devices rather than human operators on the side of the public administration. Automation of public administration's response to users' queries about information or services has been already implemented in e-government tools. The evolution of AI, in conjunction with related technologies (e.g., speech recognition), can further expand the range and ease of use of public administration's online services through interfaces that simulate human interaction.

Instances of l-government, as conceived here, are still in short supply. Some experiences based on the implementation of ubiquitous computing, however, shed some light on the potentials of next generation Internet-based tools for public administration. Oja and Schrader (2008), for example, discussed the use of fingerprint biometric identification for ticket purchases. Murakami (2003) described the shift from 'e-Japan' to 'u-Japan' policies, which include, in the field of public administration, Intelligent Traffic Systems (ITSs) and Vehicle Information and Communication Systems (VICS). Anttiroiko (2005) especially focused on Finland's strategy of incremental, user-centered approach to the development of u-government tools. In this Nordic country, examples of u-government include mobile Travel Information Service provided by the Ministry of Foreign Affairs, 'StreetChannel' service for complaining about road maintenance, various provision of information concerning vehicles, trains, flights, and public officials.

Generally, these instances do not seem to make justice of the untapped potentials of state-of-the-art technology (or, for that matter, technology that will be reasonably made available in the not-so-distant future). Imagine a sort of control panel on your mobile device: it shows the main areas in which you—as a citizen—can interact with public administration: registry, education, health, safety, environment, taxation, etc. Think of the possibility to make enquiries through the only user interface—although, in a way which goes unnoticed, information are retrieved and service is provided on the basis of resources which are stored and elaborated by different public administration bodies. Think of the possibility that, while wondering around, you are kept informed of government's services around you—e.g., that, if you have some spare time, an historical heritage site is just a few yards far—and especially of those which are more relevant to your status—e.g., that, on the basis of your family structure, you might be interest about local government's swimming pool offers for parents with kids. Think of the possibility that you could make an enquiry through your mobile device to a virtual public officers—which is actually an 'avatar-like' automated character endowed with some AI, which enables it to understand queries about where's the closest pharmacy. Today's science fiction could be tomorrow's taken for granted, even in the field of public administration.

By and large, however, fully-fledged examples of next generation Internet-based tools for public administration are still to come. It is possible that they will emerge slowly, as improvements built on existing 'conventional' e-government applications rather than radical shifts in the ways information are provided and services delivered. This may be the case in the EU, where apparently the policy and strategic discourse about e-government (and e-governance) placed relatively little attention to radical transformations of public administration's service promised under the u-government label (Anttiroiko, 2005). Other areas in the world, especially Eastern Asian countries and developing ones, however, may undertake more self-conscious efforts to pursue the implementation of next generation Internet-based tools for public administration.

CONCLUSION

The old days in which the government provided information and delivered services through 'brick and mortar' technologies are being supplanted by complex arrangements in which traditional tools of public administration are integrated or substituted by others based on ICT. This chapter aimed to trace some directions of the development and diffusion of Internet-based solution in public administration. The new trend in the use of ICT in public administration is headed towards more pervasive use of computing, integration of public administration services, user-centric approaches, and automation. One possible future scenario emerging from this trend could be one of enhanced capabilities of public administration to intertwine with daily life of citizens and businesses. If we draw an analogy with 'augmented reality'—that is, the branch of virtual reality concerning the overlapping between the physical and the virtual (digital) world, which provides the possibility to interact with digital and physical artifacts alike in a continuous and undistinguishable way—we can envision the future of e-governance as one of 'augmented public administration,' where users can feel an underlying and constant present of public services and government policies.

Such scenario may not come without any cost, however. Alongside the development of ICT and the use of Internet-based tools for public administration comes greater concern for issues of privacy, security, accountability, and democratic check and balances. Storage and tracking of electronic Ids would be a resource of significant value in the political, social, and commercial fields, and safeguards must provided to prevent undue access and exploitation. Pervasive systems of e-governance deeply embedded with citizens' lives would require considerable protection from threats posed by terrorism, crime, corruption, and collusion between system administrators and businesses. Mechanisms should be put in place to ensure fairness, equality, and integrity of public administration and conditions of access to e-services and information. Safeguards would be needed to prevent governments to bend ICT tools to serve partisan purposes and undermine the position of minority parties and groups.

The development of ICT and the use of Internet-based tools for public administration offers the promise of a fertile ground for further research. Issues are open concerning what makes e-governance policies raise up in governments' agendas, and what accounts for the particular features of e-governance public policies, e.g., whether policy-makers adopt an explicit stance towards the adoption of next generation Internet-based tools for public administration or whether they employ 'e-governance' as an umbrella concept for any ICT applications. Additional issues arise concerning what accounts for the variety of e-governance applications, and for the mechanisms which trigger the undertaking of innovative projects and the diffusion of ICT applications in public administration in the implementation stage, e.g., why certain public administration bodies are more inclined to adopt e-government tools than others, and what facilitates or hamper the transfer of innovation experiences across public administrations.

REFERENCES

Al-Kibsi, G., Boer, K., Mourshed, M., & Rea, N. (2001). Putting citizens on-line, not in line. *The McKinsey Quarterly*, *2*, 65–73.

Anttiroiko, A. (2005). Towards ubiquitous government: The case of Finland. *E-Service Journal*, *4*(1), 65–70. doi:10.2979/ESJ.2005.4.1.65

Chadwick, A. (2003). Bringing e-democracy back in. *Social Science Computer Review*, *21*(4), 443–455. doi:10.1177/0894439303256372

Ciborra, C., & Navarra, D. D. (2005). Good governance, development theory, and aid policy: Risks and challenges of e-government in Jordan. *Information Technology for Development*, *11*(2), 14–15. doi:10.1002/itdj.20008

EU Commission. (2000, March 23). E-Europe: An information society for all. *Lisbon*.

Dada, D. (2006). The failure of e-government in developing countries: A literature review. *The Electronic Journal of Information Systems in Developing Countries*, *26*(7), 1–10.

Finger, M., & Pécoud, G. (2003). *From e-government to e-governance? Towards a model of e-governance*. Paper presented at the 3rd European Conference on e-Government. Dublin, Ireland.

Fountain, J. (2001). *Building the virtual state: Information technology and institutional change*. Washington, DC: Brookings Institution.

Gant, J., & Gant, D. (2002). *Web portal functionality and state government e-service*. Paper presented at the 35th Hawaii International Conference on System Sciences. Hawaii, HI.

Heeks, R. (2000). *Reinventing government in the information age*. London, UK: Roultedge.

Ho, A. (2002). Reinventing local government and the e-government initiative. *Public Administration Review*, *62*(4), 434–444. doi:10.1111/0033-3352.00197

Hood, C. (2006). The tools of government in the information age . In Moran, M., Rein, M., & Goodin, R. E. (Eds.), *The Oxford Handbook of Public Policy* (pp. 469–481). Oxford, UK: Oxford University Press. doi:10.1093/oxfordhb/9780199548453.003.0022

International Monetary Fund. (2009). *World economic outlook database-April 2010*. Geneva, Switzerland: IMF.

Islam, M. M., & Ahmed, A. M. S. (2007). Understanding e-governance: A theoretical approach. *Asian Affairs*, *29*(4), 29–46.

Jackson, P., & Curthoys, N. (2001). *E-government: Developments in the US and UK*. Paper presented at the 12th International Workshop on Database and Expert Systems Applications. Munich,Germany.

Kakihara, M. (2003). *Emerging work practices of ICT-enabled mobile professionals*. (PhD Thesis). University of London. London, UK.

Kakihara, M., & Sørensen, C. (2002). *Mobility: An extended perspective*. Paper presented at the 35th Hawaii International Conference on System Sciences. Hawaii, HI.

Kim, H. K. (2005). *Ubiquitous government - Dreams and issues*. Paper presented at the 39th International Council for Information Technology in Government Administration (ICA) Conference. Salzburg, Austria.

Lam, J. C. Y., & Lee, M. K. O. (2006). Digital inclusiveness – Longitudinal study of internet adoption by older adults. *Journal of Management Information Systems*, *22*(4), 177–206. doi:10.2753/MIS0742-1222220407

Marche, S., & Niven, J. D. (2009). E-government and e-governance: The future isn't what it used to be. *Canadian Journal of Administrative Sciences*, *20*(1), 74–86. doi:10.1111/j.1936-4490.2003.tb00306.x

Mellor, W., & Parr, V. (2002). *Government online: An international perspective annual global report*. Retrieved from http://tnsofres.com/gostudy2002

Moon, M. (2002). The evolution of e-government among municipalities: Rhetoric or reality? *Public Administration Review*, *62*(4), 424–433. doi:10.1111/0033-3352.00196

Murakami, T. (2003). Establishing the ubiquitous network environment in Japan. *Nomura Research Institute Papers, 66*.

OECD. (2003). *The e-government imperative: Main findings*. Retrieved from http://www.oecd.org/dataoecd/60/60/2502539.pdf

Oja, M.-K., & Schrader, A. (2008). *From internet to internet of things – The evolution of u-government*. Retrieved from http://www.iadis.net/dl/final_uploads/200817C045.pdf

Pica, D., & Kakihara, M. (2003). *The duality of mobility: Understanding fluid organizations and stable interaction*. Paper presented at the 11th European Conference on Information Systems. Naples, Italy.

Prins, C. (2001). Electronic government: Variations on a concept . In Prins, J. E. J. (Ed.), *Designing E-Government* (pp. 1–5). Dordrecht, The Netherlands: Kluwer Law International.

Rannu, R., Saksing, S., & Mahlakõiv, T. (2010). *The mobile government: 2010 and beyond.* White Paper. Retrieved from http://m-gov.mobi

Ridgway, B. (2006). *Ubiquitous government: Enabling innovation in a connected world.* Tokyo, Japan: Microsoft Asia-Pacific.

Roggenkamp, K. (2004). *Development modules to unleash the potential of mobile government: Developing mobile government applications from a user perspective*. Paper presented at the 4th European Conference on e-Government. Dublin, Ireland.

Saxena, K. B. C. (2005). Towards excellence in e-governance. *International Journal of Public Sector Management*, *18*(6), 498–513. doi:10.1108/09513550510616733

Silcock, R. (2001). What is e-government. *Parliamentary Affairs*, *54*, 88–101. doi:10.1093/pa/54.1.88

Solomon, L. M. (2002). *The tools of government: A guide to the new governance*. Oxford, UK: Oxford University Press.

Song, G. (2005). Mobile technology application in city management: An illumination of project nomad in UK. *Municipal Administration and Technology*, *7*(3), 103–106.

Sørensen, C. (2003). *Research issues in mobile informatics: Classical concerns, pragmatic issues and emerging discourses*. Retrieved from http://mobility.is.lse.ac.uk/html/downloads.htm

UNDESA. (2003). *World public sector report: E-government at the crossroad.* New York, NY: UN Department of Economic and Social Affairs.

Weiser, M. (1991). The computer for the 21st century. *Scientific American*, ▪▪▪, 94–100. doi:10.1038/scientificamerican0991-94

Welch, E. W., Hinnant, C. C., & Moon, M. J. (2005). Linking citizen satisfaction with e-government and trust in government. *Journal of Public Administration: Research and Theory*, *15*(3), 371–391. doi:10.1093/jopart/mui021

West, D. (2000). *Assessing e-government: The internet, democracy and service delivery by state and federal governments*. Retrieved from http://www.insidepolitics.org/egovtreport00.html

Wong, W., & Welch, E. W. (2004). Does e-government promote accountability? A comparative analysis of website openness and government accountability. *Governance: An International Journal of Policy, Administration and Institutions*, *17*(2), 275–297. doi:10.1111/j.1468-0491.2004.00246.x

World Bank. (2007). *Introduction to e-government: What is e-government?* Retrieved from http://worldbank.org/

World Bank. (2008). *World development indicators database*. Washington, DC: World Bank.

KEY TERMS AND DEFINITIONS

E-Governance: The process through which society is directed to achieve the most desirable long-term goals by means of improved interaction between public administration, citizens, and business through the opportunities offered by state-of-the-art ICT.

E-Government: A tool (or set of tools) of government which broadly consists of the use of ICT for the provision of information and the delivery of public services.

M-Government: A sub-set of e-governance tools which are characterized by an enhanced focus on the potential offered by ICT-based solutions related to Internet mobile technology.

L-Government: A sub-set of e-governance tools that (1) makes constant and pervasive use of computing, (2) builds on deep integration between different public administrations so that users experience fragmented services as wholes, (3) focuses on the specific identity and issues that citizens and businesses face, and (4) processes information, problems, and solutions through the use of automation or, whenever made possible, Artificial Intelligence (AI).

U-Government: ('Ubiquitous Government') A sub-set of e-governance tools that enable the constant availability, context-awareness, and integration of computers in the provision of information and delivery of services of public administration to citizens and businesses.

ICT: Acronym for 'Information and Communication Technologies,' which are understood as the range of technical tools used to handle information and support communication, including computer and network hardware, communication middleware as well as necessary software.

eEurope: A plan launched in 2001 by the EU Commission which aims to stimulate Member States to undertake programs for developing ICT-based solutions in their respective public administrations.

Chapter 5
Challenges of Implementing E-Governance in a Politically Driven Environment

Manish Pokharel
Korea Aerospace University, South Korea

Jong Sou Park
Korea Aerospace University, South Korea

ABSTRACT

The authors in this chapter underline the significance of a wider environment in which e-governance has to flourish. Even though each and every country in the world has programs to promote e-governance systems to provide maximum benefits to the citizens, and huge budgets and efforts are employed in developing master plans for e-governance in the developing countries, the systems often get terminated. There are many reasons for this termination; hence, the chapter has identified some of the key issues surrounding it.

NEED OF E-GOVERNANCE

The merits and benefits of the e-governance system does not confine to the developed countries only. It is equally or in some cases it is more relevant to the under developed or least developed countries. The ultimate mission of e-governance is to fill the digital divide. Every citizen of the nation should get maximum facilities of it. It should be more citizens centric. The recent remarks made by US President Barak Obama in China indicate the importance of e-government for every citizen.

In the underdeveloped countries, physical transportation facilities are very acute; people have to walk one or two days to get anything done. E-governance could be the panacea to them to get every service nearby their location. Many nations in this world have missed many opportunities in their history. They missed agricultural revolution; they missed industrial revolution, and many more. It is said that your luck knocks your door only once. If you do not open it when it is knocked then you will be deprived by the luck.

DOI: 10.4018/978-1-4666-1909-8.ch005

They (Underdeveloped countries) have missed many opportunities in their past, like the industrial revolution and the agricultural revolution, but if they miss the current ICT revolution, then these countries will never get a chance to renovate themselves. ICT is the approach of developing these underdeveloped countries. ICT as an approach and e-governance as a solution can be used to restructure these countries.

POLITICALLY DRIVEN NATION

Politics is the activity by which trustworthy people make decisions for the betterment of people, society, country, and the entire world. It is very difficult to find any country that is not driven by politics. Almost every country is politically influenced. Right from countries likes Ethiopia, Swaziland, and Myanmar to European countries, US, Japan, Korea, every country is driven by politics. The only difference is the types of politics and nature of politicians. The intensity of involvement of politicians is different. In some countries, politicians do politics as per the mandate of people, and in some countries, they ignore the people's mandate and do their own style and try to convince innocence people for the sake of being in power for a long time.

Advantages

Politics formulate the rules. It ensures the safety and integrity of the nation. The people involved in politics are known as politicians. In a democratic term, people select the candidate to do the politics. They have the responsibility to help every citizen on their problem and help the entire nation to raise the economic condition. If anything goes wrong or astray, then people expect these politicians should sort it out. Sometimes countries cannot select the politician and dictators come and dictate. There are many examples of dictators ruling countries for a long time.

Disadvantages

Good politics make the country better, but bad politics make the country worse. The politics of any country depend upon the actors of it, i.e. politicians. If politicians are not honest, educated, and smart enough to drive the nation then the entire system gets badly affected. There are many examples in which politicians work only to draw the vote from the citizens. They give this as high priority. They are not serious about their nation. These types of attitudes do not help the nation to grow in any sector.

Implementing Issues

Implementing means putting the system into action. Here, implementing issues are more inclined towards the e-government system. It is widely accepted that e-government is the only solution for developing countries to raise the economic, social, and political condition. This is the reason why almost every country is trying to develop and use the e-government system. It is found that there are three steps followed to develop the system. Every system has three main steps, i.e. Feasibility Study, Development, and Implementation. During these steps, the following activities are carried out:

a. **Feasibility Study:** In this phase, a detailed study is done to know the present condition of the country. The need of government and types of services are identified. The outcome of this phase is the formulation of the problem.
b. **Development:** In this phase, different solutions are analyzed and tried to find the solution of the identified problem in the previous phase. There are various models and architectures in e-government systems, like service-oriented architecture, enterprise architecture, client server architecture, etc. These architectures are analyzed, and based upon the problem, one of them is selected.

c. **Implementation:** This is the final stage. In this stage, the developed system is put into action. The real time output is expected. During the first two steps, many experts are involved in it. In this step, the present scenario of the nation has to be understood, the strengths and weakness are to be analyzed, and the main problems need to be identified. Each design is analyzed deeply. Proper design is recommended to solve the problems. This is a very critical part of the system development. If anything goes wrong during this period, then entire system will be affected.

Unfortunately, when the system reaches in the final stage, after big struggle, many systems terminate from this point. System is not implemented. The effort and money that we have put in goes into ashes. Why? What could be the reasons? Is this because of our poor system design? Is this because of our poor analysis? It is not.

Implementation is not an easy task. A proper approach and certain guidelines are to be maintained. How do we implement the system? What are the parameters to be followed during implementing the system? Do we have types of implementation? The answers of these questions have to be found.

First let us explain the types of popular implementation strategies, and then we have to decide which one of the following is/are appropriate for our e-governance system.

a. **Big Bang Approach:** This is the approach of replacing the existing system with an entirely new system. The risk involvement is very high in this approach, but if it is done properly, then the entire system will have new developed flavor. It is only recommended if the nature of system is known to everyone.

b. **Parallel Approach:** In this approach, both the old and new systems work in parallel. At the point when it is realized of stopping the old system, then it is replaced with the new system. This approach is more suitable when the nature of the new system is difficult to predict. As compared to Big Bang, the risk factor is low here.

c. **Phase Wise Approach:** This approach follows incremental steps. The entire system is divided into different phases or steps, and each of the phases or steps are developed incrementally. If there is any problem or ambiguity in any phase, then it has the possibility to rectify it before moving to the next phase. This approach is considered the best approach. As drawbacks, it consumes a lot of time.

Along with the types of implementing strategies and their scope, there are many others reasons that always play the role of opposite force to terminate it. What are these reasons? These reasons are as follows:

- Lack of awareness
- Less priority
- Political instability
- Lack of commitment from decision maker (Politicians)
- Human resources
- Poor ICT infrastructure
- Risk factors

Lack of Awareness

Awareness is one of the factors that affect implementing the system. It makes the people responsive toward their responsibilities. There are two types of awareness: *awareness for citizen* and *awareness for decision makers*. In developing countries, people are not aware of their need. Once in Nepal, when the government had tried to know the view of remote people about the telephone system, they got a very surprising response. Many people in the remote areas did not like the telephone because they said they did not know anyone with whom they could talk.

In spite of development in ICT, decision makers of many countries are not aware about the benefits of the electronic governance system. Their ICT literacy level is very low. We can find many latest set of computers in the organization, but they are not being used for any real purpose. These machines are simply used in typing documents. The traditional typewriter is replaced by computers. Until the end of the second step in system development, they show their interest just to convey to international committees that they are very serious about it. When the implementation phase comes, then they drop their interest.

Occupied with Legacy System

Many decision makers in the government are preoccupied with the traditional thought. They do not want to revise or change their way of working in the office. They get scared of new technology, especially in those countries where the corruption level is high. They think the ICT reveals every corruption.

They do not rely on the new approach of doing work in the office. It is not only their fault; it is human nature that they rarely want to be changed. This type of trend is experienced not only in developing countries but also in developed countries. When there was a need to move to object-oriented programming in making software, many countries in the world were in favor of a structured programming approach. Even today, many programmers do not want to be changed in developed countries.

No Priority

Priority depends upon the need of the country. If the country is economically sound, then they can be more flexible in priority, but if they are not, they have to be very rigid in priority. Economic condition is one of the prime factors in deciding about the priority, but along with this, there are many other factors, like social factor, political factor.

In the country Nepal, where the per capita income of citizen is very little, more than 60% of people live in remote mountains. They do not have electricity. They have to walk for a day or more to come to a district headquarters to buy basic needs. Road transportation is not there. There are many villages in Nepal, where people have to go for 2 to 3 hours to fetch water. Many parents are compelled to not sending their kids to school because of no school nearby. Many people die because of no availability of medicine and doctors. Hospitals are not there. Schools, drinking water, transportation, hospital, electricity, etc., are the basic needs of the citizen. Government has to provide all these needs to their citizen. For Nepalese government, these are the first priority rather than buying equipment and investing huge amount in e-governance.

Political Instability

Political structure or political scenario is also important in implementing an e-governance system. It is found political instability is more in developing countries. In every one year, the political team gets changed. If one team is replaced by another, then all the decisions made by the previous team will also be changed. These types of government behavior affect a lot in any system, especially in the e-government system. The e-government system has a very typical nature. There is not a final stage or final phase in the e-government system. It is cyclic in nature. It is a continuous process.

Let's take an example. The government has decided to establish a data center. The data center is a place where all important government data are kept and disseminated to any government body whenever they need. It is a very expensive and sensitive project. It takes at least a couple of years to build it. These are the steps in developing or building a data center:

a. Good building is to be built.
b. Technical infrastructure is to be made. Technical infrastructure includes good network system, good reliable software, hardware components, etc.
c. Experts are to be selected to run the data center. There is a need for a set of different experts like domain expert, network expert, software expert, project manager, etc.
d. Collect the data from all government organizations. Convert it into usable format.
e. Protect the data from hackers, and provide data to authorized entities.
f. As the time goes, new data gets generated, collect these data and remove the data that does not make sense.
g. And many more.

These above process should continue for a long time. If it gets break at any junction of this process then there will be great loss to the nation in terms of finance, manpower, time, effort, security, etc.

Commitment from Leader

Leaders in any nation make rules, policy, and govern the country. They should be committed in implementing the e-government system. There are many good examples about the strong commitment from leaders in different countries. Singapore is doing well in e-governance because of strong and committed leader Lee Quan Yew. India is progressing in ICT because of committed leader Rajiv Gandhi during mid-80s and Chandra Babu Naidu during the 2000s. In South Korea, there are many committed leaders, who are very committed in using ICT in the e-governance system. They have a very good vision and can visualize the scope of e-governance system in the coming 5 to 10 years.

This is not the same scenario in others countries. Leaders in many other countries are not committed. Many leaders are scared of the ICT. They think if manual voting system converts into e-voting then there will not be enough room for them to manipulate during their election. That is why it is found that many leaders, in spite of supporting or giving the commitment, create the problems in implementing it.

Human Resource

In e-governance system, human resource is very important. Different levels of human resources are required in e-government system. We have to be very careful in selecting human resources. Let us take the example of the data center again. Here, we need different level of human resources. They are as follows:

- **Low-level human resources:** People who are experts in collecting data, creating database, maintaining network system, installation of new hardware or software.
- **Medium-level human resources:** People are experts in writing software, maintaining database, providing security, exploring new technology, etc.
- **Top-level human resources:** People, who can manage project, visualize the future, recruit the junior, and make a good decision.

In the developed countries, it may be possible to find such human resources, but in the case of developing countries it is not. Universities and colleges are the producers of such manpower. Even in developing countries, there are many good international universities. In Nepal, there is an international standard university like Kathmandu University, which provides very good manpower to the market. There are other universities as well. In spite of this, developing countries always face a scarcity of good human resources. It is because of a trend called "Brain Drain." It is the trend in which local experts explore the good opportunity in developed countries and move there. It is a serious problem for the country like Nepal, where in spite of having many experts, they always face scarcity.

Besides these, need in human resources is to be analyzed. A country has to produce the human resources itself to support e-government system. Before that, it has to identify the status of a country in human resources. It can be identified from following parameters:

- **Literacy Rate:** It helps to identify how many citizens in the country can read and write.
- **Enrollment in academic institutes:** It helps to identify how many children from a family enroll in the different levels of educations such as in secondary school, in high school, and in the university.

ICT Infrastructure

ICT is the backbone of the e-government system. Simple processing units and servers are not enough to operate the system. The dissemination and distribution of information from one place to another are equally important. It is expected in e-government that citizens from remote area need not to come to the district head office to do any official job. Information should be available everywhere in the country. Each service should be affordable to every citizen irrespective of their race, location, gender, etc. These are only possible if a country has good ICT infrastructure. Good infrastructure means a reliable system that is affordable to every citizen.

In developing countries, ICT infrastructure is very poor. In Nepal itself, out of 3940 VDCs, more than 1400 VDCs do not have connectivity at all. Out of 100 people in Nepal, only 7 people have telephone facility. Internet penetration is very low in developing countries.

Need of ICT Infrastructure in E-Government

A country has to be very serious in developing ICT infrastructures to promote and smoothing e-government services. The status of a country in ICT infrastructure depends upon the following parameters:

- **PC Penetration Rate:** It is the rate of the citizen using PC in their daily activities. It can be calculated like "how many of them are using PC out of 100 citizens?"
- **Internet Penetration Rate:** It is the rate of the citizen using Internet. It is also calculated among 100 citizens.
- **Mobile Penetration Rate:** It is the rate of using mobile phone among 100 inhabitants.
- **Fixed Line Telephone Penetration Rate:** It is the rate of using fixed line telephone among 100 inhabitants.

Online Services

A country provides the e-services through the online. A country has to evaluate the strength or capability of providing online services to the citizens. We recommend an approach of evaluating the strength of a country in terms of providing e-services.

Online service is very popular with four level models and also one of the parameters in evaluating country's rank as per UN index. The human capital index, infrastructure index, and online service index are the parameters to find out the country's rank as per UN survey. Online service maturity is one of the core issues among the three indexes. The Figure 1 shows the online service maturity graph. There are four levels and each level shows the maturity in providing online service. The Figure 1 helps the nation to identify its own position and gives the guidelines to improve.

There are four main levels in Figure 1. Each one of them is explained below:

Figure 1. Maturity level of online service

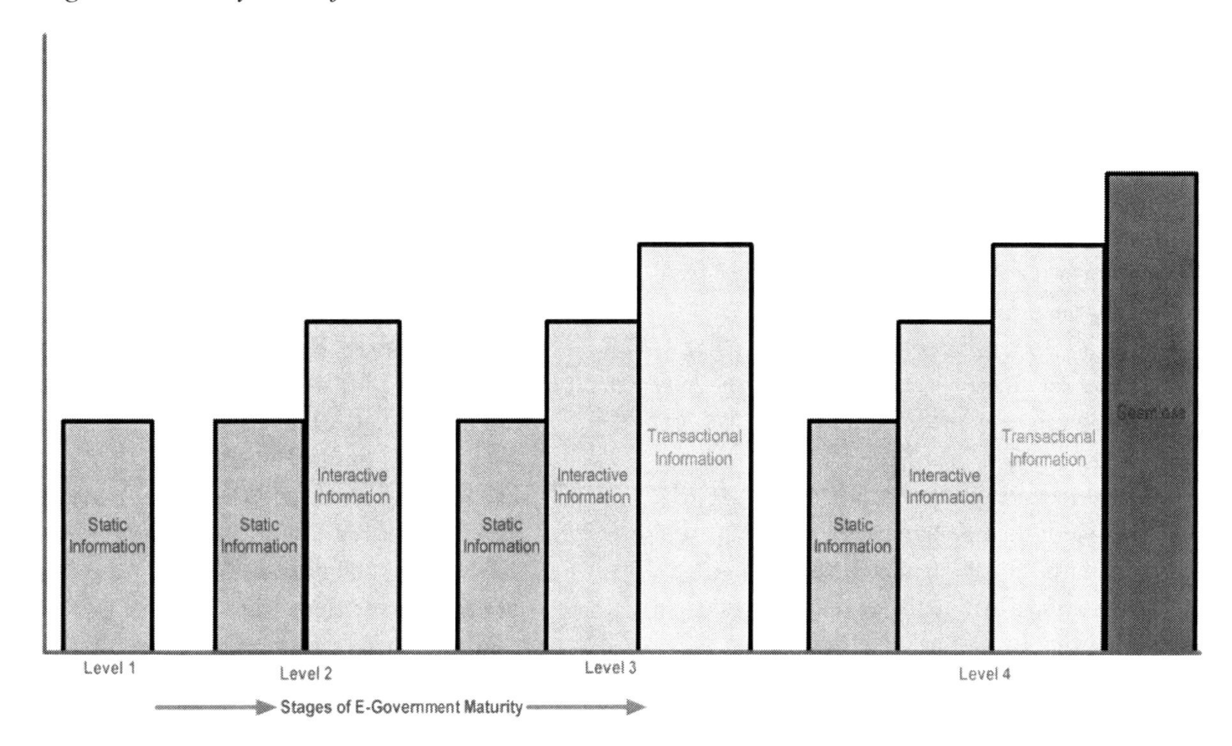

Level 1: Static Information (The information is published in government website but it is static). In this level, only required information is published in government website. Citizens can obtain the government information.

Level 2: Interactive Information (Level 1 + Interactive, i.e. citizen can access the information and interact). This level is little matured as compare to level 1 since it has information with interactive natures.

Level 3: Transactional (Level 1 + Level 2 + citizen's ability to enter secure information and engage in transactions with the organization). The level 3 provides the facilities to conduct the transactions with the organization. The features of previous levels are also included.

Level 4: Seamless (Level 1 + Level 2 + Level 3 + organization's ability to share data and information with other organization, as per the law and with the users consent).The level 4 is a matured level. It provides the entire government services to the citizens with maximum interoperability and information sharing among the government organizations.

Risk Factors

Risk is defined as a product of probability of an accident and the consequence because of the accident. Risk is also considered as a source of danger that makes the chances of loss and damage. It is also said as the probability of becoming infected and its results in a system. Risk is the probability that a hazard will turn into a disaster. The combination of vulnerability and hazards become a risk or, in other words, the probability that a disaster will happen.

Risk also refers to any factor (or threat) that may affect adversely the successful completion of the project in terms of delivery of its outputs or adverse effects on resourcing, time, cost, and quality. These factors/threats include risks to the project's business environment that may prevent the project's outcomes/benefits from being realized fully.

In e-Government, various types of risks are encountered. The risk is very high, especially in politically driven countries. Risks such as technical risk, political risk, financial risk, human resource risk, quality risk, etc., are the most probable risks. Some risks are independent and some of them are dependent. Dependent risk depends upon the existence of independent risks. Predicting possible threats during the e-Government projects together with measures for managing threats is needed to ensure that the project will be successful. In general, there are two types of threats in e-Government. One is managerial, and the other one is technical. The managerial threats involve the availability of qualified human resources to carry out project activities and completing them on time as expected where as technical threats involve responses to technical complexity and technical volatility.

How Do We Tackle These Challenges?

The mentioned challenges can be addressed if we get maximum commitment from political leaders, their wish, and determination. As we know, when there is a wish there is a way. Once we get commitment then the other challenges like awareness, priority, and infrastructure can easily be addressed. In developing countries, sometimes even if we have desire to do, we cannot because of many reasons. Developing countries could not provide the basic needs to the citizen. The government should provide so many things to the citizen.

The types of implementation strategies are to be addressed properly. Domain experts are to be hired to evaluate each one of them and follow it strictly. Along with these, the other areas such as capacity building and risk managements should be emphasized.

Capacity Building

As it is said, human resource is also one of the risks in politically driven least developed countries, there should be a mechanism to mitigate such risks. Capacity building is one of the solutions to mitigate the human resources risk.

The capacity building identifies the existing human resources status of the country. The main purpose of capacity building is to ensure that the qualified human resources or eligible workforces are identified and recruit them as per the need of local context. In order to work out in capacity building, there is a need of four main tasks and four sub-tasks in capacity building. The details of these tasks and sub-tasks are given in Table 1.

In order to identify such taskforces and the types of training, the recruitment process or capacity building process is divided into three areas (High Level, Technical Level, and Management Level) based upon the level of employees and their works.

Table 1 shows the three main areas in the government. The officials in each area are the main actors in e-government systems. The policy makers are the officials who develop the

Table 1. Capacity building

Phase	Tasks(T)	Sub Tasks(ST)
Capacity Building	Need Assessment	Identify the Need
	High Level Training	Training/Workshop for Policy Makers
	Technical Training	Training/Workshop for Technical Officials
	Management Training	Training/Workshop for Managerial Officials

policy and govern the government. People such as ministers and administrators belong to policy makers. These people are required to be aware on need and significance of e-government system in their country. In the area of technical line, the officials are mostly from technical background. People such as engineer, system administrators, and network specialists belong to this area. They should be aware of the latest technology and trained in using the latest affordable technology. The chief information officers and general managers belong to the managerial people area. They are the middle level in the government; should bridge between policy makers and technical people. Without knowing the importance of e-government, information management, and business process reengineering, they cannot communicate between policy makers and technical people. Hence, they are also required to be aware on it.

Capacity building takes the responsibility in proving required training to officials of each area and equips them with the latest knowledge and information.

High Level Training (T2)

The main purpose of this task is to identify the capacity of high-level officials in the country and provide the training to them and the course is designed accordingly.

Technical Training (T3)

The main purpose of this task is to identify the technical officials in the country and equip them with the latest technology, and its use in e-government systems. The technical capability of technical officials is identified, and the course is designed based upon their skills and country's requirements.

Management Training (T4)

The main purpose of this task is to identify the managerial officials and provide the training to them. The course is designed as per the need and training is provided.

Sub Tasks

- **Identify the Need (ST1):** The main purpose of this sub task is to identify the partner country's need in capacity development.
- **Training and Workshop for Policy Makers (ST2):** The main purpose of this sub task is to conduct the training and workshop for policy makers.
- **Training and Workshop for Technical People (ST3):** The main purpose of this sub task is to conduct the training and workshop for technical officials.
- **Training and Workshop for Managerial Officer (ST4):** The main purpose of this sub task is to conduct the training and workshop for managerial officials.

Risk Management Process

Risk management process is concerned with identifying risks and drawing up plans to minimize their effect on an e-Government system. Risk management does not just help us to prevent disasters because of the risk, but it helps us to reduce the impact of risk and make us ready to carry out alternative prevention mechanism. Risk is in every step of system development. If a risk in one stage is not handled, then the consequences of the risk accumulate, and the project would be out of control and nothing can be done. It is concerned with identifying, analyzing, and responding to project risk. It consists of risk identification, risk analysis, risk evaluation, and risk treatment or mitigation.

In the case of e-Government, the process of risk management is needed to be done throughout

Table 2. Risk management need

Question	Answer
Why do we need risk management in e-Government?	To continue e-Government projects.
When do we need risk management?	Throughout project life cycle.
Who is the responsible for managing the risk?	Specific team related to risk management
How does the risk get managed?	With risk management process.

Figure 2. Risk management process

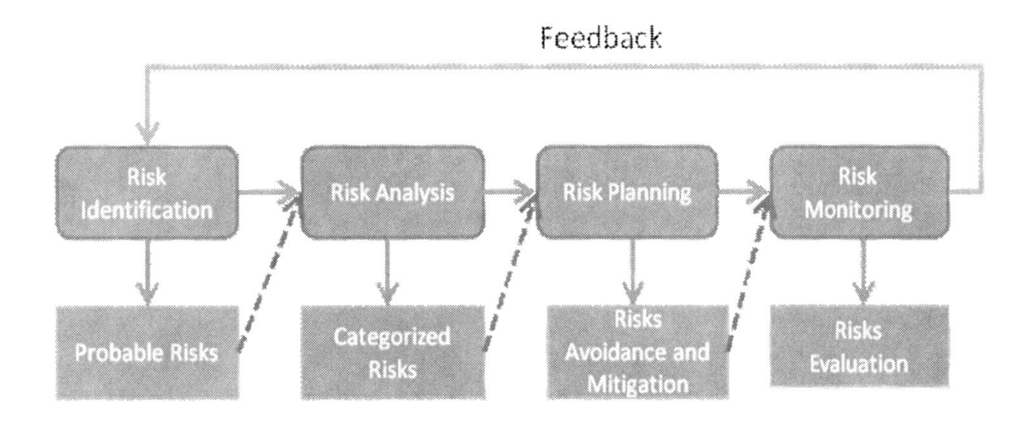

the phases. In order to know the significance of risk management in e-Government, the following questions and its answers would help to understand it better way as given in Table 2

The process of risk management is given in Figure 2.

Figure 2 is the risk management process. In general, there are four main stages. They are:

- Risk Identification
- Risk Analysis
- Risk Planning
- Risk Monitoring

The risk management process in the Figure 2 is iterative in nature, i.e. it moves throughout the project life line. The risk management process starts from the first stage, i.e. identification of the risk. This stage does the simple research on identifying the likely risks and it produces the outcome as the document of list of probable risks. The second stage is risk analysis which is done based upon the outcome produced by the previous stage, i.e. risk identification. The third stage is the risk planning which is focused on planning the mechanism of encountering the risks. This produces the document that covers the risk avoidance and mitigation mechanisms for identified risks. Finally, the monitoring on the risk is done and it also provides the document on risks assessment as outcome. At the final stage, the presence of risk is identified. If risk still exists, then again it moves to the risk identification stage to redo the procedures.

CONCLUSION

Implementing e-governance does means putting it into action. Implementing the e-government system is not easy. The different types of implementing strategies are to be analyzed thoroughly.

The best strategies are to be used, as it differs from one system to another. Economic factors of the country is one, but it is not everything. There are so many reasons behind it. In a country where political turmoil is maximum, their chances of implementing e-government master plan is very low. The commitment of leader is required to implement it. If the proposed system is not implemented, then the effort would be in dire straits. That could be finance, manpower, expertise, time, etc., and sometimes it has possibility to make the country bankrupt or failure state.

Risk is also identified as one of the main challenges in implementing e-government projects. It has to be identified during the project life cycle. The risk management process is required to be followed to mitigate the existence of risks in the e-government systems.

Chapter 6
Marketing E-Government to Citizens

María de Miguel Molina
Universitat Politècnica de València (UPV), Spain

Carlos Ripoll Soler
Universitat Politècnica de València (UPV), Spain

ABSTRACT

This chapter explores the different literature that analyses the application of marketing strategies in e-Government to help government managers promote citizen communication and participation. On occasion, the application of marketing strategies to encourage citizens to use G2C (Government to Citizen) tools is poor. Providing citizens with information is the first step whilst the final step of governments should be to reach citizen participation in decisions. To achieve this final step, governments will need to carry out a network analysis in order to find out more about their citizens and their needs. They will then need to communicate effectively how they can cover these needs through e-Government. Based on several experiences, the authors aim to study the gaps that governments at any administrative level should cover to enhance their communication with citizens and so increase the use of e-Government. In this chapter, the authors propose the use of Web 2.0 tools as a new way of communication.

INTRODUCTION

Nowadays, the majority of governments are using e-Government tools as a way to improve their relations with citizens, as well as to reduce bureaucracy and expenses. However, e-Government is sometimes implemented without having carried out a previous marketing analysis to find out what citizens' needs are, to develop citizen segmentation of certain services and, subsequently, to prepare a promotion plan to show what the advantages of e-Government are and how to make it accessible to a wide range of citizens. Moreover, in this day and age, new communication tools are likely to be a good instrument for governments to increase citizen participation.

DOI: 10.4018/978-1-4666-1909-8.ch006

Thus, the main three objectives of this chapter are:

- To provide an easy conceptual framework for public managers who would like to apply marketing techniques.
- To analyze the different steps in citizen e-Government, from information through to participation.
- To learn from different experiences which are the most suitable tools to enhance the use of e-Government by citizens.

BACKGROUND

From a managerial point of view and more specifically from a marketing background, the importance of knowing who our target audience is—to be able to cater for its needs—is well recognized. However, in Public Administration the citizen focus is not as recent and was in fact not present at the beginning of the New Public Management (NPM) paradigm, which, during the 90s, focused on the effectiveness and efficiency of governments. In fact, according to Criado (2009), e-Government was a marginal item in many NPM studies until works by significant authors such as Margetts and Hood came along in the new millennium. However, at present all government levels should take e-Government into account because "due to their multidimensionality, the management of the innovation processes entailed by e-Government requires the availability of specialized competencies in various domains, including technology, marketing, communication, policy making and project management..." (Castelnovo, 2009). Kotler and Lee (2006) remarked that ICT could be analyzed as tools for marketing purposes but that e-Government is not a specific field of Public Management *per se*. In fact, they point out that e-Government refers to using the Web as a communication or even as a distribution channel, i.e. what some call e-Governance.

However, some authors think that Governance represents a new way of governing and its raison d'être lies in the need to increase all stakeholders' involvement in the management of private and public organizations (Freeman, 1994; Rhodes, 1996; Freeman, 2004; Frederickson, 2005; Rhodes, 2007), mainly by greater participation in policy-making. This is a post-NPM or Network Governance phase (Osborne, 2006) in which each public administration must evolve in accordance with its own environment. In other words, political leaders should have the skills required to introduce NPM ideas while taking the various different stakeholders into consideration (Ferlie & Steane, 2002). Governance tends to go beyond top-level policy co-ordination to where the public sector co-ordinates and co-operates with non-state actors, such as firms and voluntary organizations, to deliver policy outcomes (Acevedo & Common, 2006). Therefore, Governance is an extension of NPM but not a new paradigm. It is rather a combination of paradigms depending on the country and agency under analysis (Andresani & Ferlie, 2006). Governance may be the foundations on which to build NPM as this would be a way of working towards relational government (De Miguel, 2010). It would be a case of starting from the beginning and instead of simply governing society, introducing a way of governing with society.

In this extension of the NPM concept, Osborne (2006) speaks of the transition to a model of New Public Governance (NPG). According to this author, NPM was a transitional period on the way towards NPG, which is based on the network theory and on social capital studies, and even goes as far as relational marketing. For the OECD one of the objectives of good governance involves, among other things, improving the transparency of governments with regard to their citizens (Grönlund & Horan, 2004). However, even Osborne (2010a) does not propose NPG as a new and definitive paradigm of public service delivery but as part of a triple model where Public Administration, NPM and NPG coexist. However, even when taking into account networks not all authors have a common view. While Rhodes "viewed the new governance as traditional structures plus networks, Klijn views it as networks alone" (Lynn, 2010).

Governance networks will be "more or less stable patterns of social relationships (interactions, cognitions, and rules) between mutually dependent public, semi-public and private actors that arise and build up around complex policy issues or policy programs" (Klijn, 2010). In any case, there are still many questions to be answered about how citizens can become more involved in the decision-making process using horizontal networks, which are different to traditional institutions (Klijn, 2005). To that end, participation would be through horizontal networks in which citizens collaborate as simply another stakeholder, providing ideas, knowledge and seeking consensus as long as participation is consistent and involves individual citizens as opposed to non-governmental organizations (deLeon, 2005). Some experiences have been carried out in several countries but it seems that the network analysis is what is left of value for analyzing complex policy processes (Acevedo & Common, 2010). In fact, even when we take networks into account, equal participation is not ensured as some members of society have no access to ICT and have silent voices (Berry, et al., 2004). Thus, it is clear that new technologies may contribute towards democratization in developed countries. However, they cannot be regarded as the only means of increasing participation since not all citizens have access to them. This is particularly true of less developed countries where the 'technological divide' is greater (Kim, et al., 2005). Consequently, a variety of techniques are still required to increase participation. If not, the opportunity to participate is limited to the so-called connected elite.

Moreover, some literature warns that the separation of politics and administration is unhelpful for a marketing model, such as the continued separation of the fields of e-Government and e-Democracy (Collins & Butler, 2003). It seems that e-Government is about service delivery and e-Democracy is about representation and participation, but both are part of the political system. Thus, if we apply the market model while forgetting that the development of public policy and the delivery of public services are not only a production but also a political process, we may miss compulsory access, parity, and deliberation. Market solutions are not a substitute for government planning and management. Thus, managers should balance technical and political concerns to secure public value (Hefetz & Warner, 2004).

In the case of public services, we believe that there is no separation between e-Government and e-Democracy but rather different phases of e-Government that demonstrate their evolution, such as the levels of participation in Arnstein's so-called ladder of participation which dates from 1971 (Borge, 2005). This consists of five levels: (1) information, (2) communication, (3) consultation, (4) deliberation, and (5) participation in decisions and elections.

When using ICT, we can compare Hagen and Snellen's models. According to Hagen (1997), there are four dimensions of political participation: (A) information, (B) active political discussion, (C) voting, and (D) political activity. From his point of view (p. 9-10), e-Democratisation is linked to representative democracy (dimensions A, B, C) while Cyberdemocracy is related to direct democracy (dimension D).

We believe that Snellen's approach (2005) is more relevant as it points out that the implementation of e-Government in public administrations goes through various phases, although few administrations have yet reached the last phase, which would be electronic (participatory) democracy. In general terms, these phases follow this order: (a) improving internal processes with the assistance of new technologies, (b) providing information via a website, (c) interacting with citizens, (d) carrying out virtual transactions, and, finally, (e) setting up a virtual democratic administration, as opposed to a representative democracy. To reach each step in these e-Government phases, it is important to get to know our citizens in order to give them not only relevant information but also to increase electronic transactions and, if we have resources, to reach virtual democracy. Some attempts have been developed in various countries using ICT, for example introducing e-voting systems (Romero & Téllez, 2010), but we must bear in mind that even if there is globalization regarding governance and relational values, the challenges may not be the same in all countries (Osborne, 2010b). Even in some developed countries, citizens only visit government websites to obtain information, rather than interacting or transacting with them (Kumar, et al., 2007; De Miguel, 2010).

Thus, the first premise should be a network analysis to get to know our citizens. According to Osborne and Plastrik (2003), public managers should analyze not only their primary target (the citizens that are directly related to their decisions) but also their secondary target (the citizens that may be indirectly benefited by these decisions). To develop an e-Government vision and to plan the strategy of electronic service development more efficiently, state agencies need to improve their understanding of their target market, its size and expectations (Golubeva & Merkuryeva, 2006). Market research enables the electronic services which are most demanded by consumers to be identified. This means efforts can be concentrated on their development and also on defining factors constraining the expansion of e-Government so as to determine the most promising ways of increasing demand. Some channels are better suited to particular types of services (Pieterson, et al., 2008) so that not all services will be suitable for e-Government.

At this step, as Kotler and Lee (2006) point out, we are moving from information (as a communication channel) to service delivery (as a distribution channel), in order to receive services, order products, and perform transactions. When citizens are prepared, then ICT can be used as a participation tool. Thus, we cannot think of citizens just as individual consumers, such as in a private organization (transactional relation), but as social citizens in a network relation (Osborne, Laughlin, & Chew, 2010). Some public services can operate in a market sector but the same cannot be said for all public services. Moreover, traditional marketing may not be sufficient for a Governance model, where the concept of Relationship Marketing arises to study three levels of marketing activity: micro (organization-consumer), macro (organization-organization), and meso (organization-society) (Osborne, McLaughlin, & Chew, 2010). However, we believe that these two marketing perspectives are not opposites. We can take various stakeholders into account when offering a new service. However, subsequently we may need different marketing strategies to reach them. Citizen characteristics need to be properly understood, along with other factors that generate satisfaction, before adopting an effective e-Government strategy (Kumar, et al., 2007).

For this reason, a marketing strategy position must be implemented in order to study: a) the public organization we are managing, b) our primary and secondary citizens, and c) the public service we would like to offer to them through e-Government. Therefore, a traditional marketing strategy can be useful to plan subsequent applications at an operational level. A marketing strategic plan follows different phases (Figure 1):

1. **Introduction.** Why do we need a Marketing Plan?
2. **SWOT Analysis.** What is our organization like? Who are our targets? What service we would like to offer?
3. **Objectives.** Where do we want to get to from our current position?
4. **Marketing Mix.** How are we going to achieve these objectives through specific actions?

Figure 1. Phases in a marketing strategic plan (author's own data)

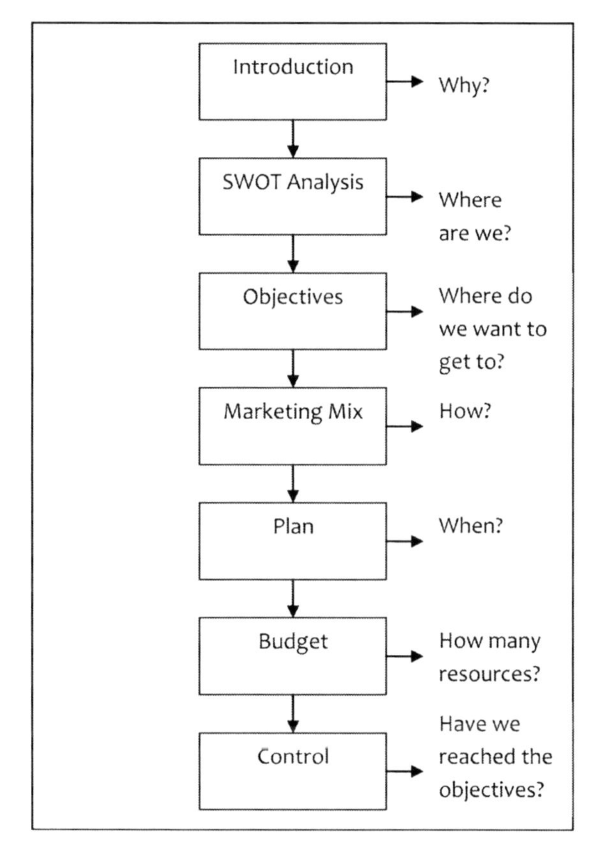

5. **Action Plan.** How can we link objectives and action in time? Who are the responsible managers?
6. **Budget.** Do we have resources to carry out the entire plan?
7. **Control.** Are our objectives being reached?

While product marketing studies four variables in the marketing mix (the 4 Ps): product, price, position, and promotion, some authors extend them to 7 Ps in the case of services, adding place, processes, and personnel (Kotler, 2004). However, Lovelock (2008) argues that these three elements can be studied alongside the service element because they are the characteristics that make services different from products.

HOW TO ENSURE CITIZENS USE E-GOVERNMENT

A Marketing Communication Plan to Enhance e-Government

It does not matter whether we consider e-Government as a distribution channel or as a communication tool, or whether we study a specific target or take a relationship focus, whatever the approach we need citizens to be aware of what our e-Government proposals are. There are a large number of so-called potential early adopters of e-government but take-up is low because awareness of e-channels is low. One solution to increase take-up is to run targeted marketing communication campaigns (Mellor, 2006).

Promotion is the key to informing citizens of our services, persuading them to use these services or reminding them of their availability. There are many communication tools but we can divide them into five groups:

- Advertisement, massive effect in the long term.
- Sales promotion, specific effect in the short term.
- Public relations, indirect effect through secondary citizens.
- Sales force or internal communication, internal effect to motivate employees.
- Direct communication, seeking personalized contact with citizens.

To coordinate all these tools, a promotion program can be created, taking into account:

- What are the services we would like to promote?
- Who is the target audience we want to reach?
- What is the promotional period?
- What is going to be the frequency of our message?
- And finally, evaluating the results.

To help managers in the use of new communication tools that could be used in their promotion programs, we will analyze Web 2.0 tools.

Specific Web 2.0 Communication Tools for Effective e-Government Marketing

Governments currently have a presence on the Internet that is mainly based on a website that performs one-way communication from government to citizen or what Reddick (2005) calls "the supply-side perspective." On the other hand, we have the demand-side perspective that considers what citizens' actual information requirements are. These real demands should be transformed into customized content so that it reaches citizens more efficiently.

From Traditional Marketing Approaches to Online Marketing

Governments usually adapt business-oriented communication solutions to their own needs. In this sense, Kotler and Lee (2006) proposed a list of several categories that government could use to establish communication channels with their citizens. Moreover, each category can be divided into offline and online tools (Table 1).

However, Leskovec, Adamic, and Huberman (2006) detected that "consumers show an increasing resistance to traditional forms of advertising such as TV or newspaper ads." This has forced organizations to find alternative communication strategies to traditional marketing. Thus, online marketing has become an alternative where the use of viral marketing has also been tried successfully. Combining both the "offline world" with the "online world" is one of the key elements that governments should consider in order to establish better communication with citizens. Traditional approaches should not be completely discarded but an intelligent move towards online alternatives to mutually reinforce elements of offline approaches will benefit in the short term. The offline world is still there and we know how it works, but the online world is becoming central to every aspect of our existence and has much potential to exploit.

Kleinberg (2008) states that "the past decade has witnessed a coming-together of the technological networks that connect computers on the Internet and the social networks that have linked humans for millennia." These networks are capturing our different styles of communication yet remain governed by our principles of human social interaction. Communication and interaction are one of the cornerstones of human existence. Thanks to online tools, these principles can now be observed, measured and quantified at amazing levels of scale never seen before. The data generated by a single online social network can easily help to give an accurate profile of every individual.

New Digital Marketing Channels to Promote Citizens' Active Participation

A new approach that promotes active citizen participation has been requested for a long time. Citizens should evolve from service consumers to service prosumers, leveraging their role towards active participation. Tapscott and Williams (2006) declared that "in the new prosumer-centric paradigm, customers want a genuine role in designing the products of the future." The same can be applied to e-Government where citizens are willing to play an active role beyond their momentary participation in any elections. For e-Government, it is an important departure point to involve citizens in designing services adapted to their real needs.

Due to the high interrelation between the offline and online world, establishing new digital marketing channels will help to better understand what and why people do things. These channels are based on the success of the Web 2.0. The main advantage for e-Government is that these channels foster effective communication with citizens. In addition, Osimo (2008) points out the following benefits of using Web 2.0 tools in e-Government (Table 2).

Table 1. Traditional marketing vs. online marketing (author's own data adapted from Kotler & Lee, 2006)

Category	Offline	Online
Advertising	Television, Radio, Newspapers, magazines, ads on backs of tickets and receipts, ads at theater using still shots and videos, billboards, busboards, bus shelter displays, subways, taxis, vinyl wrap on cars and buses, sporting events, kiosks, restroom stalls, airport billboards and signage	Internet, Adsense, Adwords, search engine marketing, online newspapers, online magazines, ipTV, ads on mobile applications
Public relations	Stories on television and radio, articles in newspapers and magazines, videos	Blogging, articles in online newspapers and magazines, YouTube videos, podcasts, Internet radio
Special events	Community meetings, demonstrations/exhibits, fairs and tours, face to face meetings with users of online communities	
Direct marketing	Mail, telemarketing, catalogs	Newsletters, customized emails, RSS feeds, Google reader
Printed materials	Forms, brochures, newsletters, flyers, calendars, posters, envelope messages, booklets, static stickers,	
Special promotional items	Clothing: T-Shirts, baseball hats, diapers, bibs Transient items: coffee sleeves, bar coasters and napkins, buttons, temporary tattoos, balloons, stickers, fortune cookies. Functional items: Key chains, flashlights, refrigerator magnets, water bottles, litterbags, pens and pencils, bookmarks, book covers, notepads, tote bags, mascots, cell phone cases.	Online discounts, free trial periods
Signage and displays	Road signs, signs and posters on government property or property regulated by government	
Personal communication channels	Face to face meetings and presentations, workshops, seminars, training sessions, word of mouth	Word of Web, blogs, chats, forums, twitter
Popular media	Public art, songs, scripts in movies, television and radio programs, comic books and comic strips, playing cards and other games	

Table 2. Benefits and applications of Web 2.0 tools to e-government (author's own data adapted from Osimo, 2008)

Benefits	Examples (UK)
Simple and user-oriented	TheStraightChoice.org allows to follow citizenship participation in general elections
Transparent and accountable	Service writetothem.org allows different stakeholders to be contacted
Participative and inclusive	Fixmystreet.org allows citizens to report, view or discuss local problems
Joined up and networked	Pledgebank.com helps get things done, especially things that require several people

Figure 2. Role of citizens and number of users (author's own data adapted from Osimo, 2008)

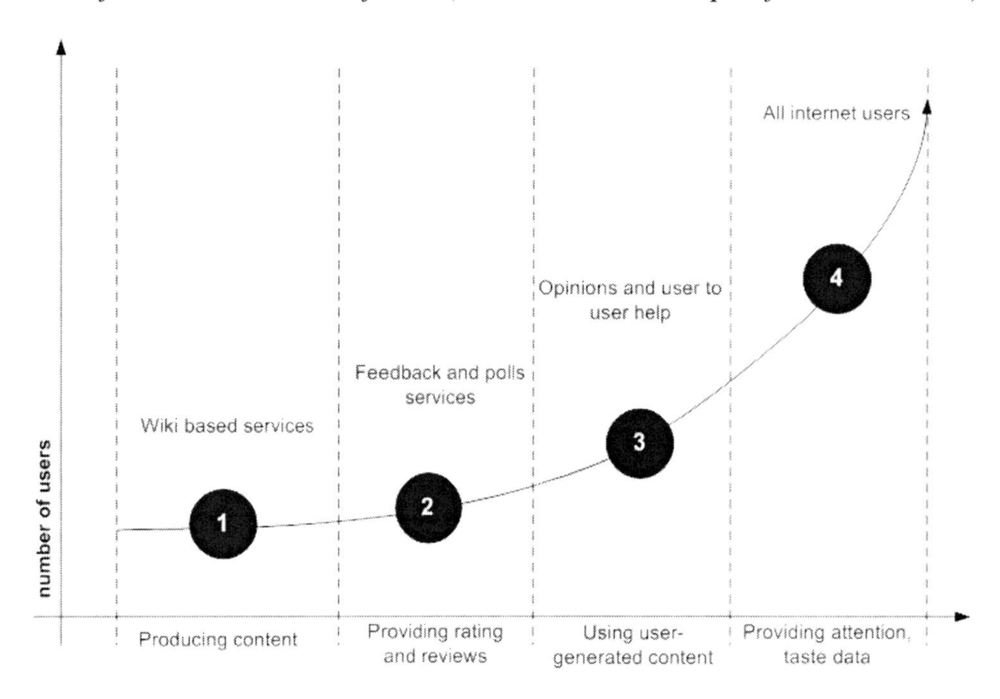

Moreover, we can distinguish different roles according the number of potential users who increase as the services become more Web 2.0 oriented (Figure 2).

In the following subheadings, we aim to analyze some of these Web 2.0 communication tools.

Audio and Video

There are many audio and video alternatives that can be used. Beyond traditional TV commercials or shows, we can use a wide variety of means that will help to improve our relationship with citizens. Thus, when talking about audio, the use of podcasts becomes one of the most relevant actors. Podcasts enables any audio content to be immediately broadcast to the Internet. The citizen can subscribe to an organization's podcasts, download them to his/her computer or iPod and play them whenever he/she wishes to.

Some governments have already decided to establish their own podcasts on the Internet. They are organized in episodes so that any updates can be sent as frequently as considered necessary. Although the most popular format is the audio podcast, video podcasts are also available. Some governments are using this kind of videos to provide a weekly address of the president to their citizens, for instance. It is particularly relevant that podcasts are consumed in different places from typical TV commercials or shows. This means that people using their iPods or mp4 players will hear an organization's message in a totally different environment, such as the underground, a bus, a taxi, or an airport.

Another interesting option is setting up a YouTube channel that contains all the available video materials. This does not just distribute content (as in the case of podcasts), but also allows people to comment or share it with other colleagues. Many universities have a YouTube channel. For example, the Universidad Politécnica de Valencia's Lifelong Learning Institute has developed its own YouTube channel containing videos that reinforce their current services to citizens (Figure 3).

Figure 3. YouTube channel: Lifelong Learning Institute, Universidad Politécnica de Valencia. Retrieved from www.youtube.com/upvcfp (2010).

Both cases enable us to gather usage level data through Web analytics, as we will analyze afterwards. We know how many people download a podcast and how many really see a podcast; how many people have reproduced a YouTube video, as well as other interesting statistics, such as geographical location.

Online Social Networks

Acquisti and Gross (2006) define online social networks at a very basic level as "an Internet community where individuals interact, often through profiles that (re)present their public persona (and their networks of connections) to others." Moreover, Boyd and Ellison (2008) define "social network sites as Web-based services that allow individuals to (1) construct a public or semi-public profile within a bounded system, (2) articulate a list of other users with whom they share a connection, and (3) view and traverse their list of connections and those made by others within the system."

Social Network Sites (SNS) can serve a dual purpose in e-government:

- Provide a better understanding of citizens' needs through data processing.

Figure 4. Universidad Politécnica de Valencia's Facebook group. Retrieved from www.facebook.com/UPolitecnicaValencia?ref=ts (2010).

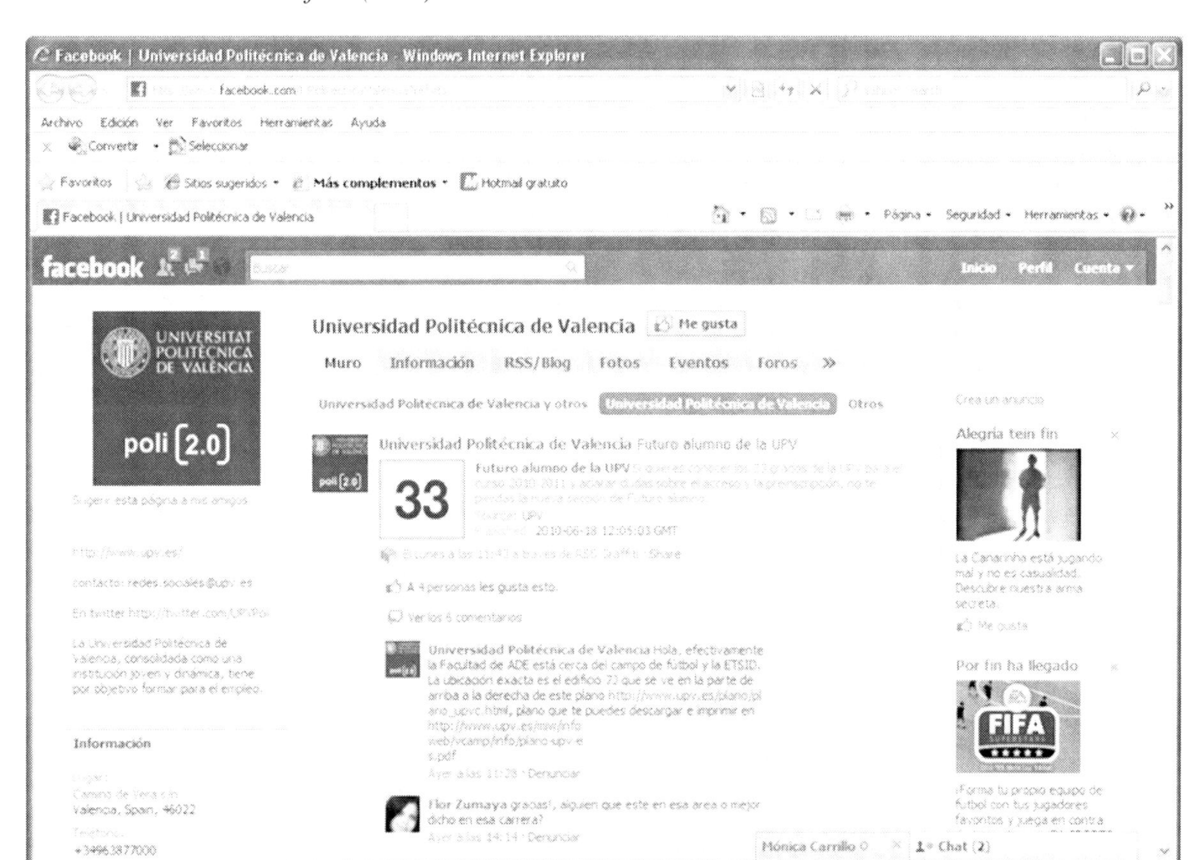

- Allow direct interaction with citizens by means of immediate interaction or by including ads.

Facebook and Twitter are two of the most relevant SNS. Ellison, Steinfield and Lampe (2007) have reviewed the evolution of Facebook from its creation in 2004. This SNS has reported over 200 million registered users. The site enables groups to be created, even by organizations. For example, a city council could create its own group and interact with its citizens in a different way than just providing a website. Facebook also allows an advertisement system that can deliver messages to very specific target groups. Ellison, Steinfield and Lampe (2007) remarked that "the typical users spend about 20 minutes a day on the site, and two thirds of users log in at least once a day."

Figure 4 shows an example of the Universidad Politécnica de Valencia's Facebook group. Students can participate in this group by asking direct questions. A group should also be encouraged by a community manager. He/she is in charge of promoting interaction between users. In this case, there is a full time employee dedicated to Web 2.0 channels.

Facebook also creates complete profiles of its users. In fact, it is the SNS that retrieves the greatest amount of information from them. Stutzman (2006) compared Facebook with 2 different major networks, such as MySpace and Friendster and clearly demonstrated that Facebook requests and gathers more relevant information from their users.

On the other hand, Twitter is a SNS based on micro-blogging to connect with counterparts, family, or coworkers. Its main characteristic is that it can instantly deliver messages (up to 140 characters) to users that are willing to receive a message from another user. According to Lenhart and Fox (2009), around 19% of Internet users actively participate in this social network. Zhao and Rosson (2009) also classify twitter as a micro-blog which provides a new communication channel for people to broadcast information that they would not likely share otherwise using existing channels (e.g., email, phone, IM, or weblogs). Micro-blogging has become popular quickly, raising its potential to serve as a new informal communication medium at work and providing a variety of impacts on teamwork (e.g., enhancing information sharing, building common ground, and sustaining a feeling of connectedness among colleagues). Users decide who they are interested in following, and are notified when one of these people has posted a new message.

Newsletters

A newsletter establishes a periodical communication channel with citizens. Information is collected regularly and then delivered to the citizen. First of all, this information has to be customized according to each citizen's defined preferences. This requires a database to store these profiles. The profiles enable customized information to be sent to each citizen, thus increasing the probability that the citizen will read that particular email. In addition to just providing information to citizens, newsletters are also an interesting way of gathering information from citizens. When a newsletter is sent we can find out how many people open the email, who opens it and which parts he/she clicks on. By way of example, Figure 5 shows the newsletter made by the Lifelong Learning Institute at the Universidad Politécnica de Valencia. This newsletter includes all the learning programs offered by this university at any particular time. The system customizes the email according to the customer preferences and records exactly what he/she does with the email. People's profiles are updated constantly with the information obtained.

Finally, we can use more diverse tools to reach and cater for citizen participation and their interests more efficiently.

Search Engine Optimization (SEO)

Nowadays there are millions of Web pages, sites, and contents. Terabytes of information circulate daily around the Internet. Thus, the main goal is to ensure that people find exactly what they are looking for. Therefore, organizational websites need to be designed in such a way that search engines understand exactly what is in them. Search engine optimization ensures that a specific government website is correctly classified and easily located. Thus, if a citizen types in a word, the government website they have searched for should appear. Search engines, such as Google, Yahoo, or Bing, are likely to be the main gateway for visitors to our websites. Citizens mainly go to a search engine, write a search term, and surf through the proposals made by the search engine. Search engines have become Internet advisors to enable people to find what they are really looking for. Otherwise, it would be impossible to know all the potential places that exist for consideration.

Web Analytics

Web analytics is a complementary tool for any of the other tools mentioned before. Web analytics can be used to monitor a website, a blog, or even a Facebook group. Kaushik (2010) presents Web analytics as a tool "to determine how to market effectively, how to truly connect with our audiences, how to improve the customer experience on our sites, how to invest our meager resources, and how to improve our return on investment, be it getting donations, increasing revenue, or winning elections!"

Currently, there are many ways of analyzing our data. One of the most famous tools is Google

Figure 5. Newsletter made by the Lifelong Learning Institute, Universidad Politécnica de Valencia. Retrieved from www.cfp.upv.es (2010).

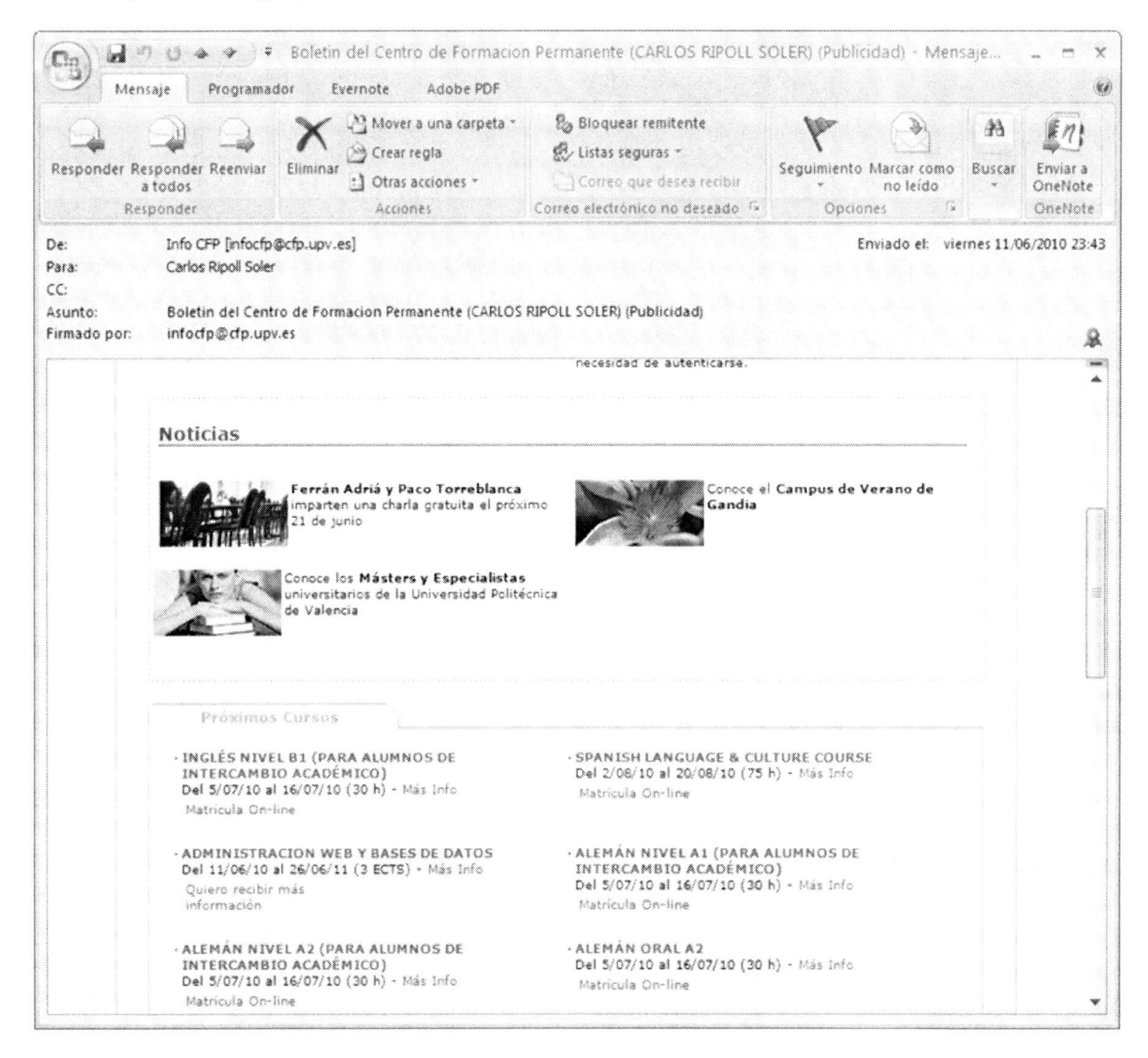

Analytics. It provides huge amounts of information and statistics that can be used to find out more about which people are visiting our sites. Figure 6 shows an example of the control panel on the analytics tool of the Lifelong Learning Institute at the Universidad Politécnica de Valencia.

Solutions and Recommendations

As we have explained, the Web 2.0 provides governments with new ways of communication, which could increase citizen participation. Moreover, they give us the chance to collect highly useful reports on website use. However, this does not mean casting aside traditional tools but instead mixing offline and online worlds to reach a wider range of citizens. We should also point out that e-Government has been developed in the online world. This means that its users can easily be reached through these 2.0 tools if search engine optimization is applied.

Figure 6. Google analytics control panel from the Lifelong Learning Institute, Universidad Politécnica de Valencia. Retrieved from www.google.com/analytics (2010).

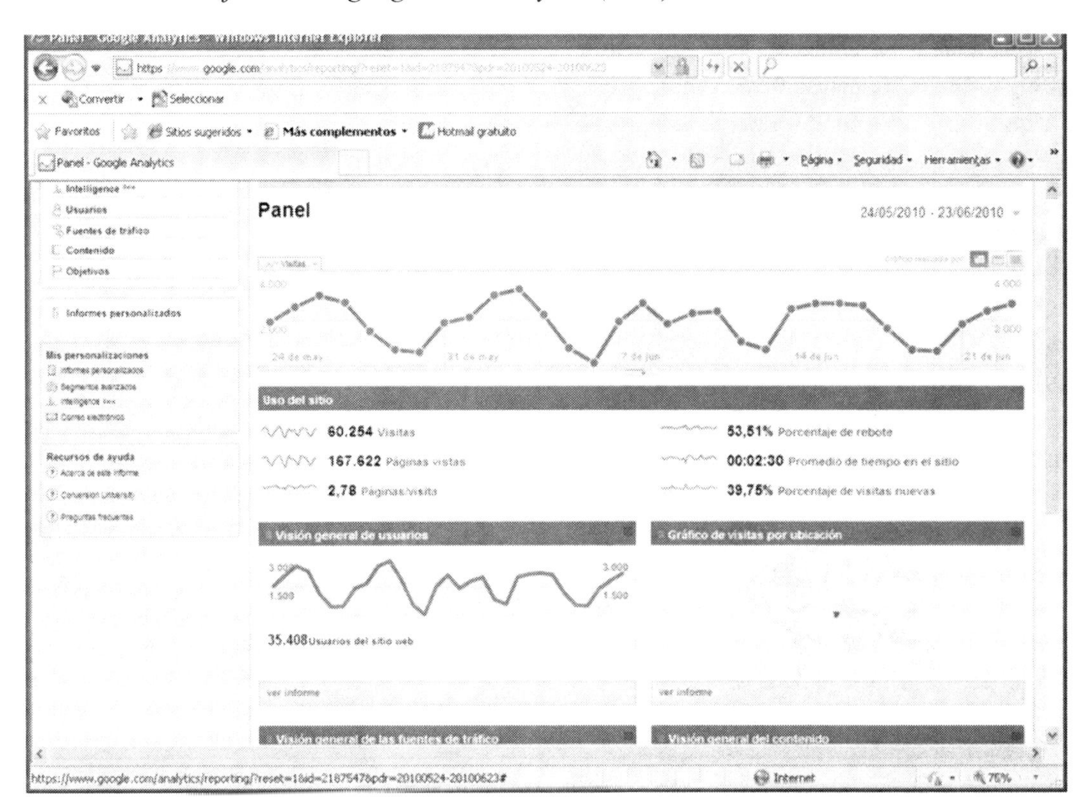

However, we should take into account the fact that in our desire to reach all our citizens there are still groups of people who do not have access to the online world or are not prepared to use it. This implies that governments also need to improve dissemination policies and other policies aimed at helping these citizens gain access to the Internet.

FUTURE RESEARCH

The trend of citizen participation is being analyzed by many scholars. However, we think these studies often have a political focus rather than a managerial and marketing approach. As we have underlined in the Background section, many governments will still need realistic marketing plans with previous network analysis in order to get to know their citizens and their needs. Without this knowledge, it will be difficult to be effective and efficient in the delivery of public services through e-Government and even more difficult to increase participation in the new e-Governance.

On the other hand, the use of Web 2.0 communication tools is still very recent. This implies that future research will be necessary to measure the efficacy and efficiency of these tools in real case scenarios. This would also allow us to compare the use of e-Government in different countries and get an idea as to how it is used in the different public sector levels of a country. The first issue will be especially interesting when comparing developed and less-developed countries, and democratic and non-democratic countries.

We believe that perspectives on the study of e-Government, apart from this chapter, which focuses on marketing to citizens, will increase in future years. It will thus be necessary to develop multidisciplinary analyses to enrich research into this field.

CONCLUSION

This chapter adds a marketing point of view to this interesting publication on Public Administration in the Digital Age, and encourages governments to use Web 2.0 communication tools in order to increase citizen participation. However, governments at any administrative level firstly have to realize that "the success of e-Government efforts depends, to a great extent, on how well the targeted users for such services, citizens in general, make use of them" (Kumar, et al., 2007). In this sense, if we want to increase the use of e-Government, a strategic perspective needs to be adopted beforehand so as to develop the services that citizens really need and to communicate them successfully.

As we suggest podcasts, videos, social network sites, and personalized newsletters are online tools that can enrich communication between governments and their citizens, and at best encourage citizen participation in order to reach the so-called "governance" status that at present seems to be the goal of many governments. However, it must be said that these tools have to form part of a planned promotion program in which search engine optimization should be used.

Furthermore, as we have pointed out in the previous section, these tools are new ways to communicate with citizens that will need more analysis in the future to test whether they have contributed to achieving government objectives. At this point in time, Web analytics is a powerful source of information available to governments so as to assess their management decisions regarding e-Government.

REFERENCES

Acevedo, B., & Common, R. (2006). Governance and the management of networks in the public sector. *Public Management Review*, *8*(3), 395–414. doi:10.1080/14719030600853188

Acevedo, B., & Common, R. (2010). Governance, networks and policy change: The case of cannabis in the United Kingdom. In Osborne, S. P. (Ed.), *The New Public Governance?* (pp. 394–412). New York, NY: Routledge.

Acquisti, A., & Gross, R. (2006). Imagined communities: Awareness, information sharing, and privacy on the Facebook. In P. Golle & G. Danezis (Eds.), *Proceedings of 6th Workshop on Privacy Enhancing Technologies,* (pp. 36–58). Cambridge, UK: Robinson College.

Andresani, G., & Ferlie, E. (2006). Studying Governance within the British public sector and without. *Public Management Review*, *8*(3), 415–431. doi:10.1080/14719030600853220

Berry, F. S., Brower, R. S., Choi, S. O., Goa, W. X., Jang, H., Kwon, M., & Word, J. (2004). Three traditions of network research: What the public management research agenda can learn from other research communities. *Public Administration Review*, *64*(5), 539–552. doi:10.1111/j.1540-6210.2004.00402.x

Borge, R. (2005). La Participación Electrónica: Estado de la Cuestión y Aproximación a su Clasificación. *IDP: Revista de Internet* [UOC]. *Derecho y Política*, *1*, 1–15.

Boyd, D., & Ellison, N. (2008). Social network sites: Definition, history and scholarship. *Journal of Computer-Mediated Communication*, *13*(1), 210–230. doi:10.1111/j.1083-6101.2007.00393.x

Castelnovo, W. (2009). Supporting innovation in small local government organizations. In *Proceedings of the 3rd European Conference on Information Management and Evaluation,* (pp. 99-106). IEEE.

Collins, N., & Butler, P. (2003). When marketing models clash with democracy. *Journal of Public Affairs*, *3*(1), 52–62. doi:10.1002/pa.133

Criado, J. I. (2009). *Entre sueños utópicos y visiones pesimistas*. Madrid, Spain: Instituto Nacional de Administración Pública.

De Miguel, M. (2010). E-government in Spain: An analysis of the right to quality e-government. *International Journal of Public Administration, 33*(1), 1–10. doi:10.1080/01900690903178454

Deleon, L. (2005). Public management, democracy, and politics. In Ferlie, Lynn Jr., & Pollit (Eds.), *The Oxford Handbook of Public Management*. Oxford, UK: Oxford University Press.

Ellison, N., Steinfield, C., & Lampe, C. (2007). The benefits of Facebook friends: Social capital and college students' use of online social network sites. *Journal of Computer-Mediated Communication, 12*(4), 1143–1168. doi:10.1111/j.1083-6101.2007.00367.x

Ferlie, E., & Steane, P. (2002). Changing developments in NPM. *International Journal of Public Administration, 25*(12), 1459–1470. doi:10.1081/PAD 120014256

Frederickson, H. G. (2005). Whatever happened to public administration? Governance, governance everywhere. In Ferlie, Lynn Jr., and Pollit (Eds.), *The Oxford Handbook of Public Management*. Oxford, UK: Oxford University Press.

Freeman, L. C. (2004). *The development of social network analysis: A study in the sociology of science*. Vancouver, Canada: Booksurge.

Freeman, R. E. (1994). The politics of stakeholder theory: Some future directions. *Business Ethics Quarterly, 4*(4), 409–421. doi:10.2307/3857340

Golubeva, A., & Merkuryeva, I. (2006). Demand for online government services: Case studies from St. Petersburg. *Information Polity: The International Journal of Government & Democracy in the Information Age, 11*(3/4), 241–254.

Grönlund, A., & Horan, T. A. (2004). Introducing e-gov: History, definitions and issues. *Communications of the Association for Information Systems, 15*, 713–729.

Hagen, M. (1997). *A typology of electronic democracy*. Retrieved 2010 from http://www.uni-giessen.de/fb03/vinci/labore/netz/hagen.htm

Hefetz, A., & Warner, M. (2004). Privatization and its reverse: Explaining the dynamics of the government contracting process. *Journal of Public Administration: Research and Theory, 14*(2), 171–190. doi:10.1093/jopart/muh012

Kaushik, A. (2010). *Web analytics 2.0*. Indianapolis, IN: Wiley Publishing.

Kim, (2005). Special report on the sixth global forum on reinventing government: Toward participatory and transparent governance. *Public Administration Review, 65*(6), 646–654. doi:10.1111/j.1540-6210.2005.00494.x

Kleinberg, J. (2008). The convergence of social and technological networks. *Communications of the ACM, 51*(11), 66–72. doi:10.1145/1400214.1400232

Klijn, E. H. (2005). Network and inter-organizational management. In Ferlie, Lynn Jr., & Pollit (Eds.), *The Oxford Handbook of Public Management*. Oxford, UK: Oxford University Press.

Klijn, E. H. (2010). Trust in governance networks: Looking for conditions for innovative solutions and outcomes. In Osborne, S. P. (Ed.), *The New Public Governance?* (pp. 303–321). New York, NY: Routledge.

Kotler, P. (2004). *El marketing de servicios profesionales*. Barcelona, Spain: Paidós Empresa.

Kotler, P., & Lee, N. (2006). *Marketing in the public sector*. Upper Saddle River, NJ: Wharton School Publishing.

Kumar, V., Mukerji, B., Butt, I., & Persaud, A. (2007). Factors for successful e-government adoption: A conceptual framework. *The Electronic. Journal of E-Government*, *5*(1), 63–76.

Lenhart, A., & Fox, S. (2009). *Twitter and status updating*. Washington, DC: Pew Internet & American Life Project.

Leskovec, J., Adamic, L., & Huberman, B. (2006). The dynamics of viral marketing. In *Proceedings of the 7th ACM Conference on Electronic Commerce,* (pp. 228-237). ACM Press.

Lovelock, C. H. (2008). *Marketing de servicios: Personal, tecnología y estrategia*. Mexico City, México: Pearson Prentice Hall.

Lynn, L. E. Jr. (2010). What endures? Public governance and the cycle of reform. In Osborne, S. P. (Ed.), *The New Public Governance?* (pp. 105–123). New York, NY: Routledge.

Mellor, N. (2006). E-citizen - Developing research-based marketing communications to increase awareness and take-up of local authority e-channels. *Aslib Proceedings*, *58*(5), 436–446. doi:10.1108/00012530610692384

Mislove, A., Marcon, M., Gummadi, K., Druschel, P., & Bhattacharjee, P. (2007). Measurement and analysis of online social networks. In *Proceedings of the 7th ACM SIGCOMM Conference on Internet Measurement,* (pp. 29-42). San Diego, CA: ACM Press.

Osborne, D., & Plastrik, P. (2003). *Herramientas para transformar el gobierno* [The reinventor's fieldbook]. Barcelona, Spain: Paidós.

Osborne, S. P. (2006). The new public governance? *Public Management Review*, *8*(3), 377–387. doi:10.1080/14719030600853022

Osborne, S. P. (2010a). The (new) public governance: A suitable case of treatment? In Osborne, S. P. (Ed.), *The New Public Governance?* (pp. 1–16). New York, NY: Routledge.

Osborne, S. P. (2010b). Public governance and public services delivey: A research agenda for the future. In Osborne, S. P. (Ed.), *The New Public Governance?* (pp. 413–428). New York, NY: Routledge.

Osborne, S. P., McLaughlin, K., & Chew, C. (2010). Relationship marketing, relational capital and the governance of public services delivery. In Osborne, S. P. (Ed.), *The New Public Governance?* (pp. 185–199). New York, NY: Routledge.

Osimo, D. (2008). *Web 2.0 in government: Why and how? JRC Scientific and Technical Reports*. Paris, France: European Commission.

Pieterson, W., Teerling, M., & Ebbers, W. (2008). Channel perceptions and usage: Beyond media richness factors. *Electronic Government Proceedings*, *5184*, 219–230. doi:10.1007/978-3-540-85204-9_19

Reddick, C. (2005). Citizen interaction with e-government: From the streets to servers? *Government Information Quarterly*, *22*, 38–57. doi:10.1016/j.giq.2004.10.003

Rhodes, R. A. W. (1996). New governance: Governing without government. *Political Studies*, *44*(4), 652–667. doi:10.1111/j.1467-9248.1996.tb01747.x

Rhodes, R. A. W. (2007). Understanding governance: Ten years on. *Organization Studies*, *28*(8), 1243–1264. doi:10.1177/0170840607076586

Romero, R., & Téllez, J. A. (2010). *Voto electrónico, derecho y otras implicaciones.* Mexico City, México: Universidad Nacional Autónoma de México (UNAM).

Snellen, I. (2005). E-government: A challenge for public management. In Ferlie, Lynn Jr., & Pollit (Eds.), *The Oxford Handbook of Public Management*. Oxford, UK: Oxford University Press.

Stutzman, F. (2006). An evaluation of identity-sharing behavior in social network communities. In *Proceedings of the 2006 iDMA and IMS Code Conference.* International Digital and Media Arts Journal.

Tapscott, D., & Williams, D. (2006). *Wikinomics: How mass collaboration changes everything.* New York, NY: Portfolio.

Zaho, D., & Rosson, M. (2009). How and why people Twitter: The role that microblogging plays in informal communication at work. In *Proceedings of the ACM International Conference on Supporting Groupworth,* (pp. 243-252). New York, NY: ACM Press.

ADDITIONAL READING

Authors, V. (2009). E-government developments. *I-Ways, 32*(3), 136–187.

Backstrom, L., Huttenlocher, D., & Kleinberg, J. (2006). Group formation in large social networks: Membership, growth and evolution. In *Proceedings of the 12th ACM SIGKDD International Conference on Knowledge Discovery and Data Mining (KDD 2006),* (pp. 44-54). ACM Press.

Baker, D. (2009). Advancing e-government performance in the United States through enhanced usability benchmarks. *Government Information Quarterly, 26*(1), 82–88. doi:10.1016/j.giq.2008.01.004

Barzelay, M. (2001). *The new public management: Improving research and policy dialogue.* Berkeley, CA: University of California Press.

Domegan, C. T. (2008). Social marketing: Implications for contemporary marketing practices classification scheme. *Journal of Business and Industrial Marketing, 23*(2), 125–141. doi:10.1108/08858620810850254

Edmiston, K. D. (2003). State and local e-government: Prospects and challenges. *American Review of Public Administration, 33*(1), 20–45. doi:10.1177/0275074002250255

Freya, K. N., & Holden, S. H. (2005). Distribution channel management in e-government: Addressing federal information policy issues. *Government Information Quarterly, 22*(4), 685–701. doi:10.1016/j.giq.2006.01.001

Haythornthwaite, C. (1996). Social network analysis: An approach and technique for the study of information exchange. *Library & Information Science Research, 18*(4), 323–342. doi:10.1016/S0740-8188(96)90003-1

Heeks, R. (2003). Most e-government-for-development projects fail: How can risks be reduced? *IDPM E-Government Working Paper, 14*, 1–19.

Heeks, R. (2006). *Implementing and managing e-government: An international text.* Thousand Oaks, CA: Sage.

Heeks, R., & Bailur, S. (2007). Analyzing e-government research: Perspectives, philosophies, theories, methods, and practice. *Government Information Quarterly, 24*, 243–265. doi:10.1016/j.giq.2006.06.005

Hood, C., & Margetts, H. (2007). *The tools of government in the digital age.* London, UK: Palgrave.

Huberman, B., Romero, D., & Wu, F. (2008). Social networks that matter: Twitter under the microscope. *First Monday Journal, 14.* Retrieved 2010 from http://ssrn.com/abstract=1313405

Kotler, P., & Levy, S. J. (1969). Broadening the concept of marketing. *Journal of Marketing, 33*(1), 10–15. doi:10.2307/1248740

Lappas, G., & Yannas, P. (2006). A framework to evaluate political party websites. In *Proceedings of EISTA 2006: 4th International Conference on Education and Information Systems: Technologies and Applicat/Soic 2006: 2nd International Conference on Social and Organizational Informatics and Cybernetics*, (vol 2), (pp. 226-231). EISTA.

Liu, H., & Maes, P. (2005). Interest map: Harvesting social network profiles for recommendations. In *Proceedings of IU Beyond Personalization: A Workshop on the Next Stage of Recommender Systems Research,* (pp. 54-59). San Diego, CA: IEEE.

Lynn, L. E., Jr. (2005). Public management: A concise history of the field. In Ferlie, Lynn Jr., & Pollit (Eds.), *The Oxford Handbook of Public Management*. Oxford, UK: Oxford University Press.

Margetts, H. (2005). Virtual organizations. In Ferlie, Lynn Jr., & Pollit (Eds.), *The Oxford Handbook of Public Management*. Oxford, UK: Oxford University Press.

OCDE. (2003). *The e-government imperative*. Paris, France: Organization for Economic Co-Operation and Development.

Osborne, S. P. (Ed.). (2010). *The new public governance?* New York, NY: Routledge.

Stankovic, M., & Jovanovic, J. (2008). Online presence in social networks. *W3C Workshop on the Future of Social Networking*. Retrieved 2010 from http://www.w3.org/2008/09/msnws/papers/w3c-workshop-opo.pdf

Steyaert, J. C. (2004). Measuring the performance of electronic government services. *Information & Management, 41*(3), 369–375. doi:10.1016/S0378-7206(03)00025-9

Taylor, J. A., & Lips, A. M. B. (2008). The citizen in the information polity: Exposing the limits of the e-government paradigm. *Information Polity: The International Journal of Government & Democracy in the Information Age, 13*(3/4), 139–152.

Torres, L., Pina, V., & Acerote, B. (2005). e-Government developments on delivering public services among EU cities. *Government Information Quarterly, 22*(2), 217–238. doi:10.1016/j.giq.2005.02.004

Vesa, P., Suomala, J., Siltala, R., & Keskinen, S. (2006). Framework to study the social innovation networks. *European Journal of Innovation Management, 9*(3), 312–326. doi:10.1108/14601060610678176

Chapter 7
Alignment of IT Projects and Investments to Community Values

Marcus Vogt
Heilbronn University, Germany

Kieth Hales
Bond University, Australia

ABSTRACT

Information Technology (IT) investments and IT management have become increasingly important for an organization's success. The principles of IT Governance, the use of IT Governance Frameworks, the application of IT Portfolio Management, and IT Value Management are proven methodologies to lead private companies to a successful and efficient implementation of their ITT investments. However, due to the structure of public organizations, they seem to have their limits. This chapter shows the issues of IT Value Management in public organizations and discusses the usability of classic IT alignment methods.

INTRODUCTION AND GENERAL RESEARCH QUESTION

Due to increasing budget pressures and public calls for transparency, governmental institutions realize that they have to justify their IT (Information Technology) projects not only by costs but also by benefits for the community (Di Maio, 2003a; Sethibe, Campbell, & McDonald, 2007). The principles of IT Governance, the application of IT Portfolio Management and IT Value Management methods like Return On Investment (ROI), Payback Period (PBP), or Net Present Value (NPV) are proven methodologies to lead private companies to a successful and efficient implementation of their IT investments. However, when applicable in public organizations they are limited in their use because of the different strategic goals and the non-commercial function of these organizations (Di Maio, 2003a, 2003b, 2003c, 2003d; Sethibe, Campbell, & McDonald, 2007). One can find a vast number of evaluation techniques for IT investments in the private sector but only a few in the public sector, because the public sector has its unique characteristics (Markov, 2006). Markov

DOI: 10.4018/978-1-4666-1909-8.ch007

(2006) states in his paper that "future evaluations … should grant the public value of IT a higher degree of interest" (p. 26). In his opinion "researchers and public administrators should rethink the value of internally-oriented financial metrics" and focus on "external benefits of deploying IT for stakeholders such as individual citizens, the business environment or society as a whole" (p. 26).

Therefore, one has to discuss which of these proven methodologies can be used and where they have to be adjusted to meet the non-commercial character of public organizations, particular in terms of their community or public value. This chapter will analyze relevant literature about IT Governance, IT Portfolio Management, and IT Value Management in terms of how they could be used and adjusted to align the IT Portfolio of a public organization to the community values.

Definition of IT Governance

IT Governance has inherited much from Corporate Governance and IT Management, but has developed into a discrete discipline (Simonsson & Johnson, 2006). There are still overlaps between the three disciplines, but as IT evolved over the time, its definition became ambiguous because of the various definitions of IT Governance in the literature. The most common ones are:

"IT Governance is the responsibility of the Board of Directors and executive management. It is an integral part of enterprise governance and consists of the leader ship and organizational structures and processes that ensure that the organization's IT sustains and extends the organization's strategy and objectives" (ITGI, 2003, p. 10).

"IT Governance is the organizational capacity exercised by the Board, executive management and IT management to control the formulation and implementation of IT strategy and in this way ensure the fusion of business and IT" (Van Grembergen, 2002, p. 7).

"IT governance: Specifying the decision rights and accountability framework to encourage desirable behavior in the use of IT" (Weill & Ross, 2004, p. 8).

These well-known definitions of IT Governance were taken to create a new working definition in the context of public organizations or governmental institutions, which will be used throughout this chapter:

IT Governance in public organizations is the responsibility of political or public representatives, executive managers, and IT managers of these institutions or political structures. It is an integrated part of their responsibility towards the society and political directives to ensure the reasonable, effective, and efficient use of IT to support public goals and interests.

According to Van Grembergen, De Haes, and Guldentops (2004) there are "two important elements of IT Governance: value delivery (which is the goal) and strategic alignment (which is the means)" (p. 18). The ITGI (2003) adds a third element "Risk Management" and a fourth element "Performance Measurement." Both parts are equally important to balance the opportunities towards threats when a decision for an IT project or investment is made, however the fourth element "Performance Measurement" has its major part while the lifecycle of the IT investment or project.

Even though slightly different, Van Grembergen, De Haes and Guldentops (2004) are stressing that these definitions have one major thing in common: The link between business and IT; so called "alignment." It is together with 'compliance' a main driver for IT Governance implementations in organizations. However, in this article compliance will not be discussed in depth since it is a too complex topic on its own.

Importance of IT Governance for public organizations

A critical factor for an organization's success is to obtain competitive advantage, which is generally the source for growth. If IT is aligned with business goals, it can deliver important strategic advantages to support organizations to become more productive (Weill & Ross, 2004). Even though public organizations do not need competitive advantage, because there is generally no alternative, they have a responsibility towards the society to improve their services and reduce cost, therefore growth can be seen as enhanced services and productivity as improved efficiency in delivering these services.

Reinhard, Sun, and Agune (2006) state that "effective ICT governance contributes to performance and e-value generation, but requires political will and determination to be implemented and enacted" (p. 1). The relationship between IT use, growth stages and governance, and IT budget and growth can be used as indicators for the effectiveness of IT Governance in governmental institutions (Reinhard, Sun, & Agune, 2006).

Other researches demonstrate the positive association between the performance of an organization, particularly profitability and growth, and IT Governance performance. However, they also show that there is not a single best IT Governance model rather all models contribute to a positive outcome (Reinhard, Sun, & Agune, 2006; Weill & Ross, 2005).

IT Governance issues vary in relation to the development stages of the organization's IT operations. Therefore the awareness of IT's strategic value has to be fostered at councils, senates or congresses as well as in their supporting administrative management and mechanisms (Plavia, Palvia, & Witworth, 2002; Moon, 2002; Reinhard, Sun, & Augune, 2006).

Strategic Alignment

As stated above, the link between IT and business it the crucial factor in IT Governance (Van Grembergen, De Haes, & Guldentops, 2004). Additionally, according to Duffy (2002) the coexistence between IT functions and non-IT functions within an organization is not enough, they have to be joined together to gain leverages and achieve previously defined strategic goals.

However, as with IT Governance there is no single definition in the literature for "Strategic Alignment." Different researches often used synonyms for their definitions. For example:

- 'Integration' is used by Weil and Broadbent (1998)
- 'Fusion' is used by Smarczny (2001)
- 'Symbiosis' is used by Duffy (2002)
- 'Harmony' is used by Luftman (2003)

According to Chan (2002), alignment is not a static status. It is rather a consistently developed process. "The 'bringing in line' of the IS function's strategy, structure, technology and processes with those of the business unit so that IS personnel and their business partners are working towards the same goals while using their respective competencies" (p. 111).

Nevertheless, all of these terminologies aim at the integration of strategies, which refer to both worlds—IT and business. In this chapter, the definition by Duffy (2002) is used in a modified way since it has the most holistic approach. According to Duffy, strategic alignment is "... the process and goal of achieving competitive advantage through developing and sustaining a symbiotic relationship between business and IT" (in Van Grembergen, De Haes, & Guldentops, 2004, p. 7).

In the context to public organizations, the term 'achieving competitive advantage' in Duffy's definition should be substituted by 'enhancing public services.'

IT Value

IT Governance is designed to enhance a company's processes and services by the alignment of IT to strategic goals and therefore add value to a product or an offered service. According to Van Grembergen, De Haes, and Guldentops (2004) there are have identified 'IT value delivery' as the goal of IT Governance.

The ITGI describes the connection between IT Value and IT Governance, as "the value that IT adds to the business is a function of the degree to which the IT organization is aligned with the business and meets the expectations of the business" (ITGI, 2003, p. 25).

A widespread and excellent approach to assess the strategic capabilities of an organization is to use Porter's value chain model (Robson, 1997). His value chain model clearly shows that all primary and support activities or investments must add value to a product or service otherwise they are pointless. One of the supporting activities he mentions in his model is technology development (Porter, 1985; Porter, 1998). If we use his model in our today's business world, we could replace 'technology development' with 'IT activities / investments.' Conclusively, all IT activities and investments must have some business value to a company and finally increase the shareholder value.

In the public sector, the equivalent to the shareholder value is the community value (Moore, 1995; Cole & Partson, 2006). In this chapter, the term community value is used interchangeably with the term public value.

IT Governance's methodology is modeled to improve the alignment between IT and business, even though it might never be completely achieved (Van Grembergen, De Haes, & Guldentops, 2004). Maturity models, IT Governance frameworks, IT Portfolio Management and IT Value Management help to realize benefits or values to an organization. However, we must keep in mind that IT "value is in the eye of the beholder" (ITGI, 2008, p. 14), and thus the value of an IT project or investment might change over time or might be worthless for another organization or in a different situation. Weil and Broadbent (1998) even discuss the different values of IT between management hierarchies and users (Weil & Broadbent, 1998; Van Grembergen, De Haes, & Guldentops, 2004).

In the context of this chapter most researches have one shared point of view that commercial enterprises have different values and indicators than public organizations and governmental institutions in particular (Van Grembergen, De Haes, & Guldentops, 2004; Weill & Broadbent, 1998; ITGI, 2003; Luftman, 2000).

In addition to this, the ITGI describes the difference as followed in their Val IT Framework: "The nature of value differs for different types of organizations. For commercial or for-profit organizations, value tends to be viewed primarily in financial terms and can be simply the increase in profit to the organization that arises from the investment. For not-for-profit organizations, including the public sector, value is more complex and is often non-financial in nature. It should be the improvement in the organization's performance against business metrics (which measure what those whom the organization exists to serve receive) and/or the net increase in income that is available to provide those services, either or both of which arise from the investment"(ITGI, 2006b, p. 13).

Importance of IT Governance for Public Organizations

'You can't manage what you can't measure,' is an old and often used quote. According to Porter (1985, 1998) all actions taken in an organization must add value or they are just wasting money. But how can we determine intangible values out of complex IT projects? Non-IT executives might not understand IT terminology or see the bigger picture of a technology. Values, benefits and risks must be explained in a way that senior executives understand how IT drives their business, they have to understand that IT is not only a cost factor but a business enabler. IT Governance might be a solution.

Peterson (2003), Van Grembergen, De Haes, and Guldentops (2004) suggest that IT Governance should be implemented by a framework of structures, processes, and relational mechanisms. According to Bhattacharjya, and Chang (2006) "a number of international standards such as Control Objectives for Information and Related Technology (COBIT), ISO17799, IT Infrastructure Library (ITIL), and Capability Maturity Model (CMM), Project Management Body of Knowledge (PMBOK), are now available to IT organizations to help them improve their accountability, governance, and management" (p. 3).

Frameworks such as COBIT, ITIL, or CMMI can help us to realize the value of IT. However, such complex frameworks cannot be implemented over night and need time to adapt to an organization's individual needs. Nevertheless, they provide a structured guidance on behalf of best practices which can lead an organization's throughout the implementation of IT Governance and the different maturity levels. Moreover, these frameworks and tools can be used conjointly to overcome their weaknesses.

Thus, an excellent way to use these tools and frameworks together is to use an even more holistic approach. The Calder-Moir IT Governance framework "is a straightforward tool for organizing and communicating IT governance issues and activities" (Calder & Moir, 2006, p. 2). Calder's and Moir's method does not come up with a new solution; it is rather a joint collection of existing frameworks as ITIL, COBIT, and ISO 27000. It organizes these tools and methodologies to support all organizational hierarchies, the board, executives, and practitioners, and places them in an end-to-end context. This provides a simple reference point to discuss all facets of IT direction and performance. The framework is built upon six segments:

1. Business Strategy
2. Business and Risk Environment
3. IT Strategy
4. Change
5. Capabilities
6. Operations

Each of them symbolizes one step in the end-to-end process and includes subsets of related methodologies. The most significant methods and frameworks in IT Governance are:

- ISO 17799 / ISO 27000
- CMMI
- ITIL
- COBIT
- Val-IT

ISO 17799 / ISO 27000 define guiding principles for the implementation of information security. These principles are based upon regulatory requirements and generally accepted best practices. Regulatory requirements are the protection of personal data, sensible organizational information and intellectual property rights whereas best practices contain information security policies, assignment of responsibility for information security, problem escalation, and business continuity management (ISACA, 2005). Even though ISO 17799 / ISO 27000 does not have direct effect on IT Value it can "secure" investments and mitigate the risk of a loss caused by a security breach.

The Capability Maturity Model Integration (CMMI) is based on the CMM. The CMM has been developed by the Software Engineering Institute (SEI) in the mid-1980s. The latest version of the CMMI is version 1.2 and has been released in August 2006. The objective of CMMI is to improve usability of maturity models for software engineering, by integrating many different models into one framework (SEI, 2008).

"CMMI® (Capability Maturity Model® Integration) is a process improvement maturity model for the development of products and services. It consists of best practices that address development and maintenance activities that cover the product lifecycle from conception through delivery and maintenance" (SEI, 2006, p. 1).

CMMI represents a set of recommended practices for key processes, which have been used to enhance software process capability. It is a collection of best practices for Project Management, Software Engineering, Process Management, and Support Processes to effectively manage software requirements, development, delivery processes and software quality. The CMMI provides a practical framework to organize evolutionary steps into five maturity levels and is compatible to the maturity levels of COBIT and ITIL.

Most benefits of CMMI come from its rigorous rating processes. With a maturity model the status of each process can be ranked from non-existent to optimized (0-5). This creates a measure for where the organization is in reference to its goals and in comparison to other companies. Maturity models also help to identify where improvements can be made (ICFI, 2004).

With regards to the IT Value, CMMI provides guidance for efficient and effective improvement across multiple process disciplines in the organization. CMMI provides "the ability to incorporate business goals within process improvement, such as increasing schedule and budget predictability, productivity, quality, customer satisfaction, employee morale, and Return On Investment (ROI)" (ICFI, 2004, p. 2).

Due to its focus on software development, CMMI has its limitations on IT Service Management components such as infrastructure or more generic processes. One of the biggest criticisms is that implementing CMMI becomes too bureaucratic and time consuming (ICFI, 2004). Therefore, its sole application to public organizations will not optimize their IT investments, but it will definitely help them to improve their processes.

ITIL has become the worldwide de-facto standard in IT Service Management. A main advantage of ITIL is its provision of a common language for the IT and business departments to get faster and better benefits out of their IT service (Van Bon, 2004; Sallé, 2004; Pengelly, 2004). Therefore, ITIL v2 is a widely used framework to improve the service delivery of IT; however, since May 2007 ITIL v3 was launched and updated much of the former domain 'Service Management.'

Even though becoming holistic in its third version, ITIL does not focus on 'WHAT' an organization should do to achieve alignment; it rather focuses on 'HOW' IT can provide services to the business.

The ITSMF defines that "a service is a means of delivering value to customers by facilitating outcomes customers want to achieve without the ownership of specific costs and risks" and "Service Management is a set of specialized organizational capabilities for providing value to customers in the form of services" (ITSMF, 2007, p. 11). Moreover, the new ITIL version v.3 has its own 'Financial Management' and 'Service Portfolio Management' domain.

Additionally to ITIL there are a few proprietary service management frameworks out on the market. They are all based on or related to ITIL, such as the Microsoft Operations Framework (MOF), Hewlett Packard's ITSM Reference Model (HPITSM), or IBM's Process Reference Model for IT (PRM-IT). Albeit they should be mentioned, they do not play such a significant role as ITIL does.

As one can see ITIL and its related frameworks clearly focus on value delivery on an operational level. They monitor and manage the exact delivery of a defined service and therefore can enhance and improve the service delivery in public organizations, particularly if it comes to e-governance where online services should be reliable and secure. Nevertheless, ITIL's biggest strength is also its biggest flaw. It strongly focuses on IT service delivery, which makes it "inaccessible" for non-IT managers. Therefore, ITIL can help to improve services but it lacks the ability to support public organizations in strategic decisions such as project portfolios and prioritization.

Controlled Objectives for Information and related Technology (COBIT) is also a well-known framework for IT Governance. It enhances, risk mitigation, IT value delivery and strategic alignment (Debraceny, 2006; Guldentops, 2004; Van Grembergen, De Haes, & Guldentops, 2004;

Holm, Kühn, & Viborg, 2006; Ridley, Young, & Caroll, 2004; Warland & Ridley, 2005).

COBIT is built on four domains (Plan and Organize, Acquire and Implement, Deliver and Support, and Monitor and Evaluate) which are split into 34 manageable processes (ISACA, 2007). Each process consists of a number of activities and control objectives. They are using different types of metrics such as Key Performance Indicators (KPI), Key Goal Indicators (KGI), and Critical Success Factors (CSF), to monitor the processes. Each process is assessed by a maturity model, which is related to CMMI. COBIT also provides a RACI chart for the different processes and activities that recommends particular positions within an organization to be responsible, accountable, consulted, and informed. COBIT is a de-facto standard in IT Governance and often used by several practitioners (Sallé & Rosenthal, 2005; Simonsson & Hultgren, 2005; Simonsson, Johnson, & Wijkström, 2006). Furthermore, COBIT is Sarbanes Oxley compliant (ITGI, 2006).

COBIT itself has a view control objectives and activities throughout its domains to realize the IT value and manage IT portfolios, e.g.: PO5 "Manage IT Investments," DS6 "Identify and allocate costs," etc. (ISACA, 2007a). COBIT is strongly related and geared to the Val IT framework (ISACA, 2007b), which specializes on IT investment management.

Hence, one can say that COBIT covers most aspects in IT Governance and in contrast to ITIL it focuses on "WHAT" an organization should do to maximize the benefit of IT investments. However, it does need additional frameworks such as Val IT and ITIL to transform strategic IT decision in operational processes.

Val IT can be seen as an extension framework to COBIT; it "focuses on the investment decision … and the realization of benefits..., while COBIT focuses on the execution" (ITGI, 2006b, p. 7). Even though its roots are in COBIT, a mature IT Governance framework, literature or case studies about the Val IT are rare yet (Symons, 2006). However, the ITGI and other researchers are constantly developing the framework and document related work to show the effectiveness of Val IT.

The ITGI has published three additional papers to their framework: 'The Business Case' (ITGI, 2006c), 'The ING Case Study' (ITGI, 2006d), and 'Value Governance—Police Case Study' (ITGI, 2007). The latter paper demonstrated the successful applicability of the Val IT framework in public organizations by mapping and weighting IT projects in relation to community values. However, it must be mentioned that this case study was done in a mid-sized police service, with a rather simple complexity. Therefore it must be discusses if the procedures used in this case, are applicable for larger, more complex public organizations.

The close relation to COBIT is also reflected in Val IT's structure, however, it can be used without a prior implementation of COBIT. The framework consists of three processes (Value Governance, Portfolio Management, and Investment Management).

The ITGI (2006b) describes these processes as:

- The goal of "Value Governance" is to optimize the value of an organization's IT-enabled investments
- The goal of "Portfolio Management" is to ensure that an organization's overall portfolio of IT-enabled investments is aligned with and contributing optimal value to the organization's strategic objectives
- The goal of "Investment Management" is to ensure that an organization's individual IT-enabled investment programs deliver optimal value at an affordable cost with a known and acceptable level of risk

The framework provides the board and executive managers a good guideline and supporting practices how they should control and monitor their IT investments to realize business value for their organization. The main goals of Val IT are (ITGI, 2006b):

- Increase the perception and transparency of costs, risks and benefits
- Increase the likelihood of selecting the "right" investments with the highest possible return for the organization's objectives
- Increase the chance of a successful of execution of selected investments such that they realize or exceed the predicted benefits

Val IT is primarily targeted at IT-enabled business investments and process changes, where IT is the means and the increase of business value is the end result. Therefore, "Val IT fosters a close partnership between IT and the business, with clear and unambiguous accountabilities and measurements—another key requirement for effective governance" (ITGI, 2006b, p. 11).

Although Val IT is particularly developed to support general decisions in IT value management, it does not give exact guidance "How" to measure the IT value since different situations demand different solutions but the framework gives a very structured description of the value management process itself. Particular the decision stages provide a very useful approach to find the right investment without spending too much effort and money. Nevertheless Val IT has to be complemented by other IT value methods to prioritize these investments effectively and accurate.

IT Value Estimation Methods for IT Project / Investment Alignment

IT Value Management is an important part of IT Governance and its related frameworks, because IT investment management requires decisions, commitments, and accountability. Without a sound IT Governance approach, it is disputable if IT Value Management can be implemented successfully (McShea, 2006). High benefits however are crucial to an organization's success, thus IT Value Management can highlight key factors for the prioritization of IT projects and therefore improve the IT portfolio management process.

Most of the literature written on the value estimation and prioritization of IT projects recommends the usage of financially driven models such as Net Present Value (NPV), Internal Rate of Return (IRR), Return On Investment (ROI), PayBack Period (PBP), and Earned Value Analysis (EVA) (Archer & Ghasemzadeh, 1999; Norrie, 2006).

According to Forrester, researchers numerous organizations try to measure expected benefits by using these standard financial measures (Symons, 2006). Even though these metrics can help to determine the financial value of an investment they have significant flaws:

1. There are too many metrics so they can be interpreted in different ways which can lead to misunderstandings (Symons, 2006).
2. Metrics imply precision but are usually based on assumptions (Symons, 2006; Norrie, 2006).
3. They are unable to measure intangible benefits and do not include strategic goals (Symons, 2006; Norrie, 2006).
4. Future opportunities are not taken into account (Symons, 2006; Norrie, 2006, Weill & Broadbent, 1998).
5. They don't reflect the risk of an investment (Symons, 2006).

Therefore, some authors state that the sole use of these financial indicators is neither enough to reflect the significance and value of an IT investment or project to an organization due to too many intangibles (McShea, 2006; Di Maio, 2003; Norrie, 2006; Artto, Martinsuo, & Aalto, 2001) nor are they always appropriate for public organizations (Norrie, 2006; ITGI, 2006b; Di Maio, 2003)

To address this problem more holistic methods are recommended, which take other factors in account as strategy, alignment, opportunity, resources, and risk (Norrie, 2006, Artto, Martinsuo, & Aalto, 2001; McShea, 2006). Some of the most important methods are:

- Business Case (BC)
- Business Value Index (BVI)
- Total Economic Impact (TEI)
- Total Value of Opportunity (TVO) / Business Value of IT (BVIT)
- Balanced Score Card (BSC)
- Applied Information Economics (AIE)

Most of these holistic valuation methods, as well as the prior discussed Val IT framework, are based around the concept of a business case (Symons, 2006). It is a fundamental technique, that is highly recommended to get rational management support and agreement for business investments and projects since it provides an in-depth analysis (Flanagan & Nichols, 2007).

Business Case (BC)

The business case is one of the most valuable tools to assist management in their decision-making process whether an investment ads business value to the company or not. The quality of the business case and its utilizations to the complete economic life cycle of an investment is crucial for the value realization of an investment. The development of a business cases demands granularity and rigor, otherwise it might lead to wrong assumptions. Therefore, a BC should include opportunities and risks for current and future strategies and processes. However, even the best BC is only a snapshot at a particular point in time. Hence, it should be reviewed and adapted on a regular basis, which is particularly necessary for lengthy projects since priorities and technologies can change over time. A business case must include answers to the "Four Are's" questioned in the Val IT framework (ITGI, 2006c; Symons, 2006).

Doubtlessly a business case provides great benefits and information in a decision making process. However, a thoroughly done business case, which answers these questions, is quite expensive particular for rather complex projects and investments. Hence, a complete business case for complex projects with less impact and value for an organization will consume large amounts of money and time before it turns out that the cost-benefit factor is too low or the risks are too high.

Business Value Index (BVI)

The Business Value Index BVI, developed by Intel, is a method that combines various factors to get a value for IT investments. It evaluates three factors: IT business value, impact to IT efficiency and the financial attractiveness. However, it is rather qualitative than quantitative (Symons, 2006; Sward & Lansford, 2007).

All three factors use a fixed set of weighted criteria that according to Sward and Lansford (2007) include:

- Customer need
- Business and technical risks
- Strategic fit
- Revenue potential
- Level of investment required
- The innovation and learning that an investment generates

The BVI uses multiple influential factors to assess the value of IT investments and displays the results in a matrix. The advantage of this method is, that it provides a structured process with given aspects, which limits the effort to gain reliable results. Even though this method does not need a business case necessarily, elements of it could either be reused or the given aspects could become part of a business case analysis. However, the simple set of given weighted criteria might not be of enough accuracy in particular cased which are more complex and must be seen from different perspectives which is often the case in political decisions.

Total Economic Impact (TEI)

Forrester's Total Economic Impact (TEI) method contains some elements of Intel's BVI, it calculates the costs and business benefits with traditional financial methods, valuates intangibles but additionally it implies risk factors and the value of future flexible components based on the Real Options Value method (ROV), which aims to put some quantifiable value on flexibility. This provides a more future and strategic oriented approach with a 'risk-adjusted ROI' (McShea, 2006; Symons, 2006).

According to Symons (2006) "TEI is a more rigorous methodology than BVI but can be customized based on client's specific requirements" (p. 9), but involves management commitment and effort. It is more rigor and quantitative than BVI but not as precise as AIE. This shows that TEI can also become quite expensive with an increasing degree of complexity in IT investments and projects.

Total Value of Opportunity (TVO) / Business Value of IT (BVIT)

The Total Value of Opportunity Approach (TVO) and Business Value of IT (BVIT) are quite similar and based on the five pillars of Gartner's Business Performance framework (direct financial payback, strategic alignment, process impact, architecture, and risk) also known as perspectives. Both methods are using quantitative and qualitative measures to evaluate the IT value of an investment (McShea, 2006; Apfel, 2002; Apfel & Murphy, 2007). The advantage of these methods is that it determines a business case from a predefined set of business metrics, which is similar to the BVI method. Metrics and measurements are modeled to respect the different entities in an investment decision. Even though these methods are multidimensional in their view, their disadvantage is also that they are just using a simple weighted scoring for their approach which might result in biased outcomes and does not consider the lifecycle of such investments and projects. Thus such a method might not always reflect all political facets and political changes that should be considered when prioritizing IT projects in public organizations.

Balanced Score Card (BSC)

The Balanced Score Card (BSC) is another multidimensional approach to measure the value of an investment or the corporate performance. Initially developed by Robert S. Kaplan and David P. Norton to measure traditional business areas, it is equally applicable to measure the success or value of IT as long as the perspectives are adapted (McFarlane, 2006; Saull, 2000).

The BSC, compared to all other methods explained in this chapter, aims towards strategy as a primary motive. Metrics are derived from the hypothesized cause-and-effect relationship between financial and non-financial strategic objectives. These key parameters, rather than ROI, are the focus of IT Value management in this method. Each perspective has underpinning strategic objectives, which consecutively have different metrics (McShea, 2006).

Even though reflecting the strategy of an organization, the balanced scorecard is, as the TVO/BVIT approach, only based on simple weighted scoring that does not always reflect an accurate value of each aspect. Nevertheless, due to its strategic focus the BSC is a very powerful tool for political decisions and commonly used by different consultants as an easy way to determine the value or the performance of IT.

Applied Information Economics (AIE)

Applied Information Economics (AIE) is a highly sophisticated approach to evaluate intensive and complex programs. It uses multiple elements out of economics, statistics, business intelligence, options theory, risk management and modern portfolio theory to provide an all-inclusive criteria set (McShea, 2006; Symons, 2006) that "uses a very deliberate ROI calculation approach that pays special attention to soft benefits, referred to as enhanced ROI (EROI). Hard-cash-flow based ROI is adjusted for value linking (collateral benefit to non-IT processes), value acceleration (reduced timescales efficiencies), value restructuring (improved productivity), and innovation (market-position oriented benefits)....AIE quantifies in hard financial terms all variables, no matter how seemingly intangible the benefits... it views outcomes of each variable as a probability distribution, or a range of results, each with an estimated probability. The combined result is a probability distribution for ROI" (McShea, 2006, p. 32). AIE being the most holistic, rigorous, and accurate of these solutions, it is, however, very difficult to implement in an organization and therefore not often used (McShea, 2006; Symons, 2006).

Problems with Classic Frameworks and Methods to Align IT Projects and Investments to Public Values

As one can see, there are various frameworks, methodologies, and measurements available that support the determination of value of IT projects. However, if it comes to public organizations all of these methods have their limits, particularly the financially driven tools. None of them can fully meet the requirements of non-profit organizations and public administration (Van Grembergen, De Haes, & Guldentops, 2004; Weill & Broadbent, 1998; ITGI, 2003; Luftman, 2000; ITGI, 2006b).

Nevertheless, a few of the methodologies have elements that can be used in public organizations. Particular the advanced frameworks like COBIT, ITIL; CMMI and Val IT and also some of the value estimation methods such as TEI, TVO/BVIT, BSC, and AIE, because they have a more holistic and multidimensional view on IT value and are not just financially driven like ROI, NPV, IRR, etc.

In general, ITIL, COBIT, and CMMI can give good directions on how public organizations can improve the operational IT management and parts of their strategic IT management. However, these frameworks are of limited use to align public values and political directions with IT investments.

Val IT complements these frameworks in terms of IT Value Management. It is more detailed in regards to value estimation than COBIT and ITIL and monitors the performance of IT projects and investments throughout their lifecycle. However, it is completely based on business cases, which are often very expensive. Nevertheless, its methods, as used in the police case study (ITGI, 2007), have shown that it can be also used in public organizations, but the rather simple methods must be discussed when applied in larger more complex organizations.

ROI, NPV, IRR, etc. are mostly inappropriate for public organizations because they cannot reflect the intangible strategic goals of such organizations. Other methods like BSC, TEI, TVO/BVIT, and AIE have a multidimensional view on IT value. They have in common that one of their perspectives is 'strategy' or 'alignment.' However, they do not handle this perspective in detail but this would be particularly important for public organizations in order to align their IT with community values and political goals.

None of the reviewed frameworks or methods can be directly applied without adaption or used as a single solution for public organizations, because they are not driven by non-financial strategic goals and they are particularly weak in defining the level of strategic alignment to community values.

Nonetheless, serving the community is the primary goal of public organizations. Thus, the alignment of IT investments and projects to community values and political goals should be a major issue. Hence, we analyzed the situation in an Australian local government institution to find a solution how IT investments and projects can be aligned with community values and political goals.

The Case Study

The analyzed organization already established a sound IT Governance Framework (based on COBIT and ITIL) to support their Corporate Strategy. However, there are issues with using traditional approaches to assess the business value of IT due to their non-profit nature. At the moment, the value and ranking of IT projects is assessed by Gartner's Business Value of IT (BVIT) methodology, which is not fully supporting their decision-making and prioritizing process to align the IT investments efficiently with their strategic priorities. The input variables for the BVIT tool are based on implicit knowledge and estimations, and therefore this process tends to be more subjective than objective and takes no account of the notions of community values. Thus, the BVIT rating is rather limited in its usability to this public organization, because it is not transparent and holistic enough to assist them in their decision process in which they want to determine those IT proposals that are most beneficial to the community.

According to their Corporate Plan the most important strategic goals are:

Community Goals:

1. A Safe Community
2. Advocacy
3. Community Capacity Building
4. Customer Service
5. Cultural Development
6. Community Health and Individual Well-Being
7. External Communication

Economic Goals:

8. City Transport Improvement
9. City Assets
10. Diversify and Strengthen the Economy
11. City Image

Environmental Goals:

12. Preserve and Enhance the Natural Environment
13. Land Use and Development Control

Internal Goals:

14. Leadership and Governance
15. People Management
16. Information and Knowledge Management
17. Internal Services

While analyzing the actual decision process in this public organization, two main issues were identified:

1. The first instance of admittance and prioritization is based on vague input values such as risk and technical alignment that cannot be clearly defined at this stage due to too many uncertainties. All these estimations are based on a three-page concept plan that covers only very general questions and initial ideas. Particularly large and very complex projects can't be judged without a thorough business case. Therefore, estimations at this stage can lead to a severe misjudgment of the value to the organization and the community.
2. The second issue is that there is a strong information gap between the initial Concept Plans (ideas) and Business Cases. This makes it difficult for the decision makers to separate these ideas into important or less important without consuming large amount of resources or pursuing the wrong projects based on unreliable estimations (Flanagan & Nichols, 2007; ITGI, 2006b).
3. The third issue is exposed, if we analyze the actual BVIT ranking as shown in Table 1. It becomes obvious that the BVIT tool is definitely not able to reflect the significance of IT projects in regards to the strategic goals of a public institution due to its business-

driven nature. As one can see in Table 1, the first seven projects have a straight score of 400, which leaves a huge gap of more than 200 points between them and the follow-ups. According to city council's IT Project Portfolio Manager, these projects are non-discretionary. After investigating this conspicuous issue, it became evident that a 400 score does not reflect the real BVIT value of these projects, in fact these scores are deliberately adjusted to rank these projects and bump them into the Top 10. Therefore, the scores of the BVIT tool had to be manually aligned. This underpins, that solely business driven tools such as BVIT cannot cope with non-financially driven changes or priorities (Di Maio, 2003).

Deliverables of an idea, respectively of a concept plan, must be clear from the beginning on and should be also predictable from non-technical organization units such as operations or administration. However, some questions of the actual Concept plan cannot be predicted precisely enough by non-technical personnel and without a thorough business plan.

Doubtlessly, resources, impact, financials, and risk should be considered when an IT project is prioritized. However, if one looks again at the structure of the BVIT tool it becomes evident that all of the input values are of the same importance to the final result no matter if they are vague or not. Another issue with the tool is that the rating criteria is not detailed enough, thus subjective estimations about the alignment of a project might be still too imprecise, due to the fact that it is not rated against particular strategic goals.

As said the analyzed organization is following the COBIT and ITIL approach, therefore their value estimation processes comply with these well-known and highly accepted best practices. They even fulfill some specifications of the Val-IT framework, such as the staged approach applied in the Val-IT Police Case Study (ITGI, 2007).

However, there is still some space for improvements. As highlighted the biggest issues are the unreliable input factors which lead to issue 1 and issue 2, as a result the BVIT ranking has to be tweaked manually (issue 3) to bump important projects into the project portfolio Top 10.

Since this organization is of non-profit nature, they should rather put the emphasis of their IT project rating on the strategic deliverables than on risk, finance, and resources. Particularly, they have to focus on the deliverables of the community values, which are not yet included in their IT value estimation method. Thus, they have to adjust the BVIT scores to reflect the importance of some of their projects as seen in Table 1. As a consequence, their As-Is process should be slightly adjusted and an additional methodology to measure the community value of their projects has to be implemented in the early stage of their process. This would enable them to prioritize their projects in the first stage without consulting technical staff or at least it would reduce their effort because the deliverables of a project will then be compared to the strategic goals and not to highly complex aspects such as risk and resources. It would also reduce the likelihood, that projects will be reprioritized just because the initial estimations of risk, resource, impact, and financials turned out wrong after an expensive business case was conducted.

The results confirm, that the As-Is analysis confirms the conclusions of the reviewed literature and reflects the opinion of most other researches in this area, that financially driven IT value estimation methods are inappropriate for non-financially driven organizations when solely applied (Van Grembergen, De Haes, & Guldentops, 2004; Weill & Broadbent, 1998; ITGI, 2003; Luftman, 2000; ITGI, 2006b). Nevertheless, the importance of aspects such as risk, resources, impact and financial indicators will increase with the maturity of the project, thus they should be applied in later stages of the process. At least their impact on the first ranking should be minimized to a certain degree, because as stated above their ac-

Table 1. Final BVIT ranking (excerpt)

BVIT Score	Project Name
400	Disaster Recovery
400	Security Management System - ARIES Upgrade
400	Interim Library ICT fit-out
400	Scientific Services Lab Relocation
400	WWTP – ICT fit-out
400	Main Library ICT extension
400	PC Fleet
177	Cameras for Impounded Animal Identification
176	ERP Initiation Project
176	Corporate Integration
165	Implementation of Final Component of Z39.50 Access
143	Geospatial Project
132	ECM - iSpot
130	Mobility Business Case
126	Nature Conservation Database Flora and Fauna Database Review
125	PPMIS (Project Office Software)
122	Asset Information Management Transition Project
116	PD Online Stage 2
116	Integrated Test Facility Project

tual influence on the prioritization of a project might lead to wrong expectations based on fuzzy estimations.

THE COMMUNITY VALUE ESTIMATION METHOD (CVE)

To address the difficulty of alignment previously identified as issue 1, issue 2, and issue 3, the decision process was complemented by a combination of two highly accepted decision support methods. The new process was called the Community Value Estimation Method (CVE).

1. AHP-Method (Analytical Hierarchy Process)
2. Delphi-Method

The general process of this method is explained in Figure 1.

The developed method, and process is compliant with the COBIT, ITIL, and Val-IT frameworks and complements all other value estimation methods such as BVIT, NPV, TVO, etc. It was developed particularly for non-profit organizations and governmental institutions, but can also be used by private companies, which want to focus on non-financial strategic goals.

The combination of the two methods ensures a dynamic yet accurate result, however the simpler the methods are implemented the less accurate the results will be. Nevertheless, this must be seen as an advantage of the method not

Figure 1. Methodology process

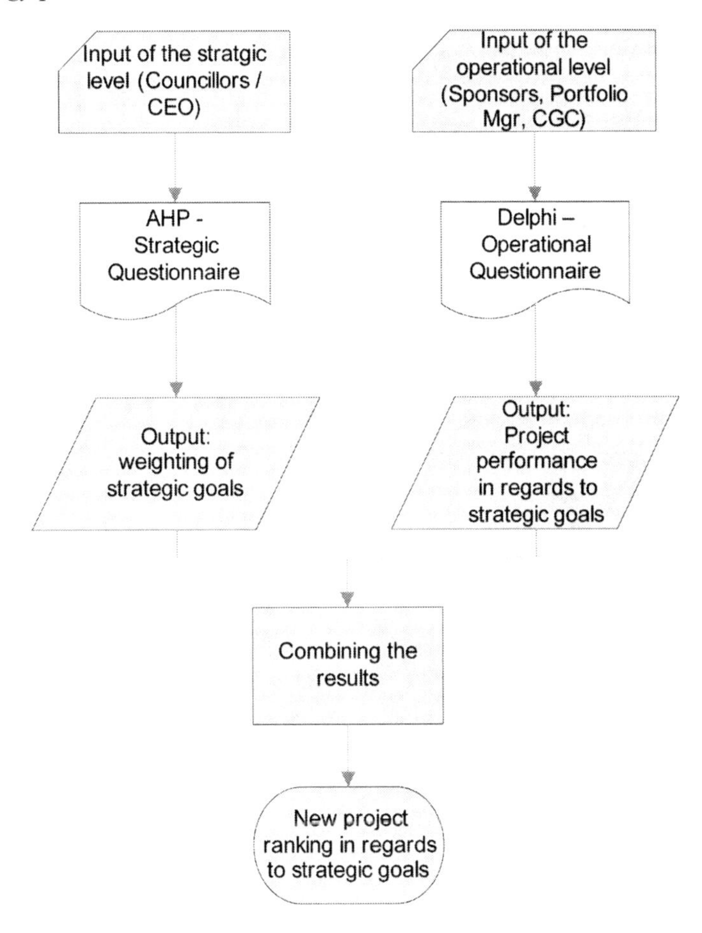

a disadvantage, because different organizations will have different demands in complexity and accuracy in regards to their IT project prioritization. Therefore, the method is suitable for small and simple organization units as well as large and complex corporations.

Since the research approach was mainly deductive, data had to be collected to test and proof the final hypothesis. For this reason, surveys, questionnaires and already existent data of the researched city council was used to feed the AHP and Delphi methods and create an alternative project portfolio.

The data collection for the surveys and the questionnaires were split into two sections:

- Strategic section (AHP)
- Operational section (Delphi)

The most suitable data collection method seemed to be the self-administrative surveys and review of existent data and documentation. Thus, all surveys and questionnaires were developed in cooperation with the CIO office of the analyzed organization to tailor them to their needs. As mentioned above each organization demands a different level of detail, granularity, and complexity. Therefore, both methods, AHP as well as the Delphi, have to be used quite flexible and thus the surveys and questionnaires, to feed the methods, must be developed accordingly.

As seen in the As-Is analysis, three major issues (difficulty to estimate financial factors in an early stage, unnecessary spending on business cases and manually adjusted BVIT scores) were identified which are addressed by the new methodology.

The different stages of project prioritization were optimized according to the Val-IT framework. This approach keeps costs and efforts low but increases the value of pursued projects to the organization. Hence, the initial prioritization of the As-Is process was substituted with the CVE-Method, which leads to the following advantages compared to the former process:

First advantage: Project deliverables can be easily predicted by non-technical personnel, without need of an expensive business case. Instead of complex technical aspects, the CVE method emphasizes on the deliverables of a project, which can easily predicted from non-technical personnel without conducting expensive business cases. Particularly the Delphi part of the methodology helps Project Sponsors and Portfolio Managers to find a consensus on the value and performance of these projects in regards to the strategic goals.

Second advantage: As a result of the initial prioritization with the CVE-Method, the strategic value of projects can be easily predicted and differentiated between strong and weak, thus this will unburden the management to deal with unimportant projects and enables them to concentrate on the overall performance of the portfolio and the management of significant projects.

Third advantage: Only projects with a high strategic value will consume resources for a business case, which is in contrast to the As-Is process, where it could turn out that due to the lack of thorough information, initial estimations of risk, cost, impact and resources were wrong. Now, business cases are only conducted for strategically important projects, thus financial factors such as risk, impact, and cost estimations are much more reliable.

Fourth advantage: The CVE-Method is completely based on the strategic deliverables. The strategic value of an IT project to the organization is now reflected in the ranking as well as traditional financial factors. Thus there should be no need to adjust the ranking values manually, which leads to better transparency and compliancy with IT Governance frameworks such as COBIT, ITIL, and Val-IT.

To proof the new concept the Top 20 BVIT projects in the analyzed organization were reviewed with the CVE-Method, thus the results of the Strategic Questionnaire and the Operational Questionnaire were combined to calculate a new project portfolio ranking, which was compared against the former BVIT ranking in Figure 2.

As described in the As-Is analysis the BVIT ranked projects 1 – 7 are non-discretionary thus their score was manually adjusted to the maximum score of 400. Therefore, this score does not reflect the original financial values of BVIT. The adjustment was a necessity to raise their importance and bump these projects into the Top 10. However, as one can see most of the projects with a manually adjusted 400 BVIT score are present in the Top 10 of the new portfolio ranking. This leads to the conclusion that the new approach is able to reflect the strategic values of the analyzed public organization.

Deviations such as the 'Security Management System—ARIES upgrade' result from the strong focus of this method on 'Community Values,' such as a 'Community Health' and 'Preserve the Environment.' Such projects do not have a direct outcome for the community and must rather be seen as part of maintenance. In contrast other projects such as 'Cameras for impounded animal identification' or 'Nature Conversation Database' are rated higher because their deliverables support the strategic goals even more.

However, one has to keep an eye on the project ranking in this method. For example in our actual case, the strategic focus is currently on 'Community Health' and 'Preserve the Environment,' so most of the Top 10 projects have generally a high influence on these factors. However, even though this is an advantage to figure out which projects deliver the highest value to the community in this moment, the strategic focus might change

Figure 2. BVIT – AHP comparison

TOP 20	Project Name / Strategic Goals	BVIT Score
1	WWTP – ICT fit-out	400
2	Scientific Services Lab Relocation	400
3	Disaster Recovery	400
4	PC Fleet	400
5	Main Library ICT extension	400
6	Interim Library ICT Fitout	400
7	Security Management System - ARIES Upgrade	400
8	Cameras for Impounded Animal Identification	177
9	ERP Initiation Project	176
10	Corporate Integration	176
11	Implementation of Final Component of Z39.50 Access	165
12	Geospatial Project	143
13	ECM - iSpot	132
14	Mobility Business Case	130
15	Nature Conservation Database Flora and Fauna Database	126
16	PPMIS (Project Office Software)	125
17	Asset Information Management Transition Project	122
18	PD Online Stage 2	116
19	Integrated Test Facility Project	116
20	HRIS Foundation Project	115

AHP / Delphi Score	Project Name / Strategic Goals
179	Cameras for Impounded Animal Identification
171	WWTP – ICT fit-out
143	Scientific Services Lab Relocation
128	Disaster Recovery
121	Nature Conservation Database Flora and Fauna Database
104	PC Fleet
95	Main Library ICT extension
94	Interim Library ICT Fitout
93	PD Online Stage 2
71	Mobility Business Case
66	Geospatial Project
56	Security Management System - ARIES Upgrade
50	ERP Initiation Project
44	Asset Information Management Transition Project
31	Corporate Integration
25	ECM - iSpot
24	Implementation of Final Component of Z39.50 Access
13	PPMIS (Project Office Software)
9	HRIS Foundation Project
8	Integrated Test Facility Project

immediately after the successful implementation of one of these Top 10 projects. So it could happen that the importance of 'Preserve the Environment' loses its high strategic value because one of the projects nearly solves a problem, and thus similar projects with identical deliverables become less important, which would change the ranking of the follow-up projects.

CONCLUSION

This article compared various IT Governance and IT Value Management methods in regards to the IT project portfolio management in public organizations. The reviewed literature and the case study highlighted the strengths and the weaknesses of important frameworks and methods, which led to the conclusion that none of the analyzed IT Governance frameworks and IT value estimation tools is adequate enough to align IT projects and investments with the strategic goals and community values in a public organization. It became clear that the sole implementation of these financially driven methods do not work properly in non-profit organizations such as the analyzed governmental institution. In particular, these methods are not able to reflect the strategic importance in a political context and thus their ability to prioritize an IT project portfolio is limited. Thus this article briefly explained a method, which was developed during a research project, how public organizations can prioritize IT projects on basis of their relevance to community values.

The described CVE-Method focuses rather on the relevance and outcomes of IT projects for the community than on financial indicators. By using the Analytical Hierarchy Process (AHP) and the Delphi-Method a decision matrix was developed which takes the importance of strategic goals into account and predicts the value of deliverables of pursued projects to the community. This led to a new IT project ranking and an optimized prioritization process that increased the transparency and compliancy of their decision making process. Due to the fact that the new method highlights the strategic value of projects it also saves cost and effort because less important projects will not be pursued at the expenses of more important projects.

Even though CVE seems to be a promising approach to align IT projects to community values, further research and implementations are necessary to proof the general usability of this method in different organizations. Additionally it would be worthwhile to consider the interdependencies between projects. As stated above a successful implementation of a project might change the importance of strategic goals since the project might solve a problem up to a satisfying level. It might also be that some projects do not have a direct impact on strategic goals but enable other technologies that do, and therefore, they must be considered as prerequisites or enablers in an IT project portfolio.

REFERENCES

Apfel, A. L. (2002). *The total value of opportunity approach*. Washington, DC: Gartner Research.

Apfel, A. L., & Murphy, T. (2007). *Five perspectives beyond ROI: A process for scoring and prioritizing projects and program*. Washington, DC: Gartner Research.

Archer, N. P., & Ghasemzadeh, F. (1999). An integrated framework for project portfolio selection. *International Journal of Project Management*, *17*(4), 207–216. doi:10.1016/S0263-7863(98)00032-5

Artto, K. A., Martinsuo, M., & Aalto, T. (2001). *Project portfolio management: Strategic business management through projects*. Helsinki, Finland: Project Management Association Finland.

Bhattacharjya, J., & Chang, V. (2006). *An Exploration of the Implementation and Effectiveness of IT Governance Processes in Institutions of Higher Education in Australia*. Australia: Curtin University of Technology.

Calder, A., & Moir, W. (2006). *The Calder-Moir IT governance framework*. Retrieved March 28. 2008, from http://www.itgovernance.co.uk

Chan, Y. E. (2002). Why haven't we mastered alignment? The importance of the informal organizational structure. *MIS Quarterly Executive*, *1*(2), 97–112.

Cole, M., & Partson, G. (2006). *Unlocking public value: A new model for achieving high performance in public service organizations*. Hoboken, NJ: Wiley Publishing.

Debraceny, R. S. (2006). Re-engineering IT internal controls - Applying capability maturity models to the evaluation of IT controls. In *Proceedings of the 39th Hawaii International Conference on System Sciences*. Hawaii, HI: IEEE.

Di Maio, A. (2003a). *Value for money is not enough in public sector IT projects*. Washington, DC: Gartner Research.

Di Maio, A. (2003b). *Traditional ROI measures will fail in government*. Washington, DC: Gartner Research.

Di Maio, A. (2003c). *New performance framework measures public value of IT*. Washington, DC: Gartner Research.

Di Maio, A. (2003d). *How to measure the public value of IT*. Washington, DC: Gartner Research.

Duffy, J. (2002). *IT/business alignment: Is it an option or is it mandatory*. Framingham, MA: IDC.

Flanagan, J., & Nicholls, P. (2007). *Public sector business cases using the five case model: A toolkit*. Retrieved April 4, 2008, from http://www.hm-treasury.gov.uk/d/greenbook_toolkit-templates170707.pdf

Guldentops, E. (2004). Governing information technology through COBIT. In Van Grembergen, W. (Ed.), *Strategies for Information Technology Governance*. Hershey, PA: IGI Global.

Holm, L. M., Kühn P., M. & Viborg A., K. (2006). IT Governance – Reviewing 17 IT Governance Tools and Analyzing the Case of Novozymes A/S. *Proceedings of the 39th Hawaii International Conference on System Science*. Hawaii/USA.

ICF International. (2004, Fall). The business value of CMMI. *Perspective Reports.*

ISACA. (2005). *Aligning COBIT, ITIL and ISO17799 for business benefit*. Rolling Meadows, IL: ISACA.

ISACA. (2007a). *Control objectives for information and related technology*. Rolling Meadows, IL: ISACA.

ISACA. (2007b). *COBIT 4.1 excerpt*. Rolling Meadows, IL: ISACA.

ITGI. (2003). *Board briefing on IT governance* (2nd ed.). Rolling Meadows, IL: ITGI.

ITGI. (2006a). *IT control objectives for Sarbanes-Oxley* (2nd ed.). Rolling Meadows, IL: ITGI.

ITGI. (2006b). *Enterprise value: Governance of IT investments: The Val IT framework*. Rolling Meadows, IL: ITGI.

ITGI. (2006c). *Enterprise value: Governance of IT investments, the business case*. Rolling Meadows, IL: ITGI.

ITGI. (2006d). *Enterprise value: Governance of IT investments, the ING case study*. Rolling Meadows, IL: ITGI.

ITGI. (2007). *VAL IT case study: Value governance - The police case study*. Rolling Meadows, IL: ITGI.

ITGI. (2008). *Enterprise value: Governance of IT investments, the Val IT framework 2.0 extract*. Rolling Meadows, IL: ITGI.

ITSMF. (2007). *An introductory overview of ITIL® V3*. Berkshire, UK: ITSMF.

Luftman, J. (2000). Assessing business-IT alignment maturity. *Communications of AIS, 4*.

Luftman, J. (2003). *Competing in the information age: Align in the sand* (2nd ed.). Oxford, UK: Oxford University Press. doi:10.1093/0195159535.001.0001

Markov, R. (2006). Economical impact of IT investments in the public sector: The case of local electronic government. In *Proceedings of the International Business Informatics Challenge 2006*. Dublin, Ireland: Business Informatics Challenge.

McFarlane, G. (2006). *IT and the balanced scorecard*. Retrieved April 4. 2008, from http://www.deloitte.com/view/en_KZ/kz/services/enterprise-risk-services/b907e60e612fb110Vgn-VCM100000ba42f00aRCRD.htm

McShea, M. (2006). IT metrics - IT value management: Creating a balanced program. *IEEE Computer Society. IT Professional*, *8*(6), 31–37. doi:10.1109/MITP.2006.138

Moon, M. J. (2002). The evolution of e-government among municipalities: Rhetoric or reality? *Public Administration Review*, *62*(4), 424–433. doi:10.1111/0033-3352.00196

Moore, M. H. (1995). *Creating public value - Strategic management in government*. Cambridge, MA: Harvard University Press.

Norrie, J. L. (2006). *Improving results of project portfolio management in the public sector using a balanced strategic scoring model.* (DPM Thesis). Royal Melbourne Institute of Technology University. Melbourne, Australia.

Palvia, P., Palvia, S., & Whitworth, J. E. (2002). Global information technology: A meta analysis of key issues. *Information & Management*, *39*(5), 403–414. doi:10.1016/S0378-7206(01)00106-9

Pengelly, J. (2004). *ITIL foundations*. London, UK: GTS Learning.

Peterson, R. (2003). Information strategies and tactics for information technology governance. In *W. Van Grembergen (2004), Strategies for Information Technology Governance*. Hershey, PA: IGI Global. doi:10.4018/978-1-59140-140-7.ch002

Porter, M. (1985). *Competitive advantage: Creating and sustaining performance*. New York, NY: Free Press.

Porter, M. (1998). *On competition*. Boston, MA: Harvard Business School Press.

Reinhard, N., Sun, V., & Agune, R. (2006). ICT Spending in Brazilian Public Administration. *Proceedings of the 19th Bled eConference eValues*. Bled, Slovenia.

Ridley, G., Young, J., & Carroll, P. (2004). COBIT and its utilization - A framework from the literature. In *Proceedings of the 37th Hawaii International Conference on System Sciences*. Hawaii, HI: IEEE.

Robson, W. (1997). *Strategic management and information systems*. London, UK: Prentice Hall.

Sallé, M. (2004). *IT service management and IT governance: Review, comparative analysis and their impact on utility computing*. Palo Alto, CA: HP Laboratories.

Sallé, M., & Rosenthal, S. (2005). Formulating and implementing an HP IT program strategy using COBIT and HP ITSM. In *Proceedings of the 38th Hawaii International Conference on System Sciences.* Hawaii, HI: IEEE.

Saul, R. (2000). The IT balanced scorecard – A roadmap to effective governance of a shared services IT organization. *Information Systems Control Journal*, *2*, 31–38. doi:10.1023/A:1010041819361

SEI. (2006). *CMMI for development version 1.2*. Pittsburgh, PA: Carnegie Mellon University.

SEI. (2008). *CMMI history*. Retrieved March 29, 2008 from http://www.sei.cmu.edu/cmmi/faq/his-faq.html

Sethibe, T., Campbell, J., & McDonald, C. (2007). IT governance in public and private sector organisations: Examining the differences and defining future research directions. In *Proceedings of the 18th Australasian Conference on Information Systems.* Toowoomba, Australia: IEEE.

Simonsson, M., & Hultgren, E. (2005). Administrative systems and operation support systems – A comparison of IT governance maturity. In *Proceedings of the CIGRÉ International Colloquium on Telecommunications and Informatics for the Power Industry.* Cuernavaca, Mexico: CIGRE.

Simonsson, M., & Johnson, P. (2006). *Defining IT governance – A consolidation of literature*. Stockholm, Sweden: Royal Institute of Technology (KTH).

Simonsson, M., Johnson, P., & Wijkström, H. (2006). *Model-based IT governance maturity assessment with COBIT*. Stockholm, Sweden: Royal Institute of Technology (KTH).

Sward, D., & Lansford, R. (2007). *Measuring IT success at the bottom line*. Palo Alto, CA: Intel Corporation.

Symons, C. (2006). *Measuring the business value of IT*. Cambridge, MA: Forrester Research Inc.

Van Bon, J. (2004). *IT service management, an introduction based on ITIL*. Zasltbommel, The Netherlands: Van Haren Publishing.

Van Grembergen, W. (2002). Introduction to minitrack: IT governance and its mechanisms. In *Proceedings of the 35th Hawaii International Conference on System Sciences*. Hawaii, HI: IEEE.

Van Grembergen, W., De Haes, S., & Guldentops, E. (2004). Structures, processes and relational mechanisms for IT governance. In Van Grembergen, W. (Ed.), *Strategies for Information Technology Governance*. Hershey, PA: IGI Global. doi:10.4018/978-1-59140-140-7.ch001

Warland, C., & Ridley, G. (2005). Awareness of IT control frameworks in an Australian state government: A qualitative case study. In *Proceedings of the 38th Hawaii International Conference on System Sciences*. Hawaii, HI: IEEE.

Weill, P., & Broadbent, M. (1998). *Leveraging the new infrastructure: How market leaders capitalize on information technology*. Boston, MA: Harvard Business School Press.

Weill, P., & Ross, J. W. (2004). *IT governance – How top performers manage IT decision rights for superior results*. Boston, MA: Harvard Business School Press.

Weill, P., & Ross, J. W. (2005). *A matrixed approach to designing IT governance*. Retrieved March 16. 2008, from http://sloanreview.mit.edu/the-magazine/2005-winter/46208/a-matrixed-approach-to-designing-it-governance/

Chapter 8
From Town Hall to the Virtual Community:
Engaging the Public with Web 2.0 and Social Media Applications

Kathryn Kloby
Monmouth University, USA

Leila Sadeghi
Kean University, USA

ABSTRACT

Engaging the public is a vital component of the public policy process. Traditional strategies for civic engagement include town hall meetings as well as citizen surveys, 311 call systems, and more interactive meetings for public deliberation. Each of these approaches has their limitations, leading many to consider new ways of engaging the public and the role that technology can play in the process. The authors focus on a discussion of the traditional citizen engagement approaches that are widely used by government to communicate with and interact with the public. Focusing on new interactive media, they discuss what is meant by "Web 2.0" and present the capabilities and potential applications of social media in the public sector. Highlighting government programs that utilize these technologies and interviewing subject-matter experts on this new form of communication, the authors present some of the adoption concerns and implementation strategies that public administrators should consider as they adopt Web 2.0 technologies. They conclude with a discussion of the potential that these new civic engagement techniques can offer the public sector as strategies to communicate, interact, and engage the public.

INTRODUCTION

We don't have a choice on whether we do social media, the question is how well we do it (Qualman, 2010).

DOI: 10.4018/978-1-4666-1909-8.ch008

Public administration theorists and scholars have long debated how public administrators should engage the public in the policy process. Early public administrators operationalizing the orthodox principles of the day, for example, assumed the role of a technical expert who was insulated from the public. Their more contemporary counterparts working in the eras of reinvention and new public management, and with demands of accountability for results are reconsidering the ways in which they engage the citizenry. Rather than relying on traditional methods for citizen engagement such as open public meetings, some are breaking new ground as they implement strategies to determine what matters to the public and how they can improve services and deliver desired results. More recent strategies include citizen surveys, 311 call systems, public meetings with deliberative exercises to determine citizen priorities, and citizen assessment of government services with hand-held tracking devices. This chapter examines governments that are using new forms of engagement such as social media technologies.

Social media technologies such as Facebook, MySpace, Twitter, and other Web 2.0 capabilities have emerged as a widely used tool for sharing information, reconnecting with friends and family, and frequently updating friends on one's whereabouts and activities, or status. More recently, these networking tools have gained notable attention by the media and politicians. Calling on citizen journalists, for example, the media utilizes social media technologies to find more information about unexpected or catastrophic events. Professional associations use social media Web sites to capture the interest of new members and find ways for professional networking opportunities. These applications are useful in the political arena as the most recent presidential election demonstrated the ability of social media technologies to generate supporters and share information on policy positions and proposals.

This platform for communication is receiving more attention among public administrators in federal, state, and local governments. This is a new frontier for public administrators, leading practitioners and academicians to ask such questions as: How can social media technologies promote the good works of government? How can these networking mechanisms be used to inform citizens of government actions? How can social media facilitate some form of interaction between citizens and their government to inform management decision making? How can social media be applied to improve government performance and show results?

The demands for government to show results to its citizens are persistent and urgent considering its recent attempts to stimulate national, state, and local economy, generate jobs, and ultimately increase trust in government decisions and actions through results and transparency in the policy process. Some governments at the federal, state, and local levels are receiving attention for their attempts to utilize social media as a tool to improve performance and to communicate with and interact with citizens. While there is a spate of research that presents strategies to bolster government performance and results, as well as techniques for engaging the public in the process, there is little known about how social media is incorporated into government practices to improve results and interact with citizens.

The aim of this chapter is to advance our understanding of Web 2.0 social media capabilities and to explore how government is utilizing social media technologies. In particular, this research is designed to address the following questions:

- In the context of public administration, what are the potential applications of Web 2.0 and social media applications?
- In what ways can social media inform management decisions and improve government performance?
- What should public administrators consider as they adopt these technologies?

We begin with a brief discussion of the traditional approaches that are widely used by government to communicate with and interact with the public. Making the transition to new forms of communication and interaction, we define what we mean by Web 2.0 and social media.

Examining government programs that utilize these technologies and consulting subject-matter experts on this new form of communication, we present some of the issues administrators should consider as they adopt Web 2.0 capabilities. We conclude with a discussion of the great potential that these new technologies can offer the public sector as strategies to communicate, interact, and engage the public.

BACKGROUND: WIDELY USED PRACTICES FOR ENGAGING THE PUBLIC

Many researchers and scholars suggest that citizen involvement is a vital component of democratic governance that results in informed management decisions (Callahan, 2007; Berman, 2005), transparency and fairness in policy development (Lukensmeyer & Tores, 2006), capacity building (Cuthill & Fein, 2005), and increased trust in government (Keele, 2007). Despite these benefits, the challenges of involving citizens in the public policy process are significant and progress is often limited. In many instances, for example, public administrators are conceptualized as experts insulated from the public (Callahan, 2007). More often than not, bureaucratic processes are too rigid to accommodate new processes in support of citizen engagement (Timney, 1998). In some cases, government actors can simply be overwhelmed by citizen demands for results or an improved quality of life and may feel they cannot—or should not—be held accountable for all of them. As a result, elected officials and public administrators often rely on traditional participation mechanisms, such as public meetings, as the primary means to listen to and engage with the public (Adams, 2004). These encounters fall far short of the ideals of citizen involvement, as they are often sparsely attended due to citizens' work schedules, lack of interest, child-care needs, or fear of public speaking (Adams, 2004; King, Feltey, & O'Neil, 1998; Berner, 2001).

Some progress has been made to create better opportunities to gauge citizen satisfaction with government services, respond to citizen complaints or questions, and include citizens in public deliberation on policy issues. Citizen surveys, for example, provide a more systematic approach for public administrators to ask citizens what they think about service quality (need source). As a result of a provision by the Federal Communications Commission to reserve 311 calls for non-emergency calls, and increasing number of governments utilize this system to allow the public to access its government 24 hours a day, seven days a week to report, for example, a loss of water service or potholes (ICMA, 2007). Other nonprofit organizations, such as AmericaSpeaks, work with governments to provide opportunities for citizens to review policy proposals, contribute policy options and priorities government options via town hall meetings (see www.americaspeaks.org).

While these strategies hold the promise of increasing opportunities for a more open and transparent government, each approach has its limitations. Citizen surveys and 311 call systems, for example involve static and one-directional communication between government and the public. Citizens, in other words, are afforded opportunities to rate or complain about government efforts but are not likely to be able to share their ideas or preferences via dynamic exchanges. Large-scale town hall gatherings are significant undertakings that can be costly and present challenges of assembling a representative group of citizens for active deliberation (Williamson, 2007). There is little research that suggests that the results of these encounters actually influence the decisions of elected officials and public administrators. Such shortcomings are leading public administrators and scholars to consider new ways of engaging the public with technologies that foster real-time interactions, continuously updated information, and virtual space that permits more authentic and representative interactions and exchanges.

BREAKING FROM TRADITION: DEFINING WEB 1.0 AND WEB 2.0

The term "Web 2.0" has received considerable attention, and leaves some to question the meaning of Web 1.0 and what the prospects are for the new 2.0 version. In terms of communication, Web 1.0 is a media model where the power of communication and information sharing is in the hands of media producers who often serve as gatekeepers of news and information and who determine what the public will learn and when they will have access to such information. Several converging factors such as the integration of high-speed Internet capabilities, advancements in mobile technology and applications, and the release of handheld technologies like the iPad are credited with transforming communication strategies, and ultimately, the way information is delivered through the World Wide Web. Web 2.0 is used to describe a second generation of Web-based, social media applications and tools such as Facebook and Twitter, wikis and blogs, and Flickr and YouTube. These technologies support open communication, information sharing, and content production (O'Reilly, 2005). They are viewed by many as having the ability to reshape the communication landscape on a societal scale, as well as between government and its citizens (Internet.com, 2010). Analyzing the role of these social media applications, some suggest that we are in the midst of "the biggest shift since the Industrial Revolution" (Qualman, 2010).

Social media technology is having a significant impact on how the public learns about local and world events, and offers great potential for reconceptualizing the way government interacts with the public. To illustrate the revolutionary qualities of social media applications, consider that it took more than fifty years for radio and television to reach an audience of 50 million, and only seven years for the Internet and iPod to reach the same number of users (Qualman, 2010). Adding to the momentum of these technological advancements and their utilization, many argue that social media is fast becoming a mainstream media. Facebook alone is documented as adding over 200 million users in one year, while the number of Twitter users is increasing at rate of 10 million users per month since February of 2010, reaching 125 million users worldwide (Qualman, 2010). A majority of these online communities are developed with the purpose of participation and communication (Qualman, 2010).

Web 2.0 and social media are widely used by the private sector to interact with people both internally (e.g., employees) and externally (e.g., consumers) (Serrat, 2010). Such technologies are used to identify and target specific audiences to assess product satisfaction. Using social media as a customer service tool, for example, social media can be used to determine consumer preferences as they rate and describe their experiences with particular products and brands (Qualman, 2010). Private sector organizations are applying social media capabilities to human resource management and hiring processes. Using social media applications and tools to recruit, hire and retain employees is fast becoming the norm. Companies, for example, are using platforms such as LinkedIn to find talented professionals for hire. Last year, Microsoft claimed significant savings in recruitment costs by using this application (Flinders, 2010).

While the private sector is at the leading edge of integrating Web 2.0 into organizational processes and with communicating with consumers, the public sector is lagging in its adoption and application. It is recognized, however, that this technology offers the potential to change the nature of the relationship between government officials and citizens (Serrat, 2010). Commonly referred to as Gov 2.0, public employees, have the ability to use several Web 2.0 interfaces such as a community blog, a Twitter microblog, a social media profile on Facebook, and post videos on a multimedia channel such as Youtube, with relative ease and timeliness. Gov 2.0 is a concept that describes how government uses social media tools to help improve citizen engagement and collaboration between government and citizens. It involves an emphasis on more direct or rapid- fire style of communication as information can be readily made available online. It provides new media solutions and the ability for public administrators to engage the public with relatively inexpensive social media applications. Such an example is the availability of streaming video or continuous updates on White House Web site and Youtube describing the events of the Horizon Oil Spill off the Gulf Coast (Farber, 2008). The public is able to read information, make comments, and share ideas to improve government performance. Using a similar strategy, NASA uses social media sites such as Facebook and Twitter through the "NASA buzzroom" to generate citizen comments on current projects and elicit new ideas for space exploration and other technological developments.

ADVANCING WEB 2.0 IN GOVERNMENT

According to Darell West (2008), "few developments have had broader consequences for the public sector than the introduction of the Internet and digital." Unlike previous interactive mechanisms, the Internet enables government to conduct the business of government or engage its citizens in a virtual space. Social media, in particular, provides public administrators with a new set of possibilities for building relationships with the community with relative ease. These possibilities can include:

- Simplified, user-oriented, and potentially participative techniques to achieve transparency and collaboration with agency personnel and citizens.
- Accessibility of information and data for review and/or manipulation.
- Increased collaboration with citizens in agenda setting, implementation, and evaluation stages of the policy process.
- Opportunities to align government actions with citizen preferences.
- Collaboration across divisions or between levels of government
- Sharing best practices and solutions.
- Supporting the search and hiring processes for agency personnel.

Recent interest in how public managers can use social media was sparked by the Obama campaign's innovative plan to use *myBarackObama* and other social media applications during the 2008 election. This marked the first step toward a transition and increased recognition of the added value social media technologies can bring to public agencies. The Obama administration later formally embraced the adoption of new technologies in federal government with the *Open Directive* urging executive departments and agencies to "take specific actions to implement the principles of transparency, participation, and collaboration" (Orszag, 2009, p. 1). Federal agencies are now required to make data and information available and to develop and maintain a Web page linked to the *Open Government* Web site (http://www.whitehouse.gov/open). Agencies are also required to provide citizens with the ability to comment and receive feedback from agency personnel. The ultimate aim of this directive is to increase transparency and develop a collaborative relationship with the public to drive smart policy, and increase government effectiveness (Orszag, 2009).

Some constituents across government are experimenting with ways to facilitate interaction with citizens that informs management decision making. Government agencies such as the Department of Defense, the Center for Disease Control and Prevention, the U.S. Army, and the Library of Congress, are using social media to communicate information and generate conversations around special topics. More generally, though, government tends to fall short on actively using social media and continues to rely on old technologies such as telephone surveys and traditional mail to solve problems (Serrat, 2010).

On the state and local levels, government has been slower to integrate social media for the purpose of civic engagement. Across all levels of government, a majority of the Web 2.0 applications are used to provide information, rather than engaging the public for more deliberative purposes. Many local websites, for example, provide transactional services such as registrations, renewing a license, or form completion. The "communication" piece in social media is typically not developed (Scott, 2006). Despite a variety of Web 2.0 applications and platforms that can be geared towards the public sector, the development of a comprehensive social media strategy is still in its infancy. Such applications and platforms include, for example: www.polleverywhere.com—a mechanism to poll citizens on legislation, the www.apps.gov site—a one-stop storefront launched by the Federal government to lower costs and drive innovation into government agencies, the www.creativecommons.org site—a platform that offers free licensing for any user to post and share photos, videos, documents, and more. The Whitehouse.gov Web site uses this service to license intellectual property made available on the Web. And finally, governments can use www.nixle.com site—a Web site that promotes geographically-based public safety alerts and advisories delivered via SMS text messaging services to community residents.

Other applications can be used to broadcast meetings, share documents, post videos and photos of events, and host sub-groups organized around common interests that can lead to new forms of volunteerism. Despite the promising new opportunities, a majority of local governments are not tapping into this technological movement largely due to a lack of clarity on exactly how Web 2.0 applications work. According to the Fels Institute of Government, for example, "Local governments' reactions to this expansion have been mixed. Some have made these services a central part of their communications strategies with the public and press. Many others are ambivalent or concerned that social media are a distraction that they may nonetheless be asked to do something clever with" (Kingsley, 2010, p. 3).

Despite the availability of a broad spectrum of Web 2.0 applications and the positive results associated with applying these tools in organizations, as demonstrated in the private sector, government is slow to jump on the technological bandwagon. This is, however, an emerging area of discussion, leading many public administrators to question whether this is an appropriate tool, how it can be applied to public agencies, and how it can impact how it relates to the public. As government breaks new ground in this area, there are likely to be many unknowns and fears, as well as technical, security, and privacy concerns that can have a significant impact on whether government actually utilizes new media. To address these concerns, it is likely that public administrators working in jurisdictions that are taking the lead in this technological revolution and who are experimenting with such capabilities will play an important role in setting the course for integrating Web 2.0 offerings into the functions of government, and ultimately, changing the way it interacts with the public.

SOLUTIONS AND RECOMMENDATIONS: EXPLORING GOVERNMENT INITIATIVES USING WEB 2.0

To determine management solutions and derive a useful set of recommendations for public administrators, we are highlighting public programs that are at the leading edge of Web 2.0 applications. Our aim is to build awareness of these programs and to also learn from their experiences as they break new ground in using technology in more innovative ways to engage the public. After interviewing key administrators who are responsible for conceptualizing and implementing these initiatives, we were able to present illustrations and identify some of the major challenges and opportunities public managers are likely to address as they incorporate technology into their management and engagement practices. Our examination focuses on the works of the following government initiatives:

- **The State of Utah with insight provided by David Fletcher, Chief Technology Officer.** In 2009, Utah received the "Best of the Web" award from the Center for Digital Government, which was one of numerous accolades it received for its technological advancements. On average, the state Web site receives 900,000 unique visitors per month. Utah spearheaded an e-government portal overhaul as an outcome of its 2007-2009 Utah eGov Strategic Plan for the state Web site that focuses strongly on the use of social media to interact with citizens and businesses. Agencies use Twitter and Facebook, and Utah provides an umbrella Web site (see www.connect.Utah.gov) that provides citizens with a streamlined way to sort through and identify useful information. In 2009, Utah added 51 new state online services to its 891 online services in an effort to reach citizens in as many ways possible. To achieve transparency, the state recently launched a new Web site called Transparent.Utah.gov that provides the public with access to the state's financial records. Citizens can analyze and interpret the data in a variety of user-friendly ways. For more information: http://connect.utah.gov
- **The County of Morris, New Jersey, with insight provided by Carol Spencer, Web Manager**. Morris County leads NJ counties in its active and widespread use of social media. Using a variety of social media platforms, including Facebook, Twitter, YouTube, Flickr and Scribd, the public is able to connect with county level information, events, and activities, comment and share information, locate government documents uploaded through Scribd, and view interviews with key government officials on a Youtube channel. With over 800 Twitter fans and 350 Facebook friends, Morris County has recently added other tools to enhance the timeliness of information disseminated to citizens. Hootsuite is a tool for scheduling tweets in advance and news feeds provide citizens with access to updates on selected information from the County. For more information: http://co.morris.nj.us/generalHTML/socialmedia.asp
- **The City of Manor, Texas with insight provided by Dustin Haisler, Assistant City Manager.** Manor Labs is a virtual "thinklab," or mechanism, where citizens and agency personnel can submit ideas and solutions to improve policy in the Manor community. A participant can easily create an account and suggest ideas for a variety of departments like the municipal court or law enforcement. Manor Labs uses incentives such as "innobucks" that tallies points based on a ranking system. "Innobucks" can be redeemed for a variety of prizes, such as a ride-along with the Police Chief or as "mayor for the day." In a recent White House Office of Science & Technology Policy Blog article, Manor Labs was cited as an exemplar of government innovation. Manor recently joined the OpenCourseWare Consortium, a global higher education collaboration that stresses open educational content using a shared model. Manor anticipates creating Gov 2.0 and open government curriculum for other governments to learn from. In 2010, Manor received a visionary award in the Best of Texas awards program sponsored

by the Center for Digital Government. For more information: http://cityofmanor.org/wordpress/labs/

- **Township of Newton, New Jersey, with insight provided by Thomas Russo, Township Manager.** Newton, NJ redesigned its township website in 2009, adding social media tools and applications for citizens to bridge the gap with government. Under the manager's progressive leadership, in 2008, the town underwent a complete website redesign to improve the site appearance, navigation, and include social media tools to increase citizen engagement. Newton Township was recently recognized by the Center for Executive Leadership in Government at Rutgers University as a best practice model for municipalities. For more information: http://www.newtontownhall.com/

We present the insights of these public administrators as they reflect and comment on their experiences with the following areas: Web 2.0 and social media adoption, its impact on communication with the public and citizen utilization, and how it influences management decision making and agency performance. Interviewees were also asked to consider issues that other public administrators should be aware of as they navigate the complex terrain of adopting, institutionalizing, and benefiting from the features of Web 2.0 and social media. The programs highlighted in this chapter vary significantly in their size, structure, and context, but their leading administrators offer useful strategies for adoption and implementation. We present an overview of these insights in the discussion below and provide challenges public administrators are likely to encounter as they explore social media options in their respective jurisdictions.

INSIGHTS FROM PUBLIC ADMINISTRATORS ON THE CUTTING EDGE

Adoption Considerations

Social media can play a vital role in increasing organizational efficiency while bolstering communication with external audiences. In 2002, for example, the State of Utah had actively encouraged state employees to create a blog to enhance collaboration across divisions to share information and problem solve. As a way of institutionalizing these practices, Utah later developed a two-year e-government plan focusing largely on how to use social media to improve the way government services are delivered. A statewide social media policy and a set of online collaboration standards were later approved by the State's Technical Architecture Review Board to address questions by staff who were reluctant to use social media because such tools were not officially procured or sanctioned. Increasing the focus on citizens, in 2003 Utah became the first state to implement online chats as a way to respond to citizen queries regarding services 24 hours a day, seven days a week. David Fletcher, Utah's Chief Information Officer, acknowledges that these were some of the earliest e-government blogs in the country, and being at the leading edge of such technological advancements inadvertently prepared agency staff to incorporate more Web 2.0 tools such as Twitter and Slideshare as they became available because a core group of Utah employees were already primed to use them.

Social media is quickly gaining attention as a set of tools that can help government play a more proactive role in communicating its accomplishments and determining the preferences of citizens. Dustin Haisler, Assistant City Manager of Manor, Texas, notes that his jurisdiction adopted social media because, "city business was being discussed online and often inaccurately by citizens without the city's participation." Social media is a means to bring government agencies and personnel into the conversation to articulate what it is responsible for, who should be contacted for certain services, and what needs to be done from the perspective of citizens. As in the case of Manor, Texas, it can be a vital communication tool for governments in jurisdictions with limited media outlets (e.g., very little to no local television news coverage), and can engage younger populations who are already well versed in social media applications and their versatility. Reporting on her own observations of younger people, Carol Spencer, Web Manager of Morris County, New Jersey, notes that they are not likely to read newspapers or magazines, but, she adds that they are, "constantly connected to each other and to the world. It was obvious to me that unless governments adopted the use of these technologies, in a few short years, we would not be able to communicate with the very constituencies we serve."

The individuals we interviewed stressed the importance of leadership and the support of either elected or appointed officials for successful adoption. Endorsement by state-level legislators or township councils and/or public administrators seems to set the cultural bar and expectation of fostering a public organization that is what Thomas Russo, Manager of Newton Township, New Jersey, describes as "a technology leader rather than follower." Whether leaders value the internal problem-solving capabilities of social media or the range of ways to connect with external audiences, or both, having this support conveys to staff and the public that technology can enhance the way government operates and communicates with the public. Support of high-level officials eliminates ambiguity and triggers the development of institutional policies that outline expectations and address some of the unknown or grey areas of utilizing social media in government agencies. Such policy parameters are likely to provide safeguards and protections for agency personnel who are embarking on a new, more exposed approach to communicating with the public that leaves many feeling vulnerable. Carol Spencer of Morris County agrees with the need for high-level leadership noting that "without the support of my direct management, I never would have had the opportunity to present social media to the governing body. Without the support of the governing body, we wouldn't be doing what we're doing." Having the support of leadership behind the implementation of Web 2.0 technologies is a vital component to achieve success.

Agency leadership also plays a vital role in determining the strategic direction and purpose of social media applications in public organizations. Dustin Haisler highlights that the two challenges to consider are how to determine what technologies can be used to benefit the community and how to educate the public in the use of the technologies. He notes, "we have now been offered so many resources, we have to really analyze what we expect will be the potential benefit to the city before we accept or implement a proposed service. With the rapid changes to technologies, this can be a demanding process." Having an elected official or public administrator with a vision of how Web 2.0 technologies can enhance outreach and engagement efforts seems to make the task of finding and using the appropriate tools more manageable.

Communication and Citizen Utilization

We asked the public administrators of the highlighted programs to consider how citizens perceive and use government-sponsored social media. Overall, citizens are receptive and seem to actively engage with their governments online. As Dustin Haisler of Manor Texas illustrates, "There are, and will always be, individuals in the community who view this as a waste of time and money, but it appears only to be small pockets of resistance or perhaps just disinterest rather than any real opposition by broader segments of the population." He acknowledges that government agencies are likely to fear a negative response or rejection of citizens toward the introduction of social media activities, but stresses that the Manor experience suggests that citizens who actively participate in online services are overwhelmingly positive about helping to make the community better.

Web 2.0 technologies are described as fostering interactions between government and citizens. Dustin Haisler of Manor, Texas, highlights that citizens have traditionally had to be proactive in engaging with their government agencies, on the government's terms (e.g. rigid meeting schedules and formats), and using its preferred avenue of communication. He further notes that:

For the first time in history, governments are able to engage citizens on the citizen's platforms of choice, which further bridges the divide between citizens and government...We have found that the use of Web 2.0 technologies is optimal for routine questions and allows them to ask those questions at their convenience rather than only during specific business hours.

Demonstrating the instant capabilities to connect with citizens, Carol Spencer of Morris County adds "Instead of putting up meeting agendas and hoping people find they're there, now we can tweet "Freeholder meeting agenda now posted" and add a link to the agenda."

Dustin Haisler of Manor, Texas, cautions that communication using social media should not altogether replace face-to-face communications, however, practice shows that face-to-face communications using traditional public hearings or meetings seldom take place. Average citizens generally do not have the time to monitor government activities, much less regularly attend council and commission meetings. Social media creates a new space for citizen comments, questions, and input, with opportunities to submit them at the convenience of citizens. It allows agency personnel the ability to fully provide answers to citizen queries after exploring options and sources of information. As a result, these Web-based interactions are noted as decreasing the number of confrontational incidents with citizens as they allow city personnel to address citizen's frustrations more quickly and remotely.

Our conversations with these trailblazers suggest that social media and citizen engagement are initiatives that are conceptualized as the future of government that is proactive rather than reactive. The belief that technology is a means to build relationships with the public is a cultural norm within these public agencies that requires leaders and staff to continuously adapt to their environments, technological developments, and the demands of citizens. Dustin Haisler of Manor Texas describes notes:

Unlike a long car-ride that has an eventual end, we must realize in that in government we will never arrive at our final destination, implying that there is always room for improvement. Our motto is to 'create the future' for government, instead of waiting for a private-sector company to do it entirely for us.

Carol Spencer, Web Master of Morris County agrees with the need for continuous improvement using social media tools. She is currently developing an application that will utilize Twitter and Facebook for emergency information. David Fletcher, Chief Information Officer of Utah, views their social media approach as a service that is available those who wish to take advantage of it. He acknowledges that social media has increased the relevance of government to citizens and projects that the number of "followers" will continue to increase.

Decision Making and Performance

Social media is described as allowing public administrators to have a "direct and transparent dialogue" with citizens that ultimately increases organizational efficiency through idea-sharing activities. This dialogue occurs in several different ways. Public administrators, for example, can utilize social media to engage other practitioners in the field, nationally and internationally. GovLoop.com is one such platform that is considered useful for sharing ideas and gaining support for new initiatives. Social media can be used to build collaborative opportunities across agency offices and divisions, and with other levels of government or community-based or regional organizations and institutions.

Government Web sites use social media and other Web 2.0 applications to increase responsiveness. Web-based surveys, for example, determine what is important to citizens, and what government actions need to address. Request Tracker, is used by Newtown, New Jersey, to enable citizens to alert the township manager of problems of concerns. These queries are then routed to the appropriate department head for a resolution. All the while, a citizen can keep track of their query and will be contacted when the problem is solved. When discussing the value of this mechanism, Thomas Russo, Manager of Newton says:

Of course, residents can still use traditional methods of communication (meetings, phone, etc.) to facilitate this, but the website provides new, useful, cutting-edge tools to help us manage the Town better and make people see their tax dollars being used wisely in order to solve problems in a timely fashion.

Newton Township also uses the Web site and social media to bring information to citizens. A "Notify Me" feature allows site visitors to subscribe to e-mail lists including updates on local initiatives, news announcements, job postings, and bid opportunities.

Other innovations are designed to increase efficiency and instill accountability within government operations. Through the use of Web 2.0 tools and Web-based platforms like Manor Labs, the citizens of Manor, Texas, can participate directly within the governance of their community. Dustin Haisler illustrates that, "using Manor Labs, we implemented six citizen-generated ideas mid-budget year since October 27, 2009. Our citizens are truly taking responsibility for creating a better community by participating in our local government research and development." Some of the administrators interviewed noted, however, that they are exploring the linkages between citizen participation and management decisions.

RECOMMENDED SOCIAL MEDIA STRATEGIES FOR PUBLIC ADMINISTRATORS

Despite the many positive attributes of social media, the challenges of adopting this technology as part of public sector strategies to inform, communicate, and include citizens in the policy process should be explored and addressed. Careful planning can eradicate some of the most common obstacles and pitfalls including bureaucratic and structural barriers; privacy and security concerns related to information sharing and content production; policies to promote and protect information, and; policies to address coordination and staff responsibilities for managing social media. The public administrators we interviewed for this chapter shared what they think others should know as they consider using or introduce Web 2.0 applications in their agencies. Below, we highlight some key challenges and recommendations worthy of consideration as public administrators explore and adopt social media.

CHALLENGE: FEAR OF LEARNING NEW TECHNOLOGY

Recommendation: Address Fear to Avoid Paralysis

The administrators highlighted in this chapter warn that fear can have a debilitating effect on Web 2.0 adoption and implementation. It is viewed as an irrational response of some personnel that is grounded in an overall lack of understanding of the technology and its applicable uses for government. Agency leaders should openly embrace technology and leverage the resources necessary to support it. It should be promoted as a mechanism to communicate and publicize the accomplishments of government. Fear can also be tempered by professional development and training opportunities that address the mechanics and capabilities of new platforms so that they are fully integrated into agency activities. Administrators also noted that persuading elected officials, supervisors, and agency personnel to use new media approaches required concrete examples of how Web 2.0 and social media would enhance the work of government—as compared to current technologies and strategies.

CHALLENGE: BUILDING A SUSTAINABLE STRATEGY

Recommendation: A Plan is Not Just a Plan

Our interviewees stress that employing Web 2.0 and social media with the intent of engaging the public without proper planning is "a recipe for failure." Plans should articulate guiding goals and present clear objectives for properly engaging the public. Citizens can also be brought into the planning process to determine how the public should be engaged and to what end. Having a clear set of objectives for engagement and expectations for the role of personnel avoids ambiguity and ultimately helps to ensure that the right applications are selected and used. The aim should be to provide meaningful and useful social media opportunities rather than simply using glossy gadgets that generate meaningless chatter.

CHALLENGE: CHOOSING APPROPRIATE WEB 2.0 TOOLS

Recommendation: Shop Wisely and Align Applications with Objectives

There are a vast number of social media applications available, leaving some overwhelmed in the planning and development process. For example, Facebook may be used to house a profile page, pictures, and other media in one virtual space. Twitter can be used for microblogging important information such as changes to the town schedule, weather-related warnings and when information gets updated on the main website. Scribd can provide citizens with access to meeting minutes and other relevant documents and forms. A manager's blog can communicate information and interact with citizens new issues of interest, potential policy options, or informing the public of day-to-day actions. And finally, YouTube can host a community channel to share multimedia spotlighting special events and public meetings, or capturing ceremonies. As highlighted above, an important step in the process is identifying what needs to be achieved through this medium. Having a handle on the various uses of these capabilities makes the decision process easier and more manageable.

CHALLENGE: MAINTAINING A POSITIVE WEB 2.0 PRESENCE

Recommendation: Adopt New Technologies and Adapt Them to the Changing Environment

Some administrators stress that Web 2.0 and social media require frequent updating to reflect the changing environment and to signify that there are living and breathing agency personnel who are maintaining governments' virtual presence. Some government Web sites are described as "fantastic websites that are never updated, never grow, never change." These are the sites that people tend to visit once, become frustrated, and never return. While Web 2.0 and social media application are promoted as tools that are relatively low in cost, a considerable amount of effort must be invested in maintenance. Aligning key staff who are energized by the prospects of Web 2.0 can ensure continued updating and help to build an organizational culture that increases support among agency personnel. Frequent content updates must be made in order to engage the audience. One way to achieve this without expanding the number of staff is to use services that automatically update social media platforms like www.ping.fm

CHALLENGE: MANAGING CONTENT, STAFF, AND PRIVACY CONCERNS WITH SOCIAL MEDIA

Recommendation: Policies Can Inform Many of the Unknowns

Public managers are likely to grapple with the following concerns: Should employees be given full discretion to post material regarding their place of employment? How should employees schedule time for posting content? Is there a policy in place for determining the type of content to post on the Internet? Public administrators can adopt an internal social media policy to inform staff as to the appropriate utilization of these platforms. They may also establish a social media policy that addresses how comments and questions from citizens should be addressed and answered.

CHALLENGE: BUDGETARY CONCERNS RELATED TO PRODUCT COST AND OVERALL IMPLEMENTATION

Recommendation: Strategize and Shop Wisely

In light of significant budgetary constraints across all levels of government, it seems likely that public administrators would resist adopting new platforms that may be expensive or in need of added staff. Key administrators interviewed note that a majority of these social media tools and platforms are free and require minimal effort of select staff who may be required to devote a small percentage of their time and effort. In the case of Newton Township, Russo notes that the township initiated an RFP process and the least expensive vendor was chosen to redesign the main website while ensuring key social media applications were part of that redesign. Many of the Web 2.0 applications are Web-based, and are far less expensive than the purchase of software or hardware. Others may require a registration or account that is maintained via reasonable fees for licensing or maintenance. Furthermore, the administrators of the programs highlighted in this chapter generally agree that instituting a strategic plan that aligns organizational goals with social media applications will reduce the need to register for too many sites and overload agency personnel.

CONCLUSION

Public Administrators are operating in a new and complex era of digital communication. On the one hand, they are responsible for sustaining traditional mechanisms for informing and interacting with the public via open meetings, budget hearings, and media coverage by local news organizations. These approaches alone are met with a number of implementation challenges. Most notable is the often paltry level of citizen turnout and active deliberation in the public policy process. Rigid meeting times, locations, and overall structure are just a few of the factors that negatively impact citizen involvement. On the other hand, public administrators can adopt Web 2.0 and social media applications that offer a range of opportunities to inform the public, display the work of government, and actively engage citizens in the policy process. These strategies are undertaken in virtual space and can potentially enhance governments' ability to interact with the public, and ultimately, break down many of the barriers that inhibit authentic interactions associated with traditional mechanisms for participation.

Questions of whether Web 2.0 and social media applications should be used by the public sector and how it can impact the relationship between government and citizens are spurring an increasing number of conversations within the practitioner and scholarly communities. Our aim was to explore the utilization of this technology in select public organizations and to add to the discussion with useful illustrations and insights from public administrators who are spearheading initiatives that utilize this new media approach. Asking public administrators at the state and local levels of government to reflect on adoption, citizen interactions, and impact on agency performance, we generated a series of challenges and recommendations that public administrators should consider as they explore new and more innovative options to engage the public.

Among the programs spotlighted in this chapter, we found that fear of technology and openness, thoughtful planning, strategic selections, updating and overall content management are some of the concerns that public administrators may encounter when considering or adopting new communication technologies. Most notable is the demand for sound leaders, as they are the individuals within public organizations who will have a vision, who can inspire agency personnel and citizens to interact, and who will make strategic decisions that have an impact on the vitality of such initiatives. Planning a course of action that articulates the objectives of Web 2.0 and social media coordinates agency actions and provides a sense of purpose as administrators and agency personnel charter this new territory.

Web 2.0 and social media applications offer some promising techniques for rapid response to citizen queries, for posting information, for informing management decisions and engaging the public in the policy process. Questions remain, however, as to how the public should be engaged and if it is an adequate substitute for face-to-face interactions. Future research should examine the nature of Web-based interactions using social mapping and whether they are in fact meaningful and informative for public administrators and the public. Implementation strategies demand further attention as the number of governments using this technology increase. Lessons can be learned, for example, about data management and storage, strategies to determine how citizen queries and interactions inform management decisions, and leadership and other accountability concerns. We also believe that there is much to be learned about the generational dimensions of digital communication. Younger generations who socialize and gain knowledge using Web 2.0 applications are bound to have an impact on whether governments incorporate digital communication into its practices as citizens and as part of the public sector workforce. This generation, often referred to as the "Net Generation," is likely to stimulate a new model of democracy that is transparent, collaborative and engaging, leaving public administrators with the challenge of redesigning their technology infrastructure to meet these demands. Government may also need to reinvent itself from a technological standpoint to attract and retain the next generation of public sector employees. These are just a few of the unanswered questions related to adopting and evolving Web 2.0 and social media in government agencies. The future of government communicating and interacting with the public offers new and innovative options for public administrators who are up for the task of reinvigorating democracy through innovative communication techniques.

REFERENCES

Adams, B. (2004). Public meetings and the democratic process. *Public Administration Review*, *64*(1), 43–54. doi:10.1111/j.1540-6210.2004.00345.x

Berner, M. (2001, Spring). Citizen participation in local government budgeting. *Popular Government*, 23–30.

Callahan, K. (2007). *Elements of effective governance: Measurement, accountability and participation*. New York, NY: Taylor and Francis.

Cohn-Berman, B. (2005). *Listening to the public: Adding the voices of the people to government performance measurement and reporting*. New York, NY: The Fund for the City of New York.

Cuthill, M., & Fein, J. (2005). Capacity building facilitating citizen participation in local governance. *Australian Journal of Public Administration*, *64*(4), 63–80. doi:10.1111/j.1467-8500.2005.00465a.x

Farber, D. (2008, November 16). *Obama appoints YouTube (Google) as secretary of video*. Retrieved from http://news.cnet.com/obama-appoints-youtube-google-as-secretary-of-video/

Flinders, K. (2010, June 11). *Will LinkedIn reshape the recruitment sector?* Retrieved from http://www.computerweekly.com/Articles/2010/06/14/241559/Will-LinkedIn-reshape-the-recruitment-sector.htm

ICMA. (2007). *Customer service and 311/CRM technology in local government: Lessons on connecting with citizens*. Retrieved from http://icma.org/en/results/research_and_development/smart_communities/311

Internet.com. (2010). *Web 2.0*. Retrieved from http://www.webopedia.com/TERM/W/Web_2_point_0.html

Keele, L. (2007). Social capital and the dynamics of trust in government. *American Journal of Political Science*, *51*(2), 241–254. doi:10.1111/j.1540-5907.2007.00248.x

King, C. S., Feltey, K. M., & O'Neil, B. O. (1998). The question of participation: Toward authentic public participation in public administration. *Public Administration Review*, *58*(4), 317–327. doi:10.2307/977561

Kingsley, C. (2010). *Making the most of social media: 7 lessons from successful cities*. Retrieved from https://www.fels.upenn.edu/sites/www.fels.upenn.edu/files/PP3_SocialMedia.pdf

Lukensmeyer, C. J., & Torres, L. H. (2006). *Public deliberation: A manger's guide to citizen engagement*. Washington, DC: Center for the Business of Government.

O'Reilly, T. (2005). *What is web 2.0: Design patterns and business models for the next generation of software*. Retrieved from http://oreilly.com/web2/archive/what-is-web-20.html

Orszag, P. (2009, December 8). *Memorandum for the heads of executive departments and agencies: Open government directive*. Retrieved from http://www.whitehouse.gov/omb/assets/memoranda_2010/m10-06.pdf

Qualman, E. (2010, May 5). *Social media 2 (refresh)*. Retrieved from http://socialnomics.net/2010/05/05/social-media-revolution-2-refresh/

Scott, J. K. (2006). E" the people: Do U.S. municipal government web sites support public involvement? *Public Administration Review*, *66*(3), 341–353. doi:10.1111/j.1540-6210.2006.00593.x

Serrat, O. (2010, May 24). *Social media and the public sector*. Retrieved from http://www.globalknowledgeexchange.net/social-media-and-the-public-sector

Timney, M. (1998). Overcoming administrative barriers to citizen participation: Citizens as partners, not adversaries. In King, C. S., & Stivers, C. (Eds.), *Government is Us: Public Administration in an Anti-Government Era* (pp. 88–99). Thousand Oaks, CA: Sage.

West, D. M. (2008). Improving technology utilization in electronic government around the world, 2008. *Governance Studies at Brookings*. Retrieved from http://cendoc.ccddesarrollo.net/cendoc_docs/Doc%2010242%20(Improving%20Technology%20Utilization%20in%20Electronic%20Government%20around%20the%20World).pdf

Williamson, A. (2007). *Citizen participation in the unified New Orleans plan*. Unpublished Paper. Boston, MA: Harvard University.

ADDITIONAL READING

Bertot, J. C., Jaeger, P. T., & Grimes, J. M. (2010). Using ICTs to create a culture of transparency: E-government and social media as openness and anti-corruption tools for societies. *Government Information Quarterly*, *27*(3), 264–271. doi:10.1016/j.giq.2010.03.001

Chen, Y., & Thurmaier, K. (2008). Advancing e-government: Financing challenges and opportunities. *Public Administration Review*, *68*(3), 537–548. doi:10.1111/j.1540-6210.2008.00889.x

Childs, R. D., Gingrich, G., & Piller, M. (2009). The future workforce: Gen Y has arrived. *Public Management*, *38*(4), 21–23.

Cordello, A. (2007). E-government: Towards the e-bureaucratic form? *Journal of Information Technology*, *22*, 265–274. doi:10.1057/palgrave.jit.2000105

Council, C. I. O. (2009). *Guidelines for secure use of social media by federal departments and agencies*. Retrieved from http://www.cio.gov/Documents/Guidelines%5Ffor%5FSecure%5FUse%5FSocial%5FMedia%5Fv01%2D0%2Epdf

Drapeau, M., & Wells, L., II. (2009). Social software and national security: An initial net assessment. *Center for Technology and National Security Policy: National Defense University*. Retrieved from http://www.dtic.mil/cgi-bin/GetTRDoc?AD=ADA497525

Eggers, W. D. (2005). *Government 2.0: Using technology to improve education, cut red tape, reduce gridlock, and enhance democracy*. Lanham, MD: Rowman & Littlefield Publishers, Inc.

Godwin, B. (2008). *Government and social media*. [PDF version of presentation slides from Social Media for Communicators Conference]. Retrieved from http://www.usa.gov/webcontent/documents/Government_and_Social_Media.pdf

Godwin, B., Campbell, S., Levy, J., & Bounds, J. (2008, December 23). Social media and the federal government: Perceived and real barriers and potential solutions. *Federal Web Managers Council*. Retrieved from http://www.usa.gov/webcontent/documents/SocialMediaFed%20Govt_BarriersPotentialSolutions.pdf

Godwin, B., Ettner, M., & Greenfeld, S. (2009, April 1). *Open government, transparency, and social media*. Retrieved from http://www.usa.gov/webcontent/documents/Open_Government_Transparency.pdf

GSA. (2009, December 23). *How to use social media strategically in the federal government*. Retrieved from http://www.howcast.com/videos/241178-How-To-Use-Social-Media-Strategically-in-the-Federal-Government/

Kaylor, C., Deshazo, R., & Van Eck, D. (2001). Gauging e-government: A report on implementing services among American cities. *Government Information Quarterly*, *18*(4), 293–307. doi:10.1016/S0740-624X(01)00089-2

Moon, M. J. (2004). The evolution of e-government among municipalities: Rhetoric or reality? *Public Administration Review*, *64*(2), 424–433.

Mossberger, K., Tolbert, C., & McNeal, R. (2007). *Digital citizenship: The internet, society and participation*. Cambridge, MA: MIT Press.

Ressler, S. (2009). The rise of gov 2.0: From GovLoop to the White House. *Public Management*, *38*(3), 10–14.

Tspscott, D., Williams, A. D., & Herman, D. (2008). *Government 2.0: Transforming government and governance for the twenty-first century*. [Report presented as part of New Paradigm's Government 2.0: Wikinomics, Government and Democracy Program]. Retrieved from http://www.newparadigm.com/media/gov_transforminggovernment.pdf

USA.gov & GobiernoUSA.gov. (2010, April 12). *Social media style and editorial guidelines*. Retrieved from http://www.usa.gov/webcontent/documents/socmed_editorial_guidelines_041210.pdf

USGovernment. (2009, May 20). *New media across government*. Retrieved from http://www.youtube.com/watch?v=DPBqEdjYw-E

Welch, E. W., Hinnant, C. C., & Moon, M. J. (2005). Linking citizen satisfaction with e-government and trust in government. *Journal of Public Administration: Research and Theory*, *15*(3), 371–391. doi:10.1093/jopart/mui021

West, D. M. (2004). E-government and the transformation of service delivery and citizen attitudes. *Public Administration Review*, *64*(1), 15–27. doi:10.1111/j.1540-6210.2004.00343.x

Williams, A. D. (2007). *Gov 2.0: The power of collaborative communities connecting communities – Using information to drive change*. Washington, DC: The Brookings Institute.

Wooley, K. (2010). Collaboration: The new default setting. *Public Management*, *39*(1), 17–19.

Wyld, D. C. (2009). *Government in 3D: How public leaders can draw on virtual worlds*. Washington, DC: IBM Center for the Business of Government.

Chapter 9
Key Success Domains for Business-IT Alignment in Cross-Governmental Partnerships

Roberto Santana Tapia
Ministry of Security and Justice, The Netherlands

Pascal van Eck
University of Twente, The Netherlands

Maya Daneva
University of Twente, The Netherlands

Roel J. Wieringa
University of Twente, The Netherlands

ABSTRACT

Business-IT alignment is a crucial concept in the understanding of how profit-and-loss organizations use Information Technology (IT) to support their business requirements. This alignment concept becomes tangled when it is addressed in a socio-political context with non-financial goals and political agendas between independent organizations, i.e., in governmental settings. Collaborative problem-solving and coordination mechanisms are enabling government agencies to deal with such a complex alignment. In this chapter, the authors propose to consider four key domains for successful business-IT alignment in cross-governmental partnerships: partnering structure, IS architecture, process architecture, and coordination. Their choice of domains is based on three case studies carried out in cross-governmental partnerships, in Mexico, The Netherlands, and Canada, respectively. The business-IT alignment domains presented in this chapter can guide cross-governmental partnerships in their efforts to achieve alignment. Those domains are still open to further empirical confirmation or refutation. Although much more research is required on this important topic for governments, the authors hope that their study contributes to the pool of knowledge in this relevant research stream.

DOI: 10.4018/978-1-4666-1909-8.ch009

INTRODUCTION

Business-IT alignment (B-ITa), already a hard problem in businesses, takes on an additional complexity in government organizations because these address non-profit goals and political agendas. This adds non-measurable, unstated, or conflicting goals to the alignment problem, making it harder to assess the degree of B-ITa—not only in the public sector, but also in the private one. Moreover, cost/benefit trade-offs in government organizations are made differently from the way they are made in businesses driven by profit. In this chapter, we provide some guidelines for B-ITa in Cross-Governmental Partnership (CGP) settings, that is, settings in which two or more government organizations cooperate to provide services to citizens. The organizations may or may not represent the same level of government (e.g. federal, provincial, or municipal government) and their service delivery model may or may not span multiple jurisdictions. Note that we will refer to this by using the term "business"-IT alignment, even though we are talking about governments, because this is the term commonly in use.

Recently, an unprecedented number of organizations in government and the private sector have entered into partnerships. These partnerships help to deal with the increasing complexity of finding new sources to create competitive advantage in a global market. The increasing importance of partnerships came together with the trend for globalization and the advanced use of Information Technology (IT) to reduce transaction costs by using external resources without owning them (Santana Tapia, 2006a, pp. 3-8). B-ITa is a crucial concept in understanding how partnering organizations use IT to support their business requirements. Yet, alignment between business and IT in any organization is a hard problem that currently requires improved management methods, skills, and practices. With the advent of partnerships, the problem becomes more complex because in such environments, B-ITa is driven by goals of different independent organizations. In CGPs, B-ITa is commonly driven by mission. Mission-driven B-ITa is complex because of the culture based on rules and budgets that characterize government agencies in general (Osborne & Gaebler, 1992).

In this chapter, we propose to consider four key domains for successful B-ITa in CGPs: partnering structure, IS architecture, process architecture and coordination. The term 'domains' requires some explanation. A domain in a CGP is a coherent set of processes performed in that CGP. Our claim is that improvements in these domains cause more improvements of B-ITa in cooperation across government agencies than improvements in other domains do. B-ITa domains within single enterprises have been studied elsewhere (e.g., Chan, 2002; Federal Architecture Working Group, 2000; Luftman, 2003). Those studies show that domains such as skills, technology scope, partnership, governance, competency measurements, communications, informal organization, requirements, and IT architecture help to align business and IT in single enterprises. However, our claim is independent of the results of such studies since we focus our research on partnerships between independent organizations instead of on single enterprises.

Our selection of domains is based on literature review and professionals experience. We empirically validated our domains by means of three case studies conducted in (1) an inter-organizational collaboration among governmental departments of the state of Tamaulipas in Mexico, (2) a networked organization between the province of Overijssel, the municipalities of Zwolle and Enschede, the water board district Regge and Dinkel and Royal Grolsch N.V., all in the Netherlands; and (3) a network of government agencies in Canada.

In the next section, we first elaborate on B-ITa and CGPs. This serves as basis for the rest of the chapter, which is organized as follows: first, we briefly present the B-ITa domains that we identified after reviewing literature and conducting a focus group session. Then, we describe the case studies conducted to validate these domains. After that, we discuss the findings and present a cross-case analysis. Finally, we summarize our conclusions and present our immediate future work.

DEFINITIONS AND ASSUMPTIONS

Business-IT Alignment

B-ITa has been a concern in practice for several decades and has been studied by researchers for more than 15 years (see e.g. Chan & Reich, 2007; Henderson & Venkatraman, 1993). However, despite years of research, B-ITa still ranks as a major modern-day area of concern for both business practitioners and researchers. Interest in B-ITa is stimulated by cases of organizations that have successfully aligned their IT to gain competitive advantage (Kearns & Lederer, 2000; Powell, 1992) and to improve organizational performance (Floyd & Wooldridge, 1990; Sabherwal & Chan, 2001). This is reflected in the number of definitions proposed by practitioners and researchers. Table 1 presents a summary of several B-ITa definitions that can be found in the literature.

In this chapter, we define B-ITa as the process to make the services offered by IT support the requirements of the business, whether such services are offered individually by one organization or collaboratively by the network. This definition is related to the definitions given in Table 1 as

Table 1. B-ITa definitions

Author	Definition
Henderson & Venkatraman (1993)	The allocation of IT budgets such that business functions are supported in an optimal way.
Broadbent & Weill (1993)	The degree of congruence of an organization's IT strategy and IT infrastructure with the organization's strategic business objectives and infrastructure.
Reich & Benbasat (1996)	The degree to which the IT mission, objectives, and plans support and are supported by the business mission, objectives and plans.
Chan, Huff, Barclay, & Copeland (1997)	The situation that occurs when IS functions are amalgamated with the most fundamental strategies and core competencies of the organization.
Maes, Rijsenbrij, Truijens, Goedvolk (2000)	A continuous process, involving management and design sub processes, of consciously and coherently interrelating all components of the business/IT relationship to contribute to the organization's performance over the time.
Duffy (2001)	The process of achieving competitive advantage through developing and sustaining a symbiotic real relation between business and IT.
Luftman (2003)	A state where IT is applied in an appropriate and timely way, in harmony with business strategies, goals and the needs.
Senn (2004)	Ensuring that every single action performed by IT individuals is focused on building and delivering shareholder/stakeholder value by supporting business operations and/or achieving business goals.

Figure 1. Business-IT alignment framework in CNOs (Adapted from van Eck, Blanken, & Wieringa, 2004; Derzsi & Gordijn, 2006)

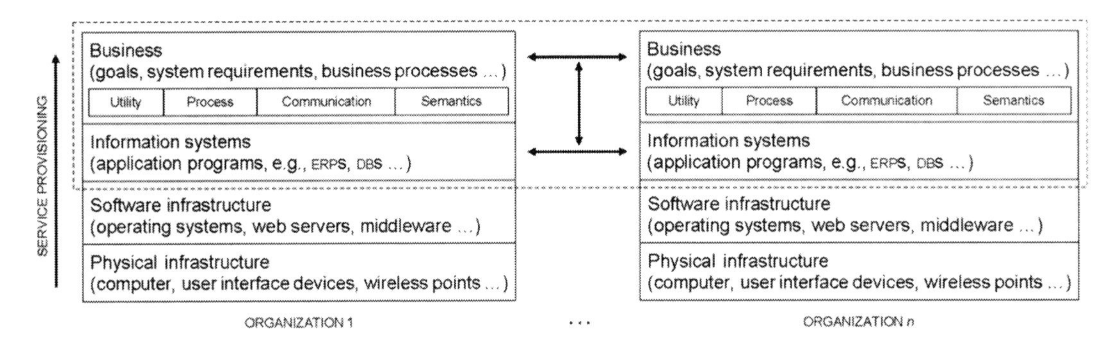

follows: first, we do not consider alignment as a steady state but as a process of continuous improvement. This is similar to those definitions that also stress that B-ITa is a process (Duffy, 2001; Henderson & Venkatraman, 1993; Maes, et al., 2000; Senn, 2004), but differs from the other half, which sees B-ITa as a desired state. As a process, B-ITa has states that can be reached, i.e., an optimal situation of B-ITa—as can be seen in some of the definitions, e.g., Broadbent and Weill (1993), Chan et al. (1997), Luftman (2003), and Reich and Benbasat (1996).

Second, our definition emphasizes the operational level of B-ITa. Many authors in the B-ITa field approach alignment purely at the strategic level, e.g. Chan et al. (1997), Luftman (2003), and Broadbent and Weill (1993). In contrast to these authors, the work of Maes et al. (2000) seems to be applicable both at the strategic as well as the operational level, while the rest of the authors presented in Table 1 seem to make no clear, explicit commitment. We make an explicit note that in our definition, the term 'services offered by IT' means only information systems as a common denominator solution to match the requirements of the business. That is, the term 'services offered by IT' does not refer to IT budgets, IT goals, or IT strategy, but it refers to software applications only. As our work is focused on CGP settings, we explore the B-ITa concept in that particular context. Thereby, the 'requirements of the business' term covers the requirements derived from analyzing the goal(s), and processes of the CGP.

We analyze the B-ITa concept in CGPs based on the scheme shown in Figure 1. This framework was originally developed for Collaborative Networked Organizations (CNOs) pursuing commercial business ideas. However, the framework is generic enough to apply it to partnerships composed of public sector organizations. The horizontal layers classify entities in a service provisioning hierarchy in an organization: physical entities provide services to a software infrastructure, which provides services to information systems, which provide services to the business processes of an organization. In the business layer, we take four views: organizations provide services that have utility, they perform processes to provide these services, they communicate with one another as part of performing these processes, and while doing that, they exchange data that has semantics. Government agencies that participate in a CGP need both to align the different entities (horizontal arrows) as well as to address B-ITa (vertical arrow). Our interest is in the upper two layers of the framework (area delimited by the dotted line), because there is where the business and IT alignment in CGPs takes place.

CROSS-GOVERNMENTAL PARTNERSHIPS

Shifting business, demographic, social, and political fronts force organizations to re-think the way they operate. Since the advent of a cheap communication infrastructure in the form of the Internet, organizations have moved much of their communication with other organizations to this infrastructure, and also have developed new ways of cooperating and communicating.

Government agencies are not an exception. Partnerships between government agencies are becoming more common and important in the current citizen-centered work environment (Cooper, Bryer, & Meek, 2006; Golden, Hughes, & Scott, 2003). These partnerships are used to remove layers of the rigid vertical structures that characterized public entities (Vincent, 1999). CGPs have been referred to in the literature in a number of ways: cross-agency services (Gortmaker, Janssen, & Wagenaar, 2005), inter-agency coordination (Serrano, 2003), whole of government (Australian Public Service Commission, 2004), joined-up government (Ling, 2002), horizontal government (Bakvis & Juillet, 2004), and coalition government (Moore & Mukherjee, 2006). What all these authors refer to is the situation in which government agencies cooperate to achieve shared goals from a citizen-centered perspective.

In our previous work (Santana Tapia, 2006a), we have defined a cross-organizational collaboration as any "mix-and-match" network of profit-and-loss responsible organizational units, or of independent organizations, connected by IT, that work together to achieve shared goals over a period of time. In a public sector context, such organizations are governmental agencies focused on improving quality of services to citizens and on increasing cost-effectiveness of their service delivery processes. We make the note that because the government agencies participating in a CGP are not-for-profit organizations (Brinkerhoff, 2000; Osborne & Gaebler, 1992), we will not use the term 'profit-and-loss responsible' in this chapter. Government agencies are independent from each other in the sense that each has its own public mission to fulfill in settings where they have some autonomy for budget allocation and service-delivery process definition. However, with the public mission also come (1) laws that must be conformed to and (2) requirements such as cost-effectiveness and accountability. A government agency improves its operations if it improves the way it satisfies these criteria and, in doing so, it has a responsibility distinct from that of the other partners in the cooperation.

In a CGP, profit is not relevant and loss consists of not fulfilling or overspending its public mission. Therefore, a CGP is a network of government agencies responsible to fulfill a shared public mission, connected by IT, which work together over a period of time. Our interest is in IT-enabled CGPs, i.e., partnerships that are made possible by IT where the participants interoperate with each other by means of information systems. We believe that IT streamlines collaboration. Similarly to businesses, public sector organizations refer to IT as to a real and lasting innovation that not only drives efficiency and effectiveness, but is also an operational asset supporting the demands of the consumers of government services in the 21st century. Regardless of the fact that government agencies do not operate in a competitive environment, changes in management ideas and disciplines lead government agencies to re-think and revise their service delivery models to better perform their mission (Gulledge & Sommers, 2002; Osborne & Gaebler, 1992). IT is one means by which government agencies can offer effective and efficient services. IT thus allows that governmental agencies focus on their core operations facilitating collaboration between agencies that have complementary competencies. An example of a successful CGP is the Washington Justice INformation Data EXchange (JINDEX) project (National Association of State Chief Information Officers, 2007). Experiences indicate that law enforcement officials in the State of Washington are saving valuable time, and gathering key information, by using JINDEX.

JINDEX is a shared integration platform designed to exchange and transfer data and information throughout the statewide justice community. Spearheaded by the Washington State Department of Information Services (DIS), this collaboration allows state and local law enforcement officers to generate electronic traffic and collision reports in their patrol cars using a scanner and laptop computer. From there, the electronic documents are uploaded to the central Statewide Electronic Collision & Ticket Online Records application (SECTOR) server and sent to the JINDEX. Once received by the JINDEX, these records are routed to the Washington State Department of Transportation (WSDOT), the Department of Licensing (DOL), and the Administrative Office of the Courts (AOC) for processing and disposition. These electronic tickets are delivered to local courts of jurisdiction within seconds, saving organizations on all sides both time and money (NASCIO, 2007, p. 1).

IT ALIGNMENT DOMAINS

From a review of literature related to business/IT processes and CNOs that we performed earlier (e.g. Halinen, Salmi, & Havila, 1999; Konsynski & McFarlan, 1990; Moulaert & Cabaret, 2006; Möller & Halinen, 1999; Santana Tapia, 2006b), we identified different domains to be considered by collaborations in their efforts for achieving B-ITa. These domains are: enterprise architecture (IS landscape), IT/business processes, workflow structure, IT governance, and coordination (see Santana Tapia, Daneva, & van Eck, 2007). We present the definition of each of these domains in turn.

- Enterprise architecture, defined as the landscape of Information Systems (IS), the interconnection relations between them, the technology infrastructure on which they run, and the way they create value for the organization.
- IT/business processes, defined as the architecture of all processes needed to reach the shared goals of the collaborating organizations. These processes are both primary business processes of the collaboration and processes needed for information exchange among the collaborating organizations.
- Workflow structure, defined as the specification of the roles and responsibilities with respect to the IT and business processes that comprise the previous domain.
- IT governance, defined as the "leadership, organizational structures and processes that ensure that the enterprise's IT sustains and extends the organization's strategies and objectives" (Information Systems Audit and Control Association, 2006, p. 5).
- Coordination, defined as the mechanisms to manage the interaction and work among the participating organizations taking into account the dependencies and the shared resources among the IT/business processes.

In order to generate a deeper understanding of these domains and also to validate that they indeed are the relevant domains to ensure B-ITa, we conducted a focus group session. We involved professionals, who are doing research on aspects related to the B-ITa domains, in an active discussion. The goal of the focus group session was to understand how suitable and adequate these first five B-ITa domains were.

Our focus group session was facilitated by one of us (Santana Tapia) who presented the domains and the rational for their consideration in collaborative B-ITa efforts. We invited professionals to be part of a group to discuss our first findings. While we targeted professionals from different background as recommended by Kitchenham, Pfleeger, Pickard, Jones, Hoaglin,

Table 2. Distribution of participants' expertise in the focus group session

Field of expertise	A	B	C	D	Total
Business-IT alignment	1	1	1		3
Software systems		1	2	1	4
Security		1	1		2
Workflows		1	1		2
Cooperative work			1		1
Requirements engineering			2		2
Enterprise architecture	2			3	5
Coordination processes	1				1
Total	4	4	8	4	20

Legend
A = PhD candidate
B = Postdoc researcher
C = Professor
D = Business partner

Emam, and Rosenberg (2002), it turned out that the professionals' availability was the key factor determining their attendance to the focus group session. The professionals were drawn from the Information Systems Group of the University of Twente (http://www.utwente.nl) and the business partners participating in our research project. Table 2 classifies the focus group participants based on their expertise.

We present in turn the results of the focus group session:

1. 9 out of 14 professionals found that in general the term 'enterprise architecture' covers all the presented B-ITa domains.
2. Having in mind the definition of enterprise architecture that we presented, professionals thought 'IS architecture' or 'IS landscape' is a better name for this domain.
3. 57 percent of the professionals raised the question: "why to include workflow structure as a separated domain if IT governance includes already some definition of roles and responsibilities?"
4. There was a strong consensus amongst the professionals that the name of the domain workflow structure must change. It is to avoid its confusion with the most common perception of the term workflow, i.e., automation of a business process, in whole or part, during which information is passed from one participant to another based on some rules (Allen, 2001).
5. Professionals indicated that—in their experience—coordination concepts and IT/business processes could not exist in separation. They found that some coordination mechanisms were always used to glue cross-organizational process fragments together and make them run smoothly.

These recommendations were used to make decisions on what to change in the domains so that its adequacy and suitability were improved. Our decisions concern both the choice of names for the B-ITa domains and their definitions. Figure 2 shows the relationships among the new B-ITa domains. These new domains are partnering structure, IS architecture, process architecture,

Figure 2. B-ITa domains

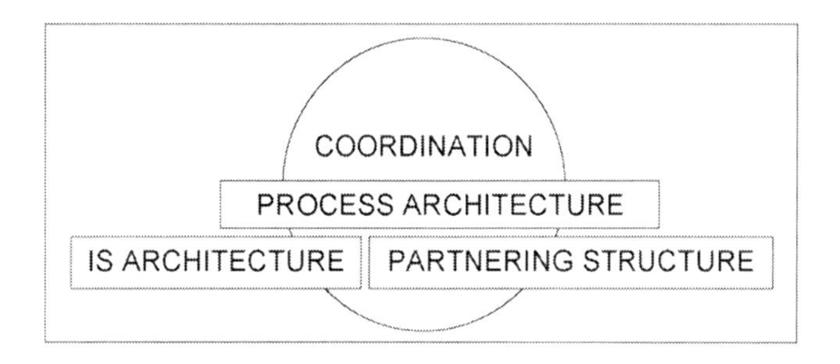

and coordination. The analysis of the results of the focus group session supported us to take the following immediate actions:

- we renamed the domain enterprise architecture to IS architecture to better reflect its new scope. This scope is apparent in the new definition, which we present below.
- we renamed the domain workflow structure to partnering structure to emphasize its contents of how cross-organizational work gets done and who is involved including the definition of roles and responsibilities, and organizational structures.
- we renamed the domain IT/business processes to process architecture to involve both the collaborative IT and business processes without discriminating between these two types.
- we merged the domain IT governance with the domain partnering structure, which meant that the definition of the domain partnering structure incorporates all aspects of what was earlier termed IT governance.

Below, we give a short summary of the resulting four B-ITa domains, following Figure 2 from bottom to top:

- Partnering structure, defined as the cross-organizational work division, organizational structure, and roles and responsibilities definition that indicate where and how the work gets done and who is involved.
- IS architecture, defined as the fundamental organization of the information management function of the participating government agencies embodied in the information systems, i.e., software applications, that realize this function, their relationships to each other and to the environment, and the principles guiding its design and evolution. It must be noted that, in our work, we distinguish IS architecture from IT architecture. For us, IT architecture consists of the (1) implementation platform, i.e., the collection of standard general-purpose software needed to run the IS architecture. It ranges from Operating Systems (OSs), middleware, network software to database management software; and the (2) physical network, i.e., the physical resources that run the software applications. This includes computers, cables, wireless access points, printers, and user interface devices to support the running of the IS architecture (van Eck, et al., 2004). We present a clear distinction of these two architectures in the second case study presented in the next section of this chapter.
- Process architecture, defined as the choreography of all processes needed to reach the shared goals of the participating government agencies. These processes are both primary business processes of the CGP and processes needed for information exchange.
- Coordination, defined as the mechanisms to manage the interaction and work among the participating government agencies taking into account the dependencies and the shared resources among the processes.

Figure 2 illustrates that understanding of both partnering structure and IS architecture is needed to efficiently support the process architecture of a CGP. Public organizations involved in cross-organizational IT alignment can (re)design the partnering structure and IS architecture separately, however, they need to understand both in order to create and maintain a solid basis for the processes required to achieve shared goals and to exchange information in the CGP. Coordination, then, comes next to manage the dependencies among the collaborative activities.

We claim that partnering structure, IS architecture, process architecture and coordination are the domains to should be considered when dealing with CGPs so that value is created for the participating government agencies and B-ITa is achieved. The next section presents three case studies we conducted to confirm these domains in real-life settings.

EMPIRICAL VALIDATION

This section describes the case studies conducted in three CGPs to test the plausibility of the claim: in any successful BITA improvement process, improvements in partnering structure, IS architecture, process architecture and coordination cause more improvements of B-ITa than improvements in other domains do. Using these case studies, we investigate whether the domains mentioned in such a claim are necessary B-ITa domains to take into account by CGPs, or whether it is required to consider more domains in CGP B-ITa efforts. Specifically, we wanted to identify both important information concerning each of the four domains and new valuable topics characteristic to B-ITa attempts in CGPs that could be considered as candidates for forming new B-ITa domains.

Sites and Timeline

Our main criterion for selecting the studied CGPs was the collaborative network perspective, which they take in their efforts towards achieving B-ITa. Once this criterion was met, the only other two requirements were that the CGPs explicitly had a B-ITa project and that they were willing to grant us access. We did not intend to conduct a comparative study across CGPs, but rather to enrich our understanding by bringing different insights from each CGP. In this sense, it is important to consider the particular context of each CGP when interpreting the data. Therefore, we chose a hermeneutic approach (Klein & Myers, 1999) to collect data but we followed the realistic approach of Pawson and Tilley (1997) to analyze the data. In our particular case, a hermeneutic approach helps to obtain results from analyzing the information sources, the sites, and their organizational contexts altogether. The realist approach then helps identifying underlying mechanisms for what we observed and to relate this to the context of the case. The data were collected during April 2006 and October 2008. The CGPs' background is presented in the next paragraphs.

Tamaulipas State Government

The first CGP we studied was a network of more than a hundred departments of the state of Tamaulipas in Mexico (hereinafter referred to as Tamaulipas CGP). The United Mexican States (Mexico) is a federal constitutional republic, i.e., a federation of thirty-one free and sovereign states and a Federal District. Tamaulipas is one of these self-governing states. Since the beginning of the administration 1999-2004, the organization of the Tamaulipas state government has not changed considerably. The government is divided in 12 secretaries. In average, each of these secretaries has 4 divisions and each division consist of at least two departments. One of these divisions is the Division of ICT (Information and Communication Technology) that is responsible for all ICT activities in the government, including development of new systems (IS architecture) and maintaining the IT infrastructure that support these systems (IT architecture). This division is the supplier of IT services within the government of the state of Tamaulipas. It is the IT side in the B-ITa problem. The business-side (IT demand) is represented by the requirements of all other secretaries, divisions and their departments (see Figure 3). The demand side drives the identification and prioritization of systems requirements and opportunities to exploit emerging technologies. This separation of IT management issues (i.e., supply and demand sides) in the collaboration among the secretaries of the Tamaulipas state government is a situation that fits properly with the operational B-ITa we are addressing.

Figure 3. IT and business sides in the Tamaulipas state government structure

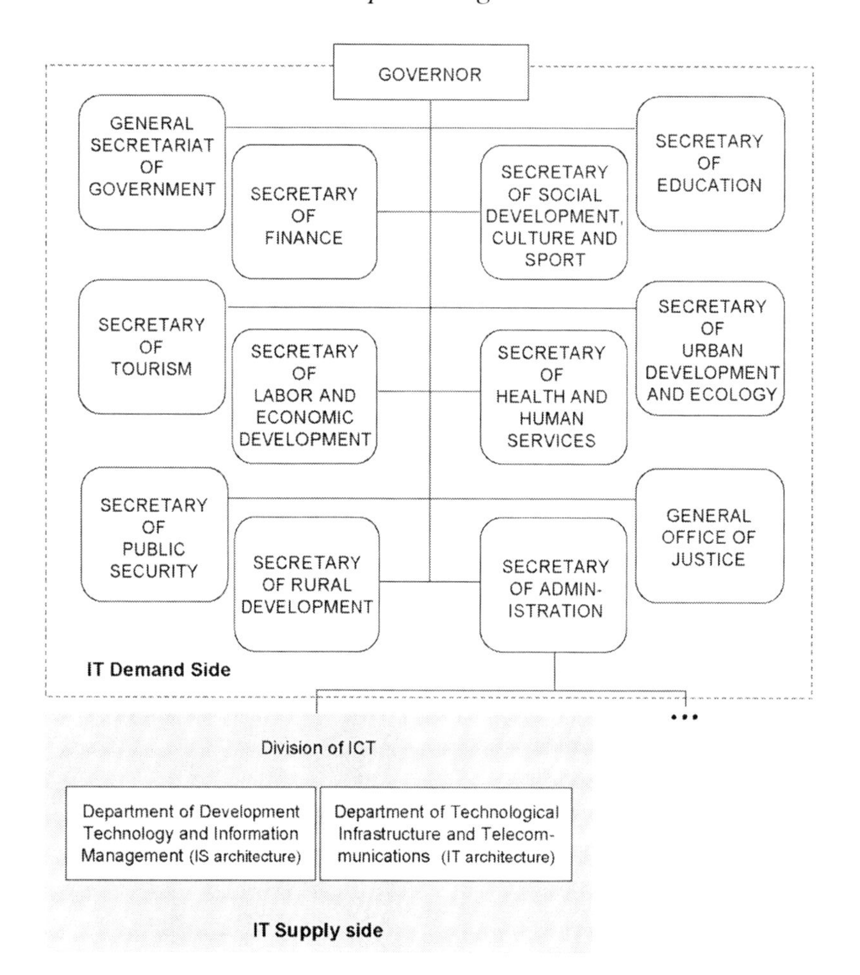

The government of the state of Tamaulipas shares the view that modern governments must be distinguished by the results that they bring, by the solutions that they generate, and by the opportunities and transparency that they offer to the society. As a response to the necessity of having a modern government administration, the government of the state of Tamaulipas implemented Domino/Notes to allow the departments to maintain fast and uninterrupted internal communication, while offering better quality of service to citizens. The goal of the project was to increase e-mail uniformity and to allow the development of collaborative systems. This goal faces the overall requirements of the government departments: to make the service-delivery process more effective and efficient, and to create a better government-citizen relation, meeting the expectations of society.

The first project under Domino/Notes was the Citizen Attention Service System (CASS). This system helps to collect all the individual requests and petitions that the citizens raise to the governor and to the government secretaries chiefs, i.e., any request concerning public services as electricity, security and the like, that the citizen wants to submit to the government of the state of Tamaulipas. The CASS project began in 2001. The initial situation in the area of service provisioning to citizens was characterized by much bureaucracy and poor response time. Only few of the departments had a system to manage the requests. Those systems were home-grown applications developed by IT sections of different departments. Each had its own application logic and data semantics and contributed in a unique way to a lack of homogeneity and communication among systems. For example, when a department received a citizen's request that involved other department(s), the official documents were sent by internal postal service and the department in charge had no longer direct control on the request. This control came back when the documents were received back. The communication among departments was primarily by means of telephone.

The new system facilitates the allocation, distribution, and communication of citizens' requests among departments, as well as all the information related to such requests, i.e., the associated preceding events, the elaboration of the official document, the current status and the answer of the departments involve in the process to satisfy the requests. This helps to have better control in each of the processes, while having a close relation with the citizens to keep them informed on the process of their requests. A visible advantage after implementing the CASS was the reduction of response time. The system records the citizens' requests and automatically sends them to the departments that are involved, thus avoiding bureaucracy and driving employees to work efficiently because of the feeling of being controlled by superiors by means of the new system.

Province Overijssel Collaboration

Our second case study was conducted in a Dutch networked organization that included agencies at the level of provincial and municipal government. The mission of this CGP was to better serve citizens who want to build, re-build, or re-use a house, factory, or barn, in the Netherlands. These citizens usually need to apply for licenses and permits regarding residency, spatial planning, and the environment. Each of these licenses and permits has their own set of criteria, procedures, administrative desks, waiting periods, fees, and staff. For both individual citizens and corporate citizens (i.e., companies), this is a complex and time consuming process that costs both applicants and the government a great deal of money. The Ministry for Housing, Spatial Planning and Environmental Management (VROM—initials in Dutch) wants to gather the different licenses together within the 'omgevingsvergunning'—the environmental permit. All aspects can then be requested from a single point and follow a single paid procedure to obtain a decision even if such decision needs the collaboration of different organizations.

The environmental permit project is part of a packet of measures that the Dutch cabinet has initiated to substantially reduce administrative charges for citizens and businesses. According to the Enviromental Licensing (General Provisions) Bill—WABO for its initials in Dutch, from January 1st 2009, municipalities, provinces and water board districts should be able to use the new process. The environmental permit is part of the modernization plan for VROM legislation, in which the ministry is reducing and improving its rules and regulations. The project includes development of an implementation plan with pilot projects and advice in different Dutch administrative regions. The second CGP we studied is one of these pilot projects.

The Netherlands is a parliamentary democratic constitutional monarchy divided into twelve administrative regions, called provinces. All provinces are divided into municipalities. The country is also subdivided in water districts, governed by a water board, each having authority in matters concerning water management. The studied CGP is a networked organization among the province of Overijssel, the municipalities of Zwolle and Enschede, the water board district Regge and Dinkel and Royal Grolsch N.V., a brewery (hereinafter referred to as Overijssel CGP), working in the WABO-ICT project. The aim of this project is to test the practical feasibility of the national online all-in-one service for environmental permits (LVO—initials in Dutch), steering the project on the services offered by IT to support the business processes.

The WABO forces the different public entities to cooperate with each other in a different way. This cooperation is possible only if it is supported by a correct provision of information. The WABO-ICT project investigates if both the process of cooperation and the support of an information system are feasible in practice. In the future situation of the Overijssel CGP, Royal Grolsch N.V. plays the role of the 'client' asking for an environmental permit to the province of Overijssel. The province of Overijssel is the competent authority that asks advice (concerning residency and spatial planning) to the municipality of Enschede, and (concerning water issues) to the water board district Regge and Dinkel as well (see Figure 4) because the province of Overijssel controls the environmental planning only. We make the note that for this particular case where Royal Grolsch N.V is physically located within the municipality of Enschede, the province of Overijssel would ask advice to this specific municipality only. The municipality of Zwolle will provide such advice just in those cases only in which the permit applicant is physically located within its territory.

Ontario's Provincial Government

The third CGP we studied was a network of provincial government agencies in Canada, namely, the Ontario Council of Agencies Serving Immigrants (www.ocasi.org), hereinafter referred to as Ontario CGP. It was formed in 1978 and today includes around 200 partnering organizations (OCASI, 2008). According to Canada's government decentralization concept, a council is a registered charity governed by a voluntary board of directors elected by the member agencies. The council reports to both the federal government of Canada and the Province of Ontario. The purpose of the council, which we study, is to act as a mediator between the immigrant-serving agencies in Ontario and the federal government, and to coordinate responses to the shared needs and concerns of the settlement sector. The 200 member agencies forming the so-called 'settlement sector' provide some services to newcomers settled in the province of Ontario. Examples of types of such agencies are: Language Instruction for Newcomers to Canada (LINC) centers (more information in http://www.eslincanada.com/linc_programs.html), the YMCA newcomers centers (http://www.ymcatoronto.org/en/newcomers/index.html), urban services organizations, francophone centers.

Figure 4. The to-be state of the Overijssel CGP: functional view

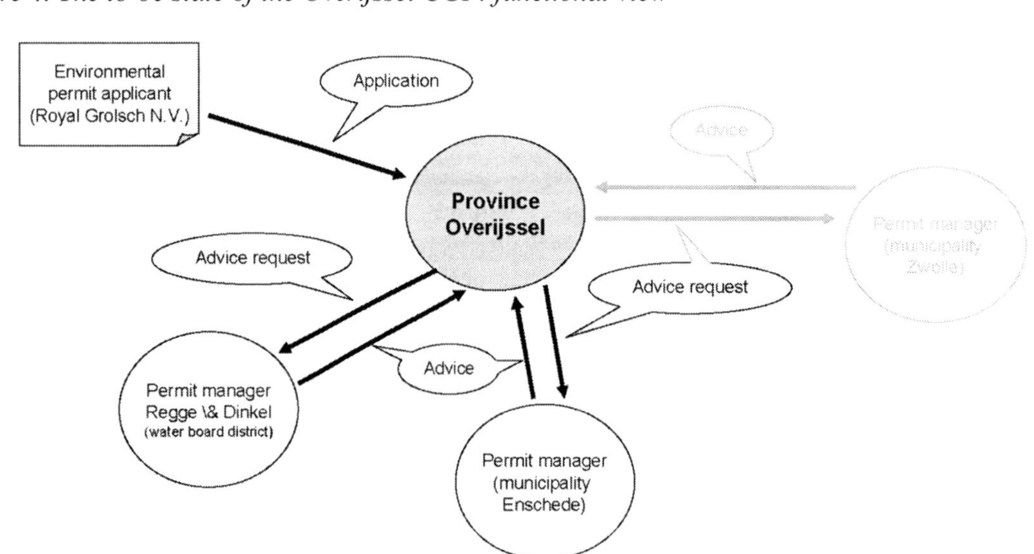

In 2004, the Ontario CGP initiated a large, settlement-sector-wide, project of aligning service-delivery-processes and IT (Commons Group, 2004). It aims at building a flexible, efficient, and responsive information management system shared by the agencies in the Ontario's settlement sector. The system, called Settlement Sector Client Administration Network (SSCAN) system, will standardize and harmonize the service-provisioning and reporting processes across the agencies within the Ontario CGP. It will be fully operational in April 2009. The service provisioning processes being automated by means of SSCAN are shared among the sector itself and its funders (those who provide money for its existence), namely, Citizenship and Immigration Canada (CIC—a ministry at federal level), Ontario's Ministry of Citizenship and Immigration (MCI), and Ontario's Women's Directorate (OWD). The shared system will enable the participating government agencies to contribute, each in its own way, to the CIC mission of "developing and implementing policies, programs and services that facilitate the arrival of persons and their integration into Canada in a way that maximizes their contribution to the country while protecting the health, safety and security of Canadians" (www.cic.gc.ca). To help the reader better understand the scope of this B-ITa initiative, we make the note that the Ontario CGP welcomes per year 50% of all immigrants to Canada (that is, 130,000 people settle in the province of Ontario and become new clients of some agencies of the settlement sector). It is also important to note that the 200 partnering agencies are diverse in terms of (1) the services they provide, (2) the clients they serve, (3) their regional contexts, and (4) their missions. For example, Ontario's MCI classifies them as 'small and large,' as 'single-service and multi-service' and as 'ethno-specific and general.' For responding to diversity of participating agencies, the shared system will support each agency in creating and managing its own service program (hence, identifying and managing its own information and reporting needs).

Prior to the start of the project, the agencies of the Ontario CGP carried out a study on the information management challenges they faced (Commons Group, 2004, pp. 31-41). They identified the following reasons for the low level of efficiency in their service delivery: tripled data-entry, desperate systems, electronic and paper record keeping, all of which added up to labor-intensive reporting to CIC, MCI, and OWD, to slowed-down service provisioning and to frustrated font-line staff.

The Ontario CGP considers the SSCAN project as the solution to let them provide better services to newcomers by using a central information repository and a common system for managing client cases (Commons Group, 2004). The expected outcome is to enhance the productivity of the settlement agencies and their capacity to deliver services

Data Collection and Analysis Techniques

The data collection technique used in each CGP was individually selected to match the particular settings. This choice was motivated by the resources at our disposal. When planning the case studies originally, we considered the use of interviews. However, at case study execution time, it turned out that we could collect data through interviews in the Tamaulipas CGP only. Professionals from the Overijssel CGP were not available to us and the documentation, which we obtained access to, was our only source of evidence in this last CGP. This documentation was carefully used and was not accepted as literal recording of information and events. Last, in the Ontario CGP case, data was collected solely based on publicly available resources published in Canadian government websites (examples are presented in Table 3). This choice was motivated by the fact that Canada's government agencies publish project documentation with level of detail suitable for a documentary case study. We judged suitability in terms of (1) the broad variety of types of documents made

Table 3. Documents studied in the Ontario CGP

Document	As-Is	To-Be
Study of settlement sector database needs (Commons Group, 2004)	-	
Results of "extensive consultation with agencies and in-depth reviews of IS" (Ibid, pp. 31-41)	-	
Current systems (Commons Group 2004, pp. 45-54)	-	
Case management system requirements (Ibid., pp. 42-44)		-
Immigration-contribution accountability measurement system (Citizenship and Immigration Canada, 2002)	-	-
Solution proposals for a shared system (Commons Group, 2004, pp. 17-30)		-
Service Mapping (Farr, 2005)		-
Business case for the shared system (Commons Group, 2004, p. 30)	-	-
Settlement Service standards – inventory of work-in-progress and future steps (Holder, 2001)		-
Evaluation of technology investments (Kerr & Simard, 2002)	-	-
Models of Settlement Service Workshop (OCASI, 2000)	-	-
Re-visioning the Newcomer settlement support system (Shakir, 2000)	-	-

publicly available to the citizens (see Table 3), (2) the broad variety of perspectives involved, e.g. those of the participating government agencies, the researchers who assist them in their B-ITa analyses, the government alignment consulting companies, (3) the broad variety of illustration examples being provided in the documents, which let us abstract high-level notions and map them to concrete instances of B-ITa phenomena relevant to our investigation. (The availability of these documentation sources is a consequence of the federal, provincial, and municipal governments' commitment to a high level of transparency in how tax-payers' money gets spent on IT and on service improvement initiatives).

In the Tamaulipas CGP we used semi-structured interviews with an average duration of 1 hour per interview. The interviews were taped to help writing the transcripts, which we used for analyzing. In this CGP, documents were supplementary sources of data (see Figure 5).

The data analysis was conducted in all three case studies by using interpretation (Klein & Myers, 1999). We bear out this decision by the following two statements: first, as we explained above, documentation was an important data source in this case study. Documents are not simply containers of meanings. They are collectively produced, exchanged, and consumed. They summarize many decisions made by more than one person for a specific purpose. Documents represent specific circumstances including different insights. Therefore, the analysis of documents requires interpretation (Finnegan, 1996).

Second, in the Tamaulipas CGP, professionals, i.e., people, were the primary data sources. In such situation, interpretation also is a suitable analysis technique. Generally, people develop and use their own understanding and observations of themselves and their environment. Therefore, it was expected that the interviewees attached their own meanings to their answers in the interviews. People interpret their world and we, as observing researchers, interpret their interpretations.

Research Approach

A high-level view of our overall research setup is presented in Figure 5. It shows the way we conducted the study through the three CGPs. We followed a replication logic (Yin, 2003) when conducting these case studies. That is, after con-

Figure 5. Research approach

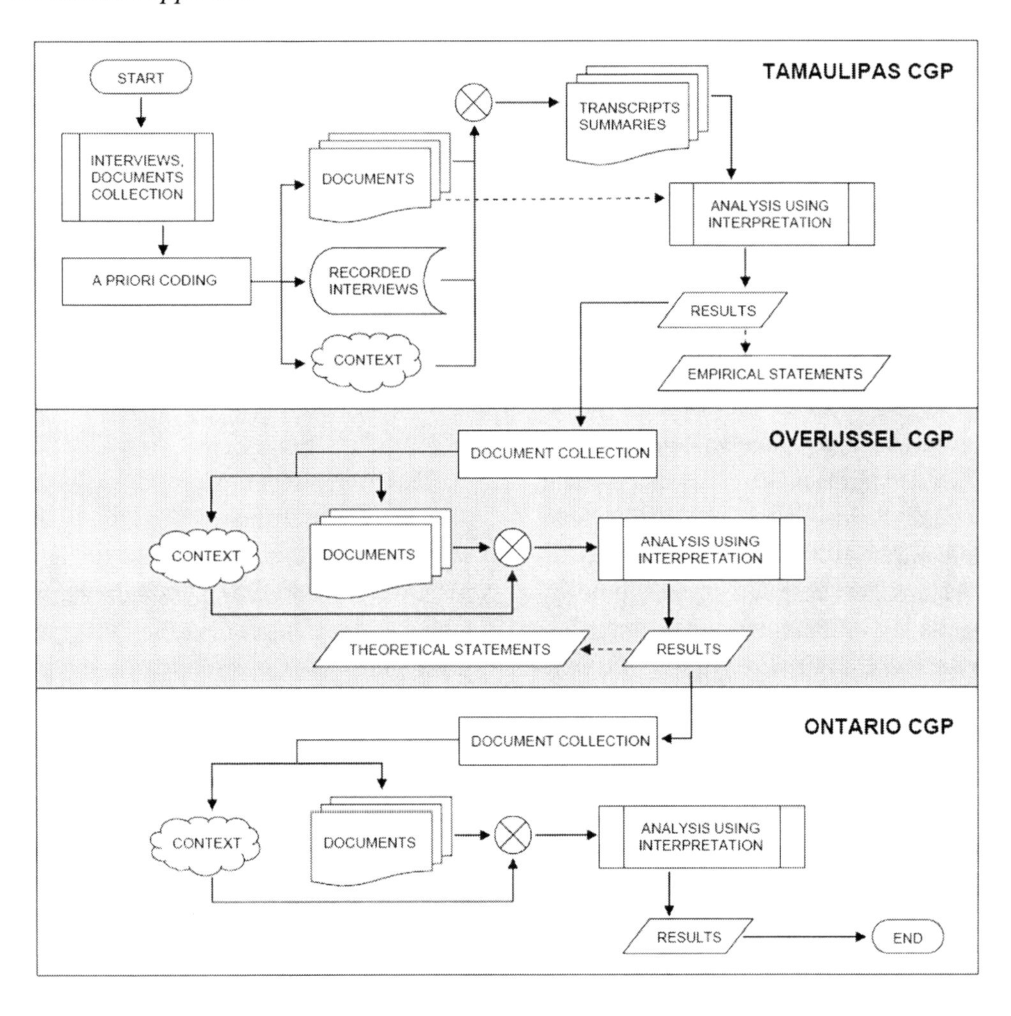

ducting the first study in the Tamaulipas CGP, we uncovered significant findings (empirical statements) that led us to conduct two more studies with replication as immediate research goal. According to Klein and Myers (1999) and Yin (2003), only with replication of findings, such findings could be robust for generalization.

The interviews in the Tamaulipas CGP were conducted with 6 persons on a one-to-one basis. It can be argued that interview data is often biased by impression management (Giacalone & Rosenfeld, 1989) and retrospective sensemaking (Weick, 1995). However, we mitigate such a bias by interviewing different and highly knowledgeable professionals who view the Tamaulipas CGP from diverse perspectives since their expertise were different.

According to literature (Eisenhardt, 1989; Yin, 2003), an "a priori codification" of expected concepts is valuable when starting an analysis process because it helps to shape the theory to be tested and built. Our previous work (Santana Tapia, et al., 2007) helped us to develop such codification based on the B-ITa domains, i.e., partnering structure, IS architecture, process architecture and coordination. This codification is later used in the analysis of the interviews transcripts (see Figure 5). If the codes prove important in the results of our case, then we consider that we have a firmer empirical ground for the findings. Individual transcripts summaries were created from each participant's recorded interview. This process helped to develop a clearer picture of the answers of the participants to the questions of the interview. Specifically, the transcriptions were useful to carry out a 'within-case' analysis (Eisenhardt, 1989), i.e., the interviews write-ups helped to generate insights and to cope with secondary data of the Tamaulipas CGP. As the interview data was, in the end, summarized in transcripts, we chose a hermeneutic approach to interpret it. This approach observes that we can only understand the meaning of a text as a whole by understanding the meanings of its parts and their interrelationships, but that to understand the meaning of the parts, we must understand the meaning of a text as a whole (Klein & Myers, 1999). For example, we can understand the interviews with stakeholders after we know what their entire set of terms means and vice versa. In practice, this means that we iterate between reflecting on the whole text and reflecting on its parts. In our particular case, having as basis the help of stakeholders and domain experts (i.e. the focus group session), a hermeneutic approach helps to obtain results from analyzing the information sources, the interviewees, and the organization altogether in a specific context to establish case boundaries and use that understanding in turn to improve our understanding of the different parts of our data (Paré & Elam, 1997; Pawson & Tilley, 1997).

In addition, in the Tamaulipas CGP, analysis of secondary data was done in parallel to the analysis of the interview transcripts. The secondary data was collected from the documentation, which referred to aspects of the domains included in our model. Proceeding from the case knowledge and case study results, we made some inferences about the B-ITa domains. Therefore, the results of the Tamaulipas CGP study served as empirical statements that were input for the Overijssel CGP study.

The Overijssel CGP provided us documentation concerning its collaborative work. We found two types of documents useful for the purpose of our case study and we carefully reviewed them. These were:

- Letters and memoranda concerning a variety of B-ITa topics in their collaborative work.
- Agendas, minutes, and reports of meetings used to discuss alignment issues and solutions.

These documents were the only ones available to us. According to Yin (2003), these types of documents are useful even though they are not always accurate and, though, they can report bias, i.e., they can reflect the bias of the author. Documents must be carefully used and should not be accepted as literally recording of information and events. The documentation we obtained was mostly clear. However, there were some unexplained acronyms and figures. To avoid confusion in our interpretations, we were able to ask our contact person in the Overijssel CGP.

In the Ontario CGP, we carried out a documentary study based on publicly available information of a province-wide large initiative to build a flexible, efficient, and responsive information management system shared by all government agencies making up Ontario's settlement sector. The documents, which we deemed valuable to include in our study, fall into two categories (Table 3): those describing the initial situation (that is, the pre-B-ITa state) and those referring to the To-Be state. As Table 3 indicates, the documents are of various types, ranging from business requirements of shared applications and business cases, to qualitative research being commissioned by the government to get the initial situation clarified and understood by all stakeholders. The variety of the information sources offered a variety of perspectives on how to reason about B-ITa success domains. The documents are authored by government stakeholders, by standardization bodies, by government consulting service companies (for example the Commons Group—see www.commons.ca), and by researchers.

The documents were analyzed by comparing and reflecting on the text and its parts. This included coding and, whenever applicable, other analytical practices from Grounded Theory (Charmaz, 2007), for example, focused coding, clustering, and theoretical sampling. The last one meant that whenever we felt a need for clarifying a concept related to B-ITa, we sampled new documents. This led us to complement the project documents from Table 3 with the following resources from Canada's settlement service sector which we found relevant to the process of understanding settlement service policies, procedures and standards: documents about standards (Ontario Council of Agencies Serving Immigrant and COSTI, 1999; Canadian Council for Refugees [CCR], 2000), about the issues of the service providing agencies (Go, Inkster, & Lee, 1996), about their best practices (CCR, 1998), about specific municipalities' settlement service provisioning processes (Owen, 1999), about the development of coordinated and collaborated models of settlement services in a specific region (Integrated Settlement Planning Research Consortium, 2000). All of them were brought to the third author's attention by the OCASI, the network who initiated the shared system project.

We also took some steps to counter validity threats. As we were uncertain whether external conditions could produce different case study results, we articulated these conditions more explicitly by identifying different CGPs (i.e., CGPs from different countries and with different amount of participants) to study. With these different conditions in the sites included in our case study, we countered external validity. Construct validity was countered by data triangulation (i.e., use of multiple sources of evidence) and having our case study reports reviewed by peers and by professionals of the studied CGPs (Yin, 2003).

Case Study Results

Figure 6 presents a summary of the findings of our empirical study. Detailed information of the results of each study is presented in the next paragraphs.

Tamaulipas State Government

The findings in the Tamaulipas CGP suggested that coordination and partnering structure are indeed necessary domains that CGPs take into account in their B-ITa efforts. However, the government of the state of Tamaulipas also consider the process architecture, IS architecture and IT architecture, as domains to address when striving for B-ITa. We summarize the results of this study in turn.

Partnering structure. The state of Tamaulipas, as a public organization, has a hierarchy of authority with powers and responsibilities understood by all, a clear-cut division of work among the departments and people, and an explicit set of pro-

Figure 6. B-ITa domains found in the studied CGPs

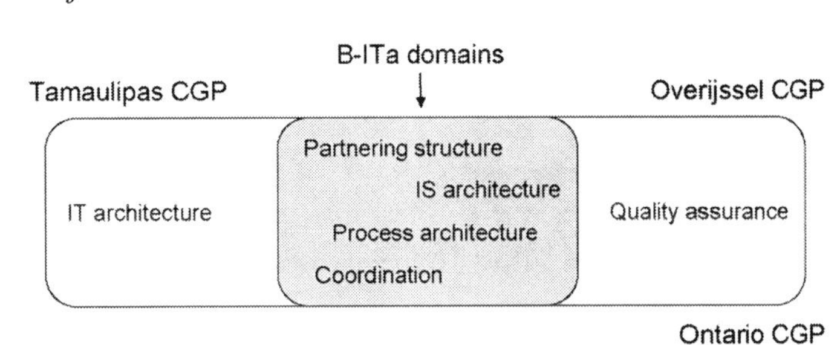

cedures for making decisions. The government of the state of Tamaulipas, every four years, upon the start of a new administration, checks and revises its "Regulations, Procedures and Organization" handbook. The complete organizational structure, the description of roles and responsibilities, the course of actions for achieving results, the norms and policies on the organizational roles, and the relationships among them, can all be found in this handbook. It means that, for the CASS project, the departments already had (1) a good definition of roles and responsibilities, and (2) an established governance structure. Although they have assigned responsibilities to the individual actors collaborating in the network, what went wrong was the inconsistent level of commitment to work effectively with a minimum waste of time and effort. In the partnership, the points of power and authority were positioned in places, which rendered it ineffective to mitigate such a situation. There was no effective use of authority to manage the work commitment program.

IS architecture. The documentation of the IS architecture of the complete CGP was indeed necessary when the CASS project began. However, it turned out that this architecture was not enough to reach the goals of the project. The IS architecture included a significant number of homegrown applications and this rendered inefficient any approach to integrating them and setting up a solid foundation for CASS. This situation led to start the development of CASS from scratch. It must be noted that for the government of the state of Tamaulipas the documentation of the architecture at the level of software applications was not enough. They also needed to define the architecture of infrastructure and technical issues (the IT architecture). For the CASS rollout team to know which departments were ready for CASS and which were not, it was important to have an inventory of the hardware and OSs of each of the participating departments. This inventory effort revealed that they needed to acquire new hardware to support the new collaborative system.

Process architecture. After implementing Domino/Notes as new platform to develop systems in the Tamaulipas state government, the definition of process architecture became an important task for the success of each project that followed. For the Tamaulipas CGP, this architecture was necessary to align the new systems, e.g., CASS, with the requirements of the CGP. The government of the state of Tamaulipas begins a project by thinking first about the processes and, then about the information systems. First, they define the processes that each participant will perform, as well as the collaborative processes. Then, they analyze what information systems could support such processes. In their case, these supporting systems are new systems as they lacked effective applications.

Coordination. In this CGP, and particularly in the CASS project, the requests and petitions management process depends on several departments (for example, Table 4 presents the departments involved in the educative credit request process—this credit is given to students who want to pursue their university studies in other Mexican states). Such a situation led to establish a considerable number of coordination mechanisms to help control the collaborative work flow. Our interviewees converged on the use of the following mechanisms: (1) coordination enabled through shared goals, (2) coordination enabled through agreements specifications, (3) definition and communication of mutual expectations, and (4) regular control meetings. All these mechanisms were used in combination to achieve concerted actions among the participating departments, i.e., to coordinate the mutual work treaties in the state of Tamaulipas.

Province Overijssel Collaboration

The results in the Overijssel CGP replicate our previous findings. This CGP considers the four B-ITa domains as key areas to work on. It must be noted that, in this case, we also could identify a

Table 4. Educative credit request process

STEP	ACTIVITY	DEPARTMENT	DAYS
1	Submit educative credit request	Student	1
2	Receive the request and check the academic situation	Division of Finance and Administration	1
3	Pursuit the request process	Department of Service of the ICEET	5
4	Verify the data of the credited and endorsement	Department of Investments and Portfolio	8
5	Approve/reject the credit	Subcommittee of the Trust of the ICEET	2
6	Elaborate the contract and the promissory note	Institute of Educative Credit	2
7	Get signatures of the credited and endorsement	Division of Finance and Administration	1

ICEET = Instituto de Crédito Educativo del Estado de Tamaulipas

Total steps	7
Total days	20
Total departments	5

domain that is distinctive from the B-ITa domains. It is quality assurance. They consider testing, verification and control as activities that need to be present in all projects. The quality team is always trying to assure quality in the B-ITa project we had access to. We summarize the Overijssel CGP results as follows.

Partnering structure. It was clear that the definition of a blueprint of the Overijssel CGP was the first activity they did to begin to collaborate in order to know what was expected from each of them and to present how they fitted together. In general, the CGP, as a networked organization in the public sector, has also a well-defined hierarchy of authority with powers and responsibilities understood by all, a clear-cut division of work among the participants and people, and an explicit set of procedures for making decisions. It was noted that increasing emphasis was being given to the balance of responsibilities. The major responsibilities in the WABO-ICT project were in the development of the technological architecture, the oversight of the information system, and the description of the collaborative processes and interfaces. Since the project began, there was a clear indication of who was responsible for achieving specific goals, and what the role of each participating organization was, which resulted in a high commitment. However, this commitment came from a specific individual goal of each participant: to be present and to contribute in a project (i.e., the WABO-ICT) that will have national impact in the future.

IS architecture. The documents analyzed revealed that the Overijssel CGP is highly dependent on information systems. We had access to different documents presenting the principles and norms they use for the development of the IS architecture, for example NORA (Kenniscentrum, 2006), SOA, ISO 17799:2005. The Dutch Government Reference Architecture (NORA—initials in Dutch) is a set of models and principles showing how e-government works. They present the way in which it is possible to collaborate, to link processes smoothly to one another and to exchange data. In the Overijssel CGP, each of the three participating public organizations (i.e., the province Overijssel and the municipalities Zwolle and Enschede) also has their own set of principles based on NORA. Such principles, together with the NORA, are the basis for the definition of specific principles and explicit requirements in the WABO-ICT project. When the project started, they created a snapshot of the existing ISs in order to have a clear view of the current situation of the CGP concerning ISs they had, and they followed the same process to define its IS to-be state. An identified problem was that the Overijssel CGP did not have any strategic innovation planning and it did not use any risk management techniques. As a result, the selection of ISs in accordance with resources, needs, and changing situations was deemed inadequate to minimize risks.

Process architecture. As in the IS architecture domain, a blueprint of the current and to-be state of the collaborative processes was clearly identified in some of the studied documents. The Overijssel CGP spent considerable time and effort in working sessions to design the choreography of all (individual and collaborative) processes needed to reach the goals of the CGP. The CGP took project management practices into account during the design of processes. For example, the allocation of human and technological resources was implicit in the process architecture design process. The results of the working sessions included (1) a workflow process that encompasses project approvals, checkpoint reviews, and periodic status reporting at project and statewide levels, (2) a plan to ensure that the work is done acceptably and that the project is in position to complete its phases successfully, and (3) documents that follow recognized best practices for project management (e.g., PMI practices as message/document-driven controls, scope statement definition and work breakdown structure).

Coordination. The Overijssel CGP tries to have highly rationalized activities resulting in simple and repetitive tasks. This situation leads to have a sharp division of work and to depend primarily on standardization of its processes and skills for coordination. Although standardization helps to coordinate the inter-dependencies among the participating organizations, differences in meanings remain creating ambiguities, which lead to conflicts. These cannot easily be handled by informal communication since this kind of communication is sometimes held back by the standardization itself. Thus, they restrict the escalation of conflicts by direct supervision. The Overijssel CGP has appointed staff members who take the responsibility for the processes and provide instructions to others to monitor their work. The coordination team tries to minimize undesirable situations by organizing social events based on informal settings to encourage the synergy and the open discussion on, and exchange of new ideas of working together. They have created an organization that was more concerned about the service to be provided to citizens and its quality than about the political games that could arise. Coordination is usually achieved with communication and active information sharing.

Ontario's Provincial Government

The findings in the Ontario CGP replicated the earlier two case studies in that partnering structure, process architecture, IS architecture and coordination are the pillars in the Ontario CGP alignment initiative. What we found unique in this case is the bold emphasis on coordination, both vertical and horizontal, and both within agencies of the same type and cross-agency coordination happening around specific service programs. The role of coordination and its specific forms are presented in more detail below.

In addition to the above four domains, the documents we reviewed revealed that Ontario CGP considered the so-called 'sectoral service standards' (Holder, 2001; CCR, 1998, 2000) as prerequisites for their alignment project success. These (settlement) service standards are a checklist against which the participating government agencies review and evaluate their programs. The standards are also used by the agencies as the basis for developing their own service manuals. We found evidence, which did suggest that the intent of standards is to enforce service quality assurance policies. For example, the following piece of text from the Ontario's government website about the goals of service standards strengthens this suggestion: "Service standards are goals that we should try to achieve. They can help us give better service because they invite feedback from the public. In addition, they can be applied to almost every service we deliver in the federal government, from issuing a license to managing an internal operation that supports front-line workers. Establishing standards for such things as quality, frequency, and cost can help improve a broad range of federal services." The evidence brought us to conclude that with respect to the importance of quality assurance, the Ontario case replicates the finding from Overijssel. Below, we present the specific results of the study.

Partnering structure. As a member of a council of autonomous community-based agencies, each agency in this CGP has a clearly defined role and responsibilities. The division of work among agencies is transparent and is traceable to the specific services, which each agency is specialized in. Within the CGP, the partner agencies are in a well-defined relationship, which includes "(1) a commitment to mutual goals, (2) a jointly developed structure and shared responsibility, (3) mutual authority and accountability for success, and (4) sharing of resources and rewards" (De Coito & Williams, 2000, p. 136).

Because the council's mission is to serve as a mediator between government institutions of different types and of different levels, decision-making processes are driven by inter-agency consensus and not by exercising hierarchical authority. Outcomes of decision-making are judged based on objective (and common) performance indicators, meaning that agencies are responsible for how the decisions they make affect their own performance. For example, in the shared system project, the adoption of the system is not mandatory for all the Ontario CGP member agencies. This means that those who would like to run their own systems should bear in mind that any service/IT misalignment issues resulting of the decision to stay with their own current systems, might have a detrimental effect over their own performance. If these agencies repeatedly do report decreased performance, then the federal/provincial funders may decide to disband them or merge them with better-performing agencies of the same type.

Based on the roles that the Ontario's CGP plays at the various committees, different information must be made easily available for performance reporting and tracking as well as management decision making (for example, the decision on launching new services or on shifting responsibilities from one type of agencies to another). To the CGP, "clarity in discussing the differing roles and responsibilities of partners is essential to the task of developing new models of settlement services and the systems which support the service provisioning processes" (OCASI, 2000). The documents, which we reviewed, indicate that in addition to the partnership's meaning as a organizational foundation for service delivering, the concept of partnership within the CGP also encompasses five other meanings: "(1) consultative (i.e. advisory), (2) contributory (i.e. managing shared funds), (3) merger/takeover (of agencies), (4) operational (i.e. delivery of work that is purchased, and (5) collaborative (i.e. real sharing of power)" (OCASI, 2000).

IS architecture. The IS architecture of the SSCAN project was analyzed by the CGP at the stage of problem analysis and business/IT requirements definition (Commons Group, 2004). This all was done by using three reference models (Farr, 2005), each referring to a specific level of government, being municipal, provincial and federal: (1) Municipal Reference Model (which at the present time has been adopted by at least 20 Canadian municipalities), (2) Public Sector Reference Model, and (3) Governments of Canada Strategic Reference Model (which is used in the Business Transformation Enablement Program [Farr, 2005] at the Treasury Board Secretariat). The latter is evolving to become a standard Pan-Canadian vocabulary for government programs and services. These reference models helped the collaborating government units define universal definitions for terms regarding their shared

Figure 7. The shared system in the Ontario CGP

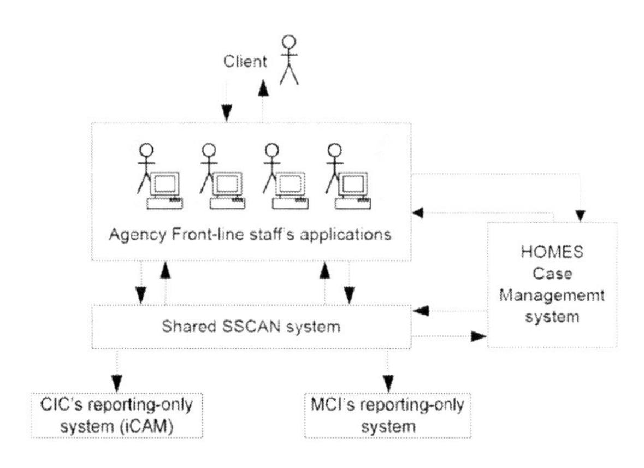

programs services, public needs, outputs, and outcomes (Farr, 2005). It was used to map and produce a picture of each service-delivery process and the applications used in its support. For example, one outcome of the analysis of the current IS architecture was the realization that within the CGP there were 11 different systems with duplicating case management functionality being used by individual member agencies. The To-Be IS architecture envisioned the system for case management to be a common solution shared by a large numbers of partnering agencies within the CGP (Figure 7).

The documents we reviewed indicate the IS architecture as key to evaluating solution implementation options. The CGP considered three different ways to implement the SSCAN system: distributed, centralized and hybrid (see Commons Group, 2004, pp. 22-25). The recommendation, which came out of comparing the three options, referred to the hybrid approach as the best way to go. It included a centralized technology and distributed control, which was aligned with the features of the partnering structure. This option included the development of two centralized Web-based systems – one to be used by agencies and one by the funders.

Process architecture. The participating government agencies in the Ontario CGP analyzed their current cross-organizational service provisioning process and the to-be processes supported by means of the SSCAN system. An important quality aspect of the process architecture in this CGP is that for some processes, multiple channels of service provisioning must exist. For example, certain services (or certain parts of them) should be delivered equally well by personal counseling, telephone counseling, and e-counseling. The new system will support the service delivery process independently of the channel through which it happens. We found that this feature of the process architecture led to the choice to keep the shared SSCAN system open and inclusive to any agency-specific application, which a CGP-partner might deem important in supporting its local (internal) process architecture. For example, the documents indicate that the SSCAN database will be customizable by agencies to meet their particular needs. In line with this, clients' data will continue to be owned and controlled by the agencies (Commons Group, 2004). The client information in the files of an agency is theirs and will remain that way. Security protocols will be established to allow for data evaluation for research and funder-reporting needs based only on aggregated information that does not identify any individual client or client group. At the same time, security and privacy measures will be put in place to ensure that only those with proper permissions within each agency can access the information in question.

The documents we reviewed emphasize that the CGP plans to allow for certain variability in terms of cross-organizational processes, which compose the shared process architecture. For example, SSCAN will not be mandatory for those agencies whose existing systems already have the capacity to meet funder requirements and exchange data with SSCAN. These agencies are welcome to keep variety in terms of service-provisioning processes and client information management processes but they should know why they do it and why their processes vary. For example, the Ontario CGP found a few agencies that have strong systems and strong databases of their own and the CGP considers that it would not make sense for them to replace these applications by SSCAN. Instead, the Ontario CGP plans to work together with them to make sure that data can be shared between SSCAN and these agency-specific systems, ensuring data integrity, security, and privacy.

Coordination. The Ontario CGP makes a clear distinction between horizontal and vertical coordination forms and is well aware of their implications on coordinated service provisioning. Horizontal coordination refers to agencies at one specific level of government (e.g. either municipal or provincial). Vertical coordination refers to coordinating service provisioning actions across institutions at all three levels: municipal, provincial and federal. Partner agencies of the CGP also make a clear distinction among "cooperation, coordination, and collaboration" (De Coito & Williams, 2000, p. 136). They acknowledge that while these three are "functionally related, they are quite different and distinct from each other" (De Coito & Williams, 2000, p. 136) in terms of the level of commitment to mutual relationships and goals, the level of formality in specifying the shared responsibility, the level of mutual authority and accountability for success, and the level of sharing of resources and rewards. To illustrate these distinct characteristics, we cite some evidence from De Coito and Williams (2000): therein, cooperation refers to "informal relationships that exist without any commonly defined mission, structure or planning effort. Information is shared as needed, and authority is retained by each organization, so that there is virtually no risk to the parties involved. Resources are separately owned as are rewards" (De Coito & Williams, 2000). In contrast to cooperation, coordination implies "more formal relationships and understanding of compatible missions. Some planning and division of roles are required, and communication channels are established. Authority still rests with the individual organizations, but there is some increased risk to all participants. Resources are available to participants and rewards are mutually acknowledged" (De Coito & Williams, 2000). Last, collaboration is characterized by "a more durable and pervasive relationship than those related to cooperation and coordination. Collaborations bring previously separated organizations into a new structure with full commitment to a common mission. Such relationships require comprehensive planning and well-defined communication channels operating on many levels. Authority is determined by the collaborative structure. Risk is much greater because each member of the collaborative contributes its own resources and reputation. Resources are pooled or jointly secured, and the products are shared" (De Coito & Williams, 2000). For more information and examples on cooperation, coordination, and collaboration, we refer interested readers to the reference of (De Coito & Williams, 2000, p. 137).

The Ontario CGP made an inventory of where coordination takes place (De Coito & Williams, 2000, p. 80) and their implications for service coordination. The inventory identified six forms: (1) "coordination within a specific service sector" (example: among agencies providing language learning services), (2) "coordination between two of more service sectors" (example: between language learning services and employment services), (3) "coordination around a particular service program" (example: around a pre-school program), (4) "coordination around a particular social problem" (example: alcoholism), (5) "coordination among organizational functions" (example: client intake, service delivery, marketing), and (6) "coordination around services for a specific client group" (example: around services for children with disabilities). Their inventory also indicates inter-agency coordination around specific service programs as their most common form of coordination.

The Ontario CGP uses their knowledge of coordination forms to design an approach to service coordination. They concluded that coordination of services to immigrants should not be approached from a technical or re-engineering perspective, but from a community development perspective. The latter places emphasis on building trust, respect, the capacity for achieving consensus and managing conflict. It also recognizes and respects the need for technical expertise in the development of services and systems. This suggests that in the community-based process for achieving service-coordination, there should be at least two types of facilitators: one should be people-oriented and the other should be technically-oriented (De Coito & Williams, 2000, p. 103).

Furthermore, the Ontario CGP is set out to supports their service coordination models by means of a Shared Interoperability Framework (SIF). SIF is about municipal-provincial-federal interoperability and is defined as "the structured set of de jure standards, de facto standards, specifications, and policies allowing computer systems to interoperate" (Coallier & Gérin-Lajoie, 2006, p. 14). It was found to be "a necessary condition for accomplishing the goals of online government, for controlling costs in certain sectors (such as health), and for providing new services to citizens."

Cross-Case Analysis

In this chapter, we propose four B-ITa domains (partnering structure, IS architecture, process architecture and coordination) to be considered by CGPs in their efforts for achieving B-ITa. In this section, we analyze data across the cases in order to identify similarities and differences in the degree of formalization of the B-ITa process. By identifying similarities and differences, we provide further insight into issues concerning the B-ITa by generalizing the case study results.

The partnering structure in the Tamaulipas CGP was well defined. However, they had a distribution of authority, which did not match the required distribution of work. This situation is an obstacle to proper B-ITa. In this case the partnering structure domain was crucial to improve B-ITa. On the basis of the Tamaulipas CGP case, we can observe that more obstacles in the partnering structure may exist in other cases. So already from this case, we can generalize that partnering structure always must be attended to when improving B-ITa because it can contain critical obstacles towards improving B-ITa. The results of the Overijssel CGP and the Ontario CGP cases support such generalization. In these cases, such a partnering structure definition was important when collaborating. We, however, found an interesting distinction between the partnering structure in the latter two settings: while the Overijssel case emphasized the importance of a well-defined hierarchy of authority with powers and responsibilities to standardize, the Ontario case emphasized "the space for diversity" (Commons Group, 2004, p. 19) and the network's philosophy and lived commitment to actively promote diversity. For example, in their experience, "using standards to actively support diversity has proven helpful" in dealing with "IT-vendor competition and in encouraging the development of innovative custom solutions for agencies with unique needs" (Commons Group, 2004, p. 19). In both cases, each CGP claimed that without the distinctive characteristic of the partnering structure, their B-ITa improvement efforts had not had positive results. The Ontario CGP says that "only when we leave space for diversity, we can be successful." We believe that this might be attributable to Canada's government decentralization concept (Citizenship and Immigration Canada, 1996).

As we explain in our definition of B-ITa, the business requirements come out from the analysis of the CGP goals and processes. In all the three case study sites we investigated, these requirements were definitely reflected in both the IS architecture and the process architecture. These observations let us conclude that, if one of these two domains (IS architecture and process architecture) is not mature enough (though with mature partnering structure and mature coordination), the B-ITa maturity of the entire CGP cannot be high. In the Tamaulipas CGP case, process architecture was attended to routinely. From a hermeneutic perspective, when analyzing the data, we can clearly see the IS architecture and the process architecture in the foreground and we consider a possible explanation for this the matter that government agencies concentrate their attention more on procedures and on applications and then on other domains.

The results of the case studies converge on that coordination is also an important domain to consider when achieving B-ITa. Commonly, government agencies depend on standardization of processes and skills for coordination. For instance, the Overijssel CGP and the Ontario CGP use the NORA and the SIF frameworks respectively to standardize their IT environment between the different participants. However, we also have found some mechanisms that can be an obstacle for proper coordination in a networked organization. For instance, the differences in meaning between the participating government agencies with respect to a situation when working together in a CGP diminish the importance of standardization. This diminishing mechanism is also an obstacle for a mature B-ITa, and leads us to be aware that more coordination obstacles can appear in CGPs. In the Tamaulipas CGP and the Overijssel CGP, both communication and information sharing have helped to maintain a proper coordination. From our studies, we can conclude that coordination then must be addressed in CGPs considering active communication and active information sharing as supporting mechanisms.

The relations between the B-ITa domains assert that (1) having a well-defined collaborative work structure and (2) knowing which parts of this structure can vary, as a basis for the definition of the architecture of information systems and of the process architecture, helps the CGPs to react promptly to the business needs. This is in a situation where coordination mechanisms are present to manage the interaction and work among the CGP participants. This understanding is reflected in Figure 8. The figure presents both a sequence of relations (namely that partnering structure and coordination precede IS architecture and process architecture—see timeline), and a dependency relation (see dotted line) between the four B-ITa domains. These two kinds of relations structure the four B-ITa domains and help explain how the CGPs we studied achieve B-ITa by considering the four domains.

Regardless of how a CGP defines its partnering structure, the IS/process architects in the CGP must (1) consider such a structure for the design of the IS architecture and process architecture, and (2) make sure that a good communication exists between the partnering structure team and the IS architecture and process architecture teams. This is necessary because the partnering structure design produces constraints that need to be met by the IS architecture and process architecture designs. This is very clearly evident in the Ontario case study, in which the CGP rejected those solution options which did not respond to the

Figure 8. Relations between the B-ITa domains in CGPs

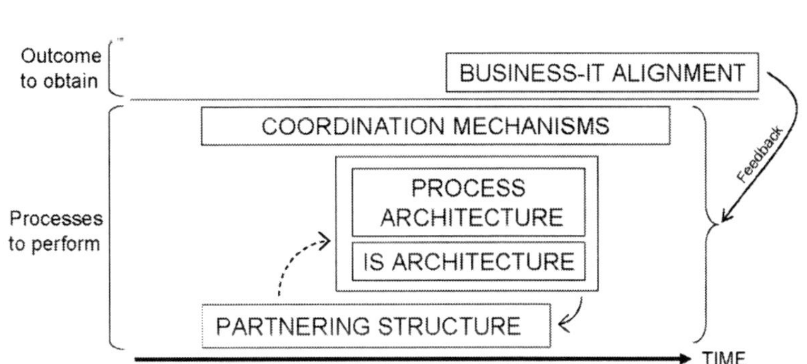

partnering structure requirement. However, we must acknowledge that the CGP's partnering structure pertains to 200 partners, while the number of partners in the other two case studies was much lower.

The small curved arrow starting from the IS architecture and process architecture domains and pointing to the partnering structure in Figure 8 asserts that, for example, it is possible to first design an IS architecture and later re-adjust the design of the partnering structure in a way that ensures the crucial fit between the structure and the IS architecture. Then, later it may be possible to check whether the constraints/requirements resulting from this partnering structure mandate some adjustment in the IS architecture. This situation represents a feedback loop that plays an important role in the enhancement of the B-ITa domains when achieving B-ITa. We must note, however, acknowledge that the Ontario CGP rejected the route to first design the IS architecture and later re-adjust the design of the partnering structure. Instead, the Ontario CGP believed that if an IS architecture is designed around a common database and a solution which fit the partnering structure, this will, later on, have a positive impact in that it will make other government agencies at federal level consider aligning to the Ontario CGP's IS architecture. The motivation for this is that when 200 provincial organizations must report data to one other agency at federal level, it will be much cheaper that this one agency aligns its reporting requirements to the reporting standard of the 200 partners, rather than the other way around.

Finally, the remaining arrow starting from the "outcome to obtain" side and pointing to the "processes to perform" side in Figure 8 represents a feedback process. When achieving B-ITa, measurement and analysis processes help to define measures (e.g., performance measures, as it is in the Ontario CGP, or financial and internal measures) and communicate the B-ITa results in order to consider them in future B-ITa projects to assure quality and real improvements. In the Ontario case, a specific performance assessment framework, namely the Contribution Accountability Framework, has been used for this purpose. This feedback process is to improve and evolve the four domains through experience.

CONCLUSION

This chapter addresses the alignment of IT services (software applications) and business requirements in CGPs. It was born of the perception that some specific set of processes (domains) cause improvements of business-IT alignment. While previous theoretical and methodological efforts have broadly dealt with the context of single entrepreneur-led organizations, few attempts have been made to identify those domains in government agencies. And even fewer, in CGPs. We focus our research on partnerships among independent government agencies. We believe that business-IT alignment in these CGPs is important since it helps to improve organizational performance when providing services to citizens.

To identify the domains, which impact on business-IT alignment, we reviewed literature and conducted a focus group session. We also presented how we have empirically validated these domains. We claim that the identified domains are the necessary domains to consider when achieving business-IT alignment in CGPs. The case study findings suggest that partnering structure, IS architecture, process architecture and coordination are these necessary domains. A new critical responsibility for CGPs involved in business-IT alignment projects is focus on these domains in order to obtain new insights and to strive for business-IT alignment maturity.

We believe that our results are not specific to networked organizations in the public sector only. As we have conducted our cases in three CGPs, we consider our results to be useful for the public sector. However, the results are not clear-cut-limited to this sector. After conducting the case studies presented in this chapter, we have found some similarities between public and private organizations. We believe that networked organizations in both the public and the private sector begin a B-ITa project with a solid mission statement that drives the strategic planning process to meet the common goal(s), and reminds the participating organizations of their work principles and respective roles in the network. As commonly the participants of entrepreneur-led networked organizations do, the participating government agencies of the studied CGPs (1) pool costs, skills, and core competences to provide world-class services that could not be provided by any of them individually; (2) use information systems to respond dynamically to meet the ever-changing customer needs and to communicate and share information among them; and (3) have a clear understanding of the common goal(s) and the functions of each of them in order to know what to expect from each other. Such characteristics are reflected in the CGPs explanation we present in the section "Definitions and Assumptions" of this chapter. So, both sectors are similar. Only their purpose could vary. In addition, this difference in intention creates the environment in which they operate and how they do it.

From the study we present in this chapter, we can also distill the next new theoretical statements that need to be validated in future work:

- The order in which the business-IT alignment domains are taken into account should not affect the results of the alignment effort. If the CGP is organization-centered, it will certainly start to concentrate on its structure first and that would not be the case of a process-centered partnership.
- Effective communication between the parties involved in designing partnering structure and those involved in the IS architecture and process architecture is necessary to create understanding of the capabilities and requirements of the domains, and to achieve business-IT alignment in a CGP.
- The four business-IT alignment domains are enhanced when measurement and analysis processes take place and a feedback process communicates the alignment project results to the involved government agencies.

REFERENCES

Allen, R. (2001). Workflow: An introduction. In Fischer, L. (Ed.), *The Workflow Handbook 2001* (pp. 15–38). Lighthouse Point, FL: Future Strategies.

Australian Public Service Commission. (2004). *Connecting government: Whole of government responses to Australia's priority challenges*. Report No. 4. Canberra, Australia: Management Advisory Committee.

Bakvis, H., & Juillet, L. (2004). *The horizontal challenge: Line departments, central agencies and leadership*. Ottawa, Canada: Canada School of Public Service.

Brinkerhoff, P. C. (2000). *Mission-based management: Leading your not-for-profit in the 21st century*. New York, NY: John Wiley & Sons.

Broadbent, M., & Weill, P. (1993). Improving business and information strategy alignment: Learning from the banking industry. *IBM Systems Journal*, *32*(1), 162–179. doi:10.1147/sj.321.0162

Canadian Council for Refugees. (1998). *Best settlement practices*. Retrieved October 2008, from http://ccrweb.ca/bpfinal.pdf

Canadian Council for Refugees. (2000). *Canadian national settlement service standards framework*. Retrieved October 2008, from http://ccrweb.ca/standards.pdf

Chan, Y., Huff, S., Barclay, D., & Copeland, D. (1997). Business strategic orientation, information systems strategic orientation, & strategic alignment. *Information Systems Research*, *8*(1), 125–150. doi:10.1287/isre.8.2.125

Chan, Y., & Reich, B. (2007). IT alignment: What have we learned? *Journal of Information Technology*, *22*, 297–315. doi:10.1057/palgrave.jit.2000109

Chan, Y. E. (2002). Why haven't we mastered alignment? The importance of the informal organization structure. *MIS Quarterly Executive*, *1*(21), 76–112.

Charmaz, K. (2007). *Constructing grounded theory*. Thousand Oaks, CA: Sage Publications.

Citizenship and Immigration Canada. (1996). *Consultations on settlement renewal: Finding a new direction for newcomer integration*. Ottawa, Canada: Citizenship and Immigration Canada.

Citizenship and Immigration Canada. (2002). *Immigration-contribution accountability measurement system: Security requirements for service provider organizations*. Retrieved October 2008, from http://integration-net.ca/english/ini/caf-cipc/doc/r300-index.htm

Coallier, F., & Gérin-Lajoie, R. (2006). *Open government architecture: The evolution of De Jure standards, consortium standards and open source software*. CIRANO Project Report Num. 2006rp-02. Montreal, Canada: CIRANO.

Commons Group. (2004). *OCASI: Settlement sector database needs study: Final report*. Toronto, Canada: Commons Group.

Cooper, T. L., Bryer, T. A., & Meek, J. W. (2006). Citizen-centered collaborative public management. *Public Administration Review*, *66*(s1), 76–88. doi:10.1111/j.1540-6210.2006.00668.x

De Coito, P., & Williams, L. (2000). *Setting the course: A framework for coordinating services for immigrants and refugees in the peel region*. Mississauga, Canada: Canadian Government Printing Office.

Derzsi, Z., & Gordijn, J. (2006). A framework for business/IT alignment in networked value constellations. In T. Latour & M. Petit (Eds.), *18th International Conference on Advanced Information Systems Engineering, CAiSE 2006*, (pp. 219-226). Namur, Belgium: Namur University Press.

Duffy, J. (2001). Maturity models: Blueprints for e-volution. *Strategy and Leadership*, *29*(6), 19–26. doi:10.1108/EUM0000000006530

Eisenhardt, K. M. (1989). Building theories from case study research. *Academy of Management Review*, *14*(4), 532–550.

Farr, D. (2005). *Service mapping, service Ontario*. Paper presented at the Provincial Government Conference. Lake Carling, Canada.

Federal Architecture Working Group. (2000). *Architecture alignment and assessment guide*. Ottawa, Canada: Federal Architecture Working Group.

Finnegan, R. (1996). Using documents. In Sapsford, R., & Jupp, V. (Eds.), *Data Collection and Analysis* (pp. 138–151). Thousand Oaks, CA: Sage Publications.

Floyd, S. W., & Wooldridge, B. (1990). Path analysis of the relationship between competitive strategy, information technology, and financial performance. *Journal of Management Information Systems*, *7*(1), 47–64.

Giacalone, R. A., & Rosenfeld, P. (1989). *Impression management in the organization*. Hillsdale, NJ: Lawrence Erlbaum Associates.

Go, A., Inkster, K., & Lee, P. (1996). *Making the road by walking it: A workbook for re-thinking settlement*. Toronto, Canada: CultureLink.

Golden, W., Hughes, M., & Scott, M. (2003). The role of process evolution in achieving citizen centered e-government. In D. Galletta & J. Ross (Eds.), *9th Americas Conference on Information Systems*, (pp. 801-810). Atlanta, GA: Association for Information Systems.

Gortmaker, J., Janssen, M., & Wagenaar, R. W. (2005). Accountability of electronic cross-agency service-delivery processes. In M. Wimmer, R. Traunmüller, Å. Grönlund, & K. V. Andersen (Eds.), *4th International Conference on Electronic Government, EGOV 2005*, (pp. 49-56). Copenhagen, Denmark: Springer.

Gulledge, T. R., & Sommers, R. A. (2002). Business process management: Public sector implications. *Business Process Management Journal*, *8*(4), 364–376. doi:10.1108/14637150210435017

Halinen, A., Salmi, A., & Havila, V. (1999). From dyadic change to changing business networks: An analytical framework. *Journal of Management Studies*, *26*(6), 779–794. doi:10.1111/1467-6486.00158

Henderson, J., & Venkatraman, H. (1993). Strategic alignment: Leveraging information technology for transforming organizations. *IBM Systems Journal*, *32*(1), 472–484. doi:10.1147/sj.382.0472

Holder, S. (2001). *Settlement service standards: An inventory of work-in-progress and future steps*. Toronto, Ontario: Ontario Council of Agencies Serving Immigrants.

Information Systems Audit and Control Association. (2006). *CobIT* (4th ed.). Ottawa, Canada: Information Systems Audit and Control Association.

Integrated Settlement Planning Research Consortium. (2000). *Revisioning the newcomer settlement support system*. Ottawa, Canada: Integrated Settlement Planning Research Consortium.

Kearns, G. S., & Lederer, A. L. (2000). The effect of strategic alignment on the use of IS-based resources for competitive advantage. *The Journal of Strategic Information Systems*, *9*(4), 265–293. doi:10.1016/S0963-8687(00)00049-4

Kenniscentrum. (2006). *Nederlandse overheid referentie architectuur: NORA 2.0 samenhang en samenwerking binnen de elektronische overheid*. The Hague, The Netherlands: Ministry of the Interior and Kingdom Relations.

Kerr, G., & Simard, L. (2002). *Powers, evaluation of the OASIS computerization project: Final report*. Retrieved October 2008, from www.realworldsystems.net

Kitchenham, B., Pfleeger, S., Pickard, L., Jones, P., Hoaglin, D., Emam, K., & Rosenberg, J. (2002). Preliminary guidelines for empirical research in software engineering. *IEEE Transactions on Software Engineering*, *28*(8), 721–734. doi:10.1109/TSE.2002.1027796

Klein, H., & Myers, M. (1999). A set of principles for conducting and evaluating interpretive field studies in information systems. *Management Information Systems Quarterly*, *23*(1), 67–93. doi:10.2307/249410

Konsynski, B., & McFarlan, W. F. (1990). Information partnerships - Shared data, shared scale. *Harvard Business Review*, *68*(5), 114–120.

Ling, T. (2002). Delivering joined-up government in the UK: Dimensions, issues and problems. *Public Administration*, *80*(4), 615–642. doi:10.1111/1467-9299.00321

Luftman, J. N. (2003). Assessing IT-business alignment. *Information Systems Management*, *20*(4), 9–15. doi:10.1201/1078/43647.20.4.20030901/77287.2

Maes, R., Rijsenbrij, D., Truijens, O., & Goedvolk, H. (2000). *Redefining business-IT alignment through a unified framework*. White Paper. Amsterdam, The Netherlands: University of Amsterdam.

Möller, K. K., & Halinen, A. (1999). Business relationships and networks: Managerial challenge of network era. *Industrial Marketing Management*, *28*(1), 413–427.

Moore, W. H., & Mukherjee, B. (2006). Coalition government formation and foreign exchange markets: Theory and evidence from Europe. *International Studies Quarterly*, *50*(1), 93–118. doi:10.1111/j.1468-2478.2006.00394.x

Moulaert, F., & Cabaret, K. (2006). Planning, networks and power relations: Is democratic planning under capitalism possible? *Planning Theory*, *5*(1), 51–70. doi:10.1177/1473095206061021

National Association of State Chief Information Officers. (2007). *Connecting state and local government: Collaboration through trust and leadership*. Lexington, KY: NASCIO.

OCASI. (2000). *Models of settlement service workshop, co-located with the OCASI annual conference Oct 20, 2000, Geneva Park, Ontario, Canada*. Retrieved October 2008, from http://atwork.settlement.org/downloads/atwork/Models_of_Settlement_Service.pdf

OCASI. (2008). *Online directory of member agencies*. Retrieved October 2008, from http://www.ocasi.org/membership/OCASI_Online_Directory.pdf

OCASI & COSTI. (1999). *The development of service and sectoral standards for the immigrant services sector: Discussion document*. Toronto, Canada: OCASI and COSTI.

Osborne, D., & Gaebler, T. (1992). *Reinventing government: How the entrepreneurial spirit is transforming the public sector*. Reading, MA: Addison-Wesley.

Owen, T. (1999). *The view from Toronto: Settlement services in the late 1990s*. Paper presented at the Vancouver Metropolis Conference. Toronto, Canada.

Paré, G., & Elam, J. (1997). Using case study research to build theories of it implementation. In Lee, A., Liebunau, J., & DeGross, J. (Eds.), *Information Systems and Qualitative Research* (pp. 70–100). London, UK: Chapman and Hall.

Pawson, R., & Tilley, N. (1997). *Realistic evaluation*. London, UK: Sage Publications Ltd.

Powell, T. (1992). Organizational alignment as competitive advantage. *Strategic Management Journal*, *13*(2), 119–134. doi:10.1002/smj.4250130204

Reich, B., & Benbasat, I. (1996). Development of measures to investigate the linkage between business and information technology objectives. *Management Information Systems Quarterly*, *20*(1), 55–81. doi:10.2307/249542

Sabherwal, R., & Chan, Y. (2001). Alignment between business and IS strategies: A study of prospectors, analyzers and defenders. *Information Systems Research*, *12*(1), 11–33. doi:10.1287/isre.12.1.11.9714

Santana Tapia, R. (2006a). *What is a networked business?* Tech. Rep. TR-CTIT-06-23a. Enschede, The Netherlands: University of Twente.

Santana Tapia, R. (2006b). *IT process architectures for enterprises development: A survey from a maturity model perspective*. Tech. Rep. TR-CTIT-06-04. Enschede, The Netherlands: University of Twente.

Santana Tapia, R., Daneva, M., & van Eck, P. (2007). Validating adequacy and suitability of business-IT alignment criteria in an inter-enterprise maturity model. In S. Ceballos (Ed.), *EDOC 2007: Proceedings of the 11th IEEE International Enterprise Distributed Object Computing Conference*, (pp. 202-213). Washington, DC: IEEE Computer Society Press.

Senn, A. (2004). *Realing: Tackling business and IT alignment*. CIO Advertising Supplement. New York, NY: Deloitte Development LLC.

Serrano, R. (2003). *What makes inter-agency co-ordination work? Insights from the literature and two case studies*. Washington, DC: Inter-American Development Bank.

Shakir, U. (2000). Re-visioning the newcomer settlement support system by ISPR consortium. In *Proceedings of the OCASI Annual Conference Oct 20, 2000*. Geneva Park, Canada: OCASI. Retrieved October 2008, from http://atwork.settlement.org/downloads/newcomer_settlement_support_system.pdf

van Eck, P., Blanken, H., & Wieringa, R. (2004). Project GRAAL: Towards operational architecture alignment. *International Journal of Cooperative Information Systems*, *13*(3), 235–255. doi:10.1142/S0218843004000961

Vincent, I. (1999). Collaboration and integrated services in the NSW public sector. *Australian Journal of Public Administration*, *58*(3), 50–54. doi:10.1111/1467-8500.00105

Weick, K. E. (1995). *Sensemaking in organizations*. Thousand Oaks, CA: Sage Publications.

Yin, R. K. (2003). *Case study research: Design and methods*. Thousand Oaks, CA: Sage Publications.

Chapter 10
Information Technology Product Quality, Impact on E-Governance, Measurement, and Evaluation

Jiri Vanicek
Czech University of Life Science, Czech Republic

Ivan Vrana
Czech University of Life Science, Czech Republic

Zdeněk Struska
Czech University of Life Sciences, Czech Republic

ABSTRACT

The chapter deals with the problems of information technologies' quality aspects and their importance for the state and public administration, e-governance, and e-democracy. The concept of quality is defined, and it is demonstrated that for quality evaluation, the clarification of needs and specification of exact requirements represents a key problem. The chapter focuses on general quality management and specifically on quality of ICT-solutions. The ISO/IEC software product quality and data quality is thereafter briefly presented. The principle of quality evaluation based on the quality decomposition into several characteristics is also described along with the principles of measurement theory. Finally, this is supplemented by selected approaches for expert rating of trends for ICT-solutions having a direct impact on long-term quality prediction.

INTRODUCTION

The main objective of this chapter is to point out the importance of the quality of Information and Communication Technologies (ICT) intended for applications in state and public administration. A well-balanced choice between the required quality and the costs is usually a fundamental problem of institutions responsible for their selection and purchase. Today it is not feasible for public institutions to develop ICT applications using their own human resources. They do not have sufficiently qualified personnel for such specialised work

DOI: 10.4018/978-1-4666-1909-8.ch010

available. To employ such highly qualified staff would not be effective in the long-term. Hence, public administration authorities depend on purchasing the necessary products and services on the market, or on ordering the development of a custom-made solution. The amount of money coming from public sources to accommodate these ICT demands is vast. It is thus crucial to allocate these funds efficiently and responsibly, for the credibility of state and public institutions.

To achieve the needed balance between the contribution of ICT facilities and acquisition costs, it is necessary to define the actual needs as accurately as possible, to assess how the evaluated ICT products meet these requirements, and to what extent. It actually requires evaluating the quality of these products. Only then, is it possible to compare the product quality to the anticipated acquisition costs. While the costs or expenses can be determined unambiguously, the assessment of quality is a much more difficult task. To evaluate the quality of complex intangible products such as computer software can be particularly complicated, and always somewhat disputable. The quality assessment is a complex problem for all interest groups such as product developers, acquirers including end users, or system integrators. It is a common practice in the developed countries that this quality evaluation is performed by independent third parties, who have sufficient reputation and adequate experience for such complex activities. Although considerably difficult, it is very important to ensure the utmost objectiveness and mutual comparability of the evaluation process.

There are two approaches towards quality evaluation. The first approach tries to find the exact definition of product quality and the evaluation process. It is based on separating the product requirements into individual quantifiable product characteristics called *attributes*. Numerical representation of these *indicators*, which should be achieved, is then compared to the actual value of the evaluated product attribute; the *measure*. Finding this value is based on a transparent method called *measurement*. Then, the quality evaluation lies in a complete comparison of deviations of instantaneous values from required indicators, taking into account the relevance of individual attributes for the user. In applying such methods, it is crucial that they are performed uniformly so that the results are comparable. In this chapter, we introduce the models and methods defined in international technical standards, which are being developed by the international standardisation authorities, ISO and IEC (International Electromechanical Committee). We also present the current situation of standardisation, the expected advancements, and possible problems in this field.

The other approach towards quality evaluation is based on previous experience and on an appraisal of existing progress in the field that is on expert analysis of existing trends. We present these methods in this chapter.

BACKGROUND

The international standards for quality management of the ISO 9000 series are well known and respected in the community of developers, suppliers, acquirers, and users of all product categories. However, these standards are general, and cannot reflect specifics of individual products including software products. These standards are primarily focused on the implementation of quality management systems for product developers, producers, suppliers and service providers, but only marginally for product acquirers and end users. They obviously cannot directly apply to the development process, and have often only limited information about it. The conformance of the development process with the requirements of the ISO 9000 series provides a certain level of product quality, but it cannot assure that all product attributes fit to particular user requirements. Accordingly, it is necessary to extend the general standard of the ISO 9000 series by standards for product quality evaluation based on specifics of individual types of products.

In the area of information technology, there operate two main worldwide standardisation organizations: ISO and IEC. To avoid duplication of the standardization effort, the ISO and IEC have created the joint standardisation authority called the "ISO/IEC Joint technical committee for information technology" – "ISO/IEC JTC1 – Information technology." The ISO/IEC JTC1 outputs are automatically assumed to be ISO and IEC documents and have the ISO/IEC label.

The issues of evaluation and measurement of system and software quality are controlled within the ISO/IEC JTC1 subcommittee SC7 "System and software engineering," primarily in its working group WG6 "Software product evaluation and product and processes measurement."

Today software product quality is covered by several standards. The standard ISO/IEC 9126-1 (2001) covers the software product quality model, and the following technical reports ISO/IEC TR 9126-2 (2003), ISO/IEC TR 9126-2 (2004), ISO/IEC TR 9126-3 (2004), and ISO/IEC TR 9126-4 (2004), propose certain quality attributes and measures ("metrics" in the earlier terminology). Another series of six standards, the ISO/IEC 14598-1 through ISO/IEC 14598-6 (1998, 1999, 2000, 2001) describes the quality of the evaluation process from different perspectives. Finally, the isolated standard ISO/IEC 12119 (1994) is focused on off-the-shelf distributed software (or software packages).

These standards are not fully mutually consistent, the nomenclature is not transparent, and the included terminology does not fully conform to the standard ISO/IEC 15939 (2007) for the measurement process. The current standards also do not contain sufficient and well-defined lists of attributes and measures considered as suitable for product quality evaluation. In view of these facts, the ISO/IEC JTC1 decided to start the research project SQuaRE (**S**oftware Quality Requirements and Evaluation) with the goal of creating the series of standards ISO/IEC 250xx that would consistently cover the software products and systems quality evaluation. Several norms of this new series have already been approved and published. Other standards are in different stages of the preparation and ratification process by international voting procedures. The actual concept of the ISO/IEC 250xx series structure and the status of the individual documents as at November 2008 are briefly presented in the separate paragraph of this section.

Besides the above-introduced principles of quality evaluation, also other alternative methods have proven useful. These alternative approaches are based at expert evaluation. The well-known consultancy agency Gartner can provide methods, which enable to assess both: long-term trends along with current software products and services evaluation. Selected methods used by Gardner will be mentioned in this chapter.

THE GENERAL QUALITY CONCEPT

The concept of *quality* is usually considered as a conformance with requirements or expectations. Some authors prefer to understand the concept in the narrow sense of the word; for example, the fitness for use, a lack or lower rate of failures, or the support for achieving a goal. We adhere to a more general definition of quality, based on the conformance of requirements.

The American Society for Quality (2008) describes quality as "a subjective term" for which each person has his or her own definition. In technical usage, quality can have two meanings:

a. Characteristics of a product or service that bear on its ability to satisfy stated or implied needs;
b. A product or service free of deficiencies.

From the ISO point of view, the object of quality evaluation can be a product or process. Process is defined in ISO 9000 (2005) as a set of interrelated or interacting activities, which transform inputs into outputs.

Product is defined as a result of a process. There are four generic product categories, as follows:

- Services (e.g. transport),
- Software (e.g. computer program or dictionary),
- Hardware (e.g. hard disk or engine mechanical part),
- Processed materials (e. g. iron ingot, or fuel),

or their combination. Many products comprise elements belonging to different product categories. Typical products are operation systems or commercial off-the-shelf software products, which are usually distributed along with the corresponding support services. Service is a result of at least one activity performed between the supplier and consumer and it is generally intangible.

To cover the product and process quality by the only definition, the ISO used the following quality definition: Quality is the degree to which a set of inherent characteristics fulfils requirements. The word "inherent" as opposed to "assigned," means existing in something as its permanent characteristic, not subsequently or temporarily added. Requirements are defined as needs or expectations that are stated, generally implied or obligatory.

Requirements can be generated by various parties concerned, persons or interest groups (e.g. customers, owners, employees, suppliers, bankers, or social partners) and hence they can be different for each group. Even several representatives of the same category of interest groups can have different requirements. For instance, one user of a text editor can have rather low requirements for the extent of advanced functions, but strong requirements for the ease of use. The other user needs a large amount of functional possibilities while being content with slightly inconvenient usability of a product. It is apparent that quality is a relative concept, depending on interest groups with a given set of requirements.

In the first attempts to define quality, the term "needs" was formerly used instead of "requirements." There are several reasons to prefer the term needs to requirements. One of them is the necessity to cover requirements originated not in needs but in external regulations, for example in legislation or in other obligatory documents. Another reason is that the term need can be considered as more indeterminate than the requirement, which has to be formulated objectively so that it can be verified.

The problem of the "needs into requirements" transformation is delicate and rather difficult. Only a minority of potential users hold adequate information and sufficiently qualified staff for the formulation of accurate quality requirements. The quality is costly and finding the balance between high quality and economically acceptable costs is a difficult but very important problem. If the transformation of "needs into requirements" is committed to an expected supplier, there is a risk that the supplier's interests prevail over the interests and priorities of consumer. This can be a particularly delicate problem for government and public authorities, where financial resources come from taxation and where responsible individuals do not invest their own money.

The quality can be evaluated from various perspectives depending on the position with respect to an evaluated object (process or product) and to the amount of available information. The group involved in the development process has obviously full information. The acquirer or potential user has obviously only a limited access to information needed, and it depends on information published by the developer or supplier before entering into a contract. It is crucial for acquirers to obtain as much information as possible. In the evaluation of quality, the middle layer between supplier and acquirer is represented by the position of independent evaluators (third parties). It is in the interest of suppliers to provide evaluators with complete information. Highly respected evaluators that are truly independent can provide valuable guidelines for a considered decision by the acquirer.

THE SPECIFIC CLAIMS FOR QUALITY OF ICT SYSTEMS AND SOFTWARE PRODUCTS

In the ISO terminology, the term software covers not only a computer program or data but also intangible products such as know-how, algorithms, or methodology. The specific sort of software product is computer software. Computer software comprises computer programs, procedures, and commands, including documentation and data, related to the system in which they are processed. In the following text, the term software is always considered as computer software. In our terminology, a software product is a set of programs, procedures, instructions, and rules, perhaps with documentation and data, delivered to a third party under a single label.

The main problem of software is that it cannot be used as such and consequently it cannot fulfill any needs or requirements. Software can only be used as a part of a system that usually also contains other software, computer hardware, operator services and operating procedures. Strictly speaking, the quality of a software product cannot be investigated and evaluated separately, but only within the context of complete system.

Nevertheless, the impact of software on the quality of information and communication system as a complex entity is often crucial. For that reason, we consider the software product quality, as the impact of the software product on the quality of system in which it is integrated. However, the quality of a software product has to be measured and evaluated based on external behaviour of a complete system. The quality of any software component identifiable within the software product is evaluated in the same way. Then, the quality of a system can be considered as the result of the quality of individual components and their interaction.

The present ISO/IEC quality model, which is used as a reference for our considerations, defines three different views of quality:

- Quality in use.
- External software quality.
- Internal software quality.

Quality in use is the degree to which a product used by specific users meets their needs to achieve specific goals. The software quality in use can be viewed as the capability of the software to enable quality in use in its operational environment, for carrying out specific tasks by specific users. In fact, the quality in use is not a product quality but the process quality, i.e. the quality of product exploitation. Although the alternative term *quality in operation* can be used for such a view, we follow the terminology used in the standard.

External software quality provides a "black box" view of the software and addresses properties related to the execution of the software on computer hardware and on an operating system. It can be defined as the degree to which a software product enables the behaviour of a system to satisfy stated and implied needs when the system including the software is used under specified conditions.

Internal software quality provides a "white box" view of software and addresses properties of the software products that typically are available during the development. Internal software quality is mainly related to static properties of the software. It can be defined as the degree to which a set of static attributes of a software product satisfies the stated and implied needs when the software product is used under specified conditions. Internal software quality has an impact on external software quality, which again has an impact on quality in use.

The authors of this chapter believe that the terms "quality in use," "external quality" and "internal quality" are not fully appropriate. The quality is an integral property of the product or process. The product or process fulfils the requirements and meets the needs as a self-contained entity. *Ergo*, there has to exist only one criterion for quality. The evaluation or forecasting of the product quality can certainly be based on various sources and it can be performed in different stages of the software life cycle corresponding to the three quality views introduced above. Accordingly, it is more appropriate and rigorous to speak about only the software product quality and its internal or external attributes and measures for the product quality evaluation. Correspondingly, it is convenient to consider the quality of the utilisation process of the complete system containing the software product and the "in use" attributes and measures of product quality. The representation of the role of internal, external and in use quality attributes in the quality of the software product life cycle is depicted in Figure 1.

According to ISO 9000 (2005) definitions verification means the confirmation, through the provision of objective evidence that specified requirements have been fulfilled. Validation means the confirmation, through the provision of objective evidence, that the requirements for a specific intended use or application have been fulfilled.

The phase of a product under development is the subject of internal software quality. Internal software quality requirements specify the level of the required quality from the internal point of view of the product. They include requirements derived from external software quality requirements. Internal software quality requirements are used to specify properties of intermediate software products (specifications, source code, etc.). Internal software quality requirements may also be used to specify properties of deliverable, non-executable software products such as documentation and manuals. Internal software quality requirements can be used as targets for verification at various stages of development. They can also be used for defining strategies of development and criteria for evaluation and verification during development.

The phase of a product in operation is the subject of external software quality. External software quality requirements specify the required level of quality from the external point of view. They include requirements derived from user quality

Figure 1. Software product life cycle

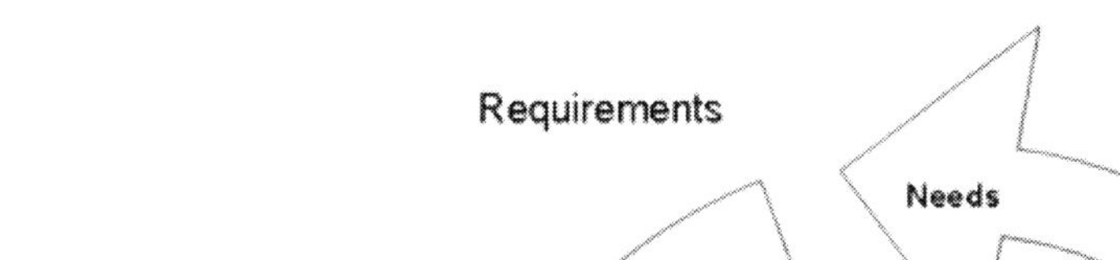
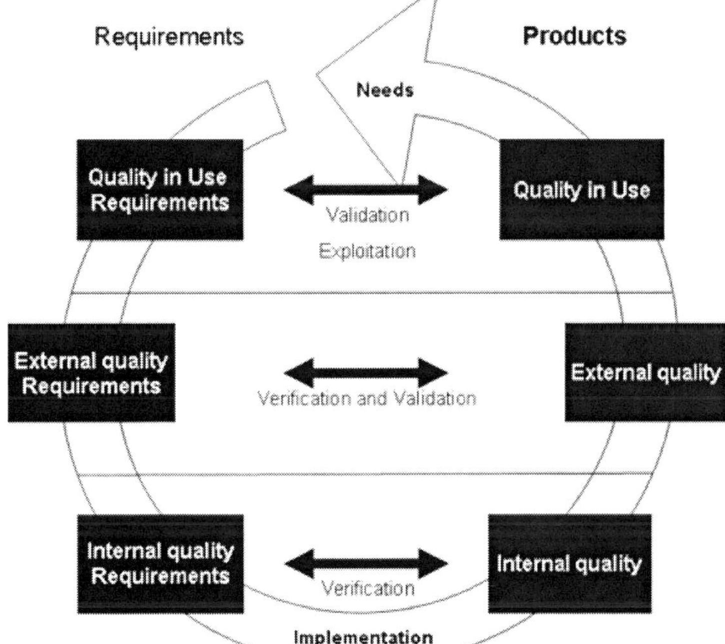

requirements, including quality in use requirements. External software quality requirements are used as the target for technical verification and validation of the software product. Requirements for external software quality characteristics should be stated quantitatively in the quality requirements specification using external measures and used as criteria when a product is evaluated.

The phase of a product in use is the subject of quality in use. Quality in use requirements specifies the required level of quality from the end user's point of view. These requirements are derived from the needs of each context of use. Quality in use requirements are used as the target for validation of the software product by the user. Requirements for quality in use characteristics should be stated in the quality requirements specification using quality in use measures and used as criteria when a product is evaluated.

In general, the sooner a possible quality problem is identified, the cheaper is its correction. The difference between the cost of corrective actions during the early stages of the development process and their cost just before distribution to the end users can be immense. On the other hand, the sooner the quality estimation is performed during the software life cycle, the less accurate and reliable will be the quality estimate of the final product obtained. The attributes of intermediate products can be regarded more as quality predictors rather than the attributes of the final product quality. The state-of-the-art software development methodologies are not able to answer the question, as to which attributes of the specification documents or the project design and development documents can affect the quality of software or system, and to what extent.

ISO / IEC QUALITY MODEL

Quality of the software product depends on their properties. Some software properties are inherent; some are assigned to the software product. The capabilities of a software product are determined by its inherent properties. Inherent properties are often classified as either functional properties or quality properties. Functional properties determine what the software is able to do. Quality properties determine how well the software performs. In other words, the quality properties show the degree to which the software is able to provide and maintain its specified services. The functional properties, however, also meet a need and satisfy the interested parties' requirements. Therefore, also according our quality definition, the functionality has to be considered as a part of product quality. Quality is inherent to a software product. Assigned properties are not considered to be the quality of the software components, since these can be changed without changing the software.

Inherent software properties which can be classified (measured) quantitatively or qualitatively, are called attributes. Quality attribute is any inherent property of the software that contributes to quality. This contribution can be direct or indirect. For example, the failure rate during the test runtime of the product is a direct contribution to product quality rating. The volume of the source code *per se* says nothing about the product quality, but together with the number of folds in the program code, can substantially contribute to the product quality, therefore shall be regarded as a quality attribute.

Quality attributes are measured by applying a measurement method. A measurement method is a logical sequence of operations used to quantify an attribute by some number. The result of applying a measurement method is called "Base measure" or "Direct measure." Base measures of quality attributes are also sometimes called "Quality measure elements." The *base measure* is obtained using observation of the respective entity (product or process). *Derived measure* can be obtained from base measures or other derived measures by applying a *measurement function*. A measurement function is some formula or an algorithm used to combine base or derived measures. The result of applying a measurement function to selected measures of quality attribute is called a "Software quality measure."

As has been highlighted before, quality is relative depending on the individual interested parties or their groups, their needs and requirements. It will not, therefore, be useful to formulate the requirements of software quality as one indivisible complex and the quality evaluation thus achieved, only as total. For example, for nuclear power plant reactor control software, the requirements for reliability and safety will be overriding, however usability can be rather complicated, and interoperability and portability can be out of the question, because it will be probably an isolated system for this unique application. For software for an information kiosk on a railway station, ease of use will be first-rated. Also for one software product, such as the text editor, various users have various requirements. For example the requirements of the layman, scientific workers, book publishers and broadsheet makers can be very different. Therefore, it is reasonable to divide the quality into several categories for which the quality requirements are recommended to be formulated separately, and the respective quality attributes also measured separately, to achieve the evaluation. Such categories are called *quality characteristics*.

Such a division is probably reasonable for quality evaluation of all products, not only for software products or ICT systems. We know from experience that the optimal number of all clear divisions is usually estimated as 7, or some number in the range between 5 and 10. If more detailed division is required, multi-level division is more appropriate for clarity. The categories on the third level of such multilevel divisions are called quality *subcharacteristics*. The structure of the division of quality into categories is shown in Figure 2.

On the ISO/IEC quality model the ideal situation is anticipated, that each quality attribute can be allocated to one and only one quality characteristic, and in the third level of the quality division into one and only one quality subcharacteristic. Unfortunately, life is not so easy. The authors of this chapter are sure that such a mutually disjointed tree division of quality cannot be developed. Every time some quality attribute will influence several categories in the division. For example, for a software product, it is beyond doubt that reliability; maintainability and portability should be different quality characteristics. The requirements of various users or interested parties for these characteristics can be different. But it is out of question that such attributes of the development process, such as object oriented design, structure oriented design or unregulated design,

Figure 2. Division of quality into categories

and such attributes of the source code as concrete programming language or the case tool used, or the comments density apparently affects all the above-mentioned software quality characteristics. The tree structure of the quality (quality characteristics—quality subcharacteristics—quality attributes from Figure 2) is very appropriate for normal workers, however unrealistic. The more complicated, but realistic approach based on actual relationships will be used.

Therefore, the correct solution of the characteristics, subcharacteristics and attributes, and their hierarchy of software product quality comprise the filling for the two dimensional matrix, with rows being quality characteristics and subcharacteristics, and columns the proposed or concluding quality attributes. In the fields of this matrix will be the indicators of the level in which the corresponding attribute applies to the characteristics or subcharacteristics.

In the following subchapter, we shall refer to the characteristics and subcharacteristics tentatively proposed for the software product quality evaluations from the external and internal point of view (in ISO/IEC terminology for evaluation of external and interval software product quality). In the next one, we shall refer to the tentative SQuaRE proposed characteristic and subcharacteristics for the process quality of the system contained utilization software (in ISO/IEC terminology the quality in use quality evaluation). As the objects of quality evaluation differ, it is evident that the quality divisions should also be different.

SOFTWARE PRODUCT QUALITY CHARACTERISTICS

On the SQuaRE research project, the following eight software product quality characteristics are proposed: Functional suitability; Reliability, Performance efficiency; Operability; Security; Compatibility; Maintainability; and Portability.

With respect to the ISO/IEC 9126-1 (2001) model, the number of software product quality characteristics increased from 6 to 8. Formerly security and interoperability were on the subcharacteristics level as a part of the characteristic "functionality." As a result of the underlying emphasis for these properties of information systems and software product from several application areas, including the government and public sector, it was decided to split these views, allocating them as separate quality characteristics and formulating quality requirements and quality evaluations for their performance separately. The names of some characteristics can be more precisely named: From functionality to functional suitability, from efficiency to performance efficiency, from usability to operability, and from portability to transferability.

The definitions of characteristics, tentative subcharacteristics, and some remarks follow.

Functional suitability is defined as the capability of the software product to provide functions that meet stated and implied needs when the software is used under specified conditions. This characteristic is concerned with what the software does, as well as to when and how it fulfils needs. The most important subcharacteristics of the functionality should be

- **Appropriateness:** Defined as the capability of the software product to provide an appropriate set of functions for specified tasks and user objectives.
- **Accuracy:** Defined as the capability of the software product to provide the right or specified results with the needed degree of precision.

Reliability is defined as the capability of the software product to maintain a specified level of performance when used under specified conditions. Wear or ageing does not occur in software. Limitations in reliability are due to faults in requirements, design, and implementation. Failures due to these faults depend on the way the software product is used and the program options selected, rather than on elapsed time.

In our consideration, we shall consistently distinguish between the terms "dependability" and "reliability" and also between the terms "fault" and "failure." In accordance with ISO 9000 (2005) terminology, the term *dependability* is used as a collective term to describe the availability, performance and its influencing factors: reliability performance, maintainability performance, and maintenance support performance. This term is used only for general descriptions in non-quantitative terms. *Fault* is an incidental condition that causes a functional unit to fail to perform its required functions. It is, in fact, the non-fulfillment of some requirements. *Failure* is an inability of a product to perform a required function or inability to perform it within previously specified limits. The failure is the consequence of some fault(s). The fault can incur failure(s), but not necessary so. Wear or ageing does not occur in software. Limitations in reliability are due to faults in requirements, design, and implementation. Failures due to these faults depend on the way the software product is used and the program options selected rather than on elapsed time. The tentative reliability subcharacteristics should be:

- **Availability:** Defined as the capability of the software product to be operational and available when required for use. Externally, availability can be assessed by the proportion of total time during which the software product is in an up state. Availability is therefore a combination of maturity (which governs the frequency of failure), fault tolerance and recoverability (which governs the length of down time following each failure.
- **Fault Tolerance:** Defined as the capability of the software product to maintain a specified level of performance in cases of software faults or of infringement of its specified interface. The specified level of performance may also include full fail-safe capability.
- **Recoverability:** Defined as the capability of the software product to re-establish a specified level of performance and recover the data directly affected in the case of a failure. Following a failure, a software product will sometimes be down for a certain period of time, the length of which is assessed by its recoverability.

Performance efficiency is defined as the capability of the software product to provide appropriate performance, relative to the amount of resources used, under stated conditions. Resources may include other software products, the software and hardware configuration of the system, and materials (e.g. print paper, CDs). For a system that is operated by a user, the combination of functional suitability, reliability, operability and performance efficiency can be measured externally by quality in use. The subcharacteristic should be:

- **Time Behavior:** Defined as the capability of the software product to provide appropriate response and processing times and throughput rates when performing its function, under stated conditions.
- **Resource Utilization:** Defined as the degree to which the software product uses appropriate amounts and types of resources when the software performs its function under stated conditions.

Usability or *Operability* is defined as the capability of the software product to be understood, learned, used and attractive to the user, when used under specified conditions. Users may include operators, end users, and indirect users who are under the influence of, or dependent on, the use of the software. Operability should address all of the different user environments that the software may affect, which may include preparation for usage and evaluation of results. Tentative subcharacteristics are:

- **Appropriateness Recognisability:** Defined as the capability of the software product to enable users to recognise whether the software is appropriate for their needs.
- **Learnability:** Defined as the capability of the software product to enable users to learn its application.
- **Ease of Use:** Defined as the capability of the software product to easy operation and control.
- **Helpfulness:** Defined as the capability of the software product to provide help when users need assistance. This includes help that is easy to find, comprehensive and effective.
- **Attractiveness:** Defined as the capability of the software product to be attractive to the user. This refers to attributes of the software that increase the pleasure and satisfaction of the user, such as the use of colour and the nature of the graphical design.
- **Technical Accessibility:** Defined as the degree of operability of the software product for users with specified disabilities.

Security is defined as the capability of the software product to provide the protection of system items from accidental or malicious access, use, modification, destruction, or disclosure, including data in transmission, sees ISO/IEC 15026 (1998). This product quality characteristic shall be distinguished from the quality in use system characteristic *safety* that is characteristic of quality in use. Safety does not relate to software alone, but to a whole system. *Survivability* (The degree to which the software product continues to fulfill its mission by providing essential services in a timely manner in spite of the presence of attacks) and *Immunity* (the degree to which the software product is resistant to attack) are covered by the Reliability characteristic:

- **Confidentiality:** Defined as the capability of the software product to provide protection from unauthorised disclosure of data or information, whether accidental or deliberate.
- **Integrity:** Defined as the degree to which the accuracy and completeness of assets are safeguarded.
- **Non-repudiation:** Defined as the degree to which actions or events can be proven to have taken place, so that the events or actions cannot be repudiated later.
- **Accountability:** Defined as the degree to which the actions of an entity can be traced uniquely to the entity.
- **Authenticity:** Defined as the degree to which the identity of a subject or resource can be proved to be the one claimed.

Compatibility is defined as the ability of two or more software components to exchange information and/or to perform their required functions while sharing the same hardware or software environment.

Tentative subcharacteristics are:

- **Replaceability:** Defined as the capability of the software product to be used in place of another specified software product for the same purpose in the same environment. For example, the replaceability of a new version of a software product is important to the user when upgrading.
- **Co-existence:** Defined as the degree to which the software product can co-exist with other independent software in a common environment, sharing common resources without any detrimental impact.
- **Interoperability:** Defined as the capability of the software product to be cooperatively operable with one or more other software products.

Maintainability is defined as the capability of the software product to be modified. Modifications may include corrections, improvements, or adaptation of the software to changes in environment, and in requirements and functional specifications. Tentative subcharacteristics should be:

- **Modularity:** Defined as the degree to which a system or computer program is composed of discrete components such that a change to one component has minimal impact on other components.
- **Reusability:** Defined as the degree to which an asset can be used in more than one software system, or in building other assets.
- **Analysability:** Defined as the capability of the software product to be diagnosed for deficiencies or causes of failures in the software, or for the parts to be modified to be identified. Implementation can include providing mechanisms for the software product to analyse its own faults and report on the conditions prior to a failure or other event.
- **Changeability:** Defined as the capability of the software product to enable a specified modification to be implemented. The ease with which a software product can be modified. The implementation includes coding, designing and documenting changes.
- **Modification Stability:** Defined as the capability of the software product to avoid unexpected effects from modifications of the software.
- **Testability:** Defined as the capability of the software product to enable modified software to be validated.

Transferability is defined as the capability of the software product to be transferred from one environment to another. Tentative subcharacteristics are:

- **Portability:** Defined as the ease with which a system or component can be transferred from one hardware or software environment to another.
- **Adaptability:** Defined as the capability of the software product to be adapted for different specified environments without applying actions or means other than those provided for this purpose for the software considered. Adaptability includes the scalability of internal capacity (e.g. screen fields, tables, transaction volumes, report formats, etc.).
- **Installability:** Defined as the capability of the software product to be successfully installed and uninstalled in a specified environment.

As a loophole, and as a convenient device for reaching the international voting agreement, it is the intention to add into each of these eight characteristics the special subcharacteristic compliance. That subcharacteristic should be defined as "The degree to which the software product adheres to standards, conventions, style guides, or regulations relating to a respective characteristic.

SYSTEM QUALITY IN USE CHARACTERISTICS

For *quality in use* as a quality of process utilisation and product effect, the characteristics and subcharacteristics mentioned above are not appropriate. The quality in use is defined as the capability of the software product to enable specified users to achieve specified goals with effectiveness, productivity, safety, and satisfaction in specified contexts of use. Therefore, quality in use is the user's view of the quality of environment containing software, and is measured from the results of using the software in that environment, rather than the properties of the software itself. The situation for quality in use division into different views is not so clear and subtle.

Before the product is released, quality in use can be specified and measured in a test environment for the intended users, goals, and contexts of use. Once in use, it can be measured for actual users, goals, and contexts of use. The actual needs of users may not be the same as those anticipated in the requirements, so the actual quality in use may be different from quality in use measured earlier in a test environment. Quality in use may be influenced by any of the quality characteristics, and is thus broader than usability, which is defined in terms of understandability, learnability, operability, attractiveness, and usability compliance.

Firstly, the following division of the quality in use into separate quality in use characteristics and subcharacteristics is prepared and discussed. Three qualities in use characteristics: Usability in use, flexibility in use and safety are proposed. Each should be divided into subcharacteristics. The list and tentative definitions follow:

Usability in use is defined as the degree to which specified users can achieve specified goals with effectiveness in use, efficiency in use and satisfaction in use in a specified context of use. Tentative subcharacteristics should be:

- **Effectiveness in Use:** Defined as the degree to which specified users can achieve specified goals with accuracy and completeness in a specified context of use.
- **Efficiency in Use:** Defined as the degree to which specified users expend appropriate amounts of resources in relation to the effectiveness achieved in a specified context of use. Relevant resources can include time to complete the task, materials, or the financial cost of usage.
- **Satisfaction in Use:** Defined as the degree to which users are satisfied in a specified context of use. Satisfaction includes *likability* (cognitive satisfaction), *pleasure* (emotional satisfaction), and *comfort* (physical satisfaction). Satisfaction is the user's response to interaction with the product, and includes attitudes towards use of the product.

Flexibility in Use is defined as the degree to which the product is usable in all potential contexts of use. Tentative subcharacteristics should be:

- **Context Conformity in Use:** Defined as the degree to which usability in use meets requirements in all the intended contexts of use.
- **Context Extendibility in Use:** Defined as the degree of usability in use in contexts beyond those initially intended. Context extendibility can include providing usability for additional user groups, tasks, and cultures.
- **Accessibility in Use:** Defined as the degree of usability in use for users with specified disabilities.

Safety is defined as the ability of the system to preserve the acceptable levels of risk of harm to people, business, data, software, property or the environment in the intended contexts of use. Risks are usually a result of deficiencies in the functionality (including security), reliability, usability, or maintainability. Tentative subcharacteristics should be:

- **Operator Health and Safety:** Defined as acceptable levels of risk of harm to the operator in the intended contexts of use.
- **Public Health and Safety:** Defined as acceptable levels of risk of harm to the public in the intended contexts of use.
- **Environmental harm in use:** Defined as acceptable levels of risk of harm to property or the environment in the intended contexts of use.
- **Commercial Damage in Use:** Defined as acceptable levels of risk of a failure that

would lead to commercial damage or reputation damage in the intended contexts of use.

For the same reasons as for product quality division, the subcharacteristic compliance will be inserted for each quality in use characteristic, defined as "The degree of conformance to standards, conventions, or regulations relating to respective characteristics."

The large number of characteristics and subcharacteristics for software product quality and in use quality of the system containing software cover various types of possible quality requirements. Such a large number seems to be excessive, and unclear, and inconvenient for real evaluation. Not everything, however, is relevant in every real situation. Nevertheless, the exhaustive list of quality characteristics and subcharacteristics can be very useful for avoiding the case when some very important view of quality could be overlooked. In the case of no relevance or low significance of some view, the respective requirement and consecutive evaluation can be bypassed. On each occasion, however, it can only be made by a conscious decision by the evaluator, and is its' responsibility.

The software and system quality also include a data quality for such data that can be positively allocated to a specified software product or information system. The modern tendency in data processing is not to bind data collections such as databases or data stores with one application only. With one database or data store many different information systems can communicate, perform transactions and jobs. The quality of such data collection can be evaluated separately from the software product quality or information system product quality in use. We shall report about the possible tentative SQuaRE project data quality model in the next subsection.

ISO/IEC DATA QUALITY MODEL

As has been shown at the end of the previous subsection, we also need to have a quality model for data independent data collections that are used by several data processing systems. The term data is defined as a reinterpretable representation of information in a formalised manner suitable for communication, interpretation, or processing. According to the general quality definition, the *data quality* is the property of data that bears on its ability to satisfy stated and implied needs when used under specified conditions.

To have a data quality model is useful, for example in the following three situations:

1. When a new system needs to define the requirements to be allocated to data.
2. When an existing system needs to assess data quality for data acquisition, integration, reengineering, migration, and compliance purposes;
3. To support data management and systems interoperability all through the data life cycle.

The SQuaRE data quality model as a software product quality model presumes both these two views to data quality:

- **Inherent Data Quality:** As the totality of intrinsic characteristics of data that bear on its ability to satisfy stated and implied needs when data are used under specified conditions, independently of other system's components. Inherent data quality refers to data itself, in particular to: data domain values and possible restrictions (e.g. business rules governing the currentness, accuracy, etc. required for a given application), metadata and also relationships between data.
- **Extended Data Quality:** As the totality of characteristics of data that bear on its ability to satisfy stated and implied needs

when data are used under specified conditions, utilising the capabilities of a system's components in the specific context of use. Extended data quality refers to artefacts such as: hardware devices (e.g. to make data accessible) and data definition language (e.g. to support precision of data).

For data quality in the present situation, fifteen possible characteristics are proposed. Seven of them are common to the inherent and extended point of view. There should be:

- **Accessibility:** As the property of data to be reached in a specific context of use, particularly by people who need supporting technology or special configuration because of some disability.
- **Compliance:** As the property of data to adhere to standards, conventions, or regulations in force and similar rules relating to data quality in a specific context of use. This characteristic was introduced in the model to assess the quality of data in situations in which regulations of public or private organisations impose specific requirements for data quality. It will however be useful to know and continuously monitor the conformance with individual laws and regulations, and to be able to measure the effort required to achieve such conformance.
- **Confidentiality:** As the property of data to be accessed and interpreted only by authorised users in a specific context of use. Confidentiality is an aspect of information security (together with availability, integrity). For example, data that refers to personal or confidential information like health or profit must be accessed only by authorised users or should be written in secret code.
- **Performance:** As the capability of data to be processed and provide the expected levels of performance by using the appropriate amounts and types of resources under stated conditions and in a specific context of use. For example using more space than necessary to store data can cause waste of storage, memory, and time.
- **Precision:** As the degree of the exactness or discrimination with which a quantity is stated in a specific context of use.
- **Traceability:** As the capability of data to provide an audit trail of accesses to the data and of any changes made to the data in a specific context of use. Public administration must keep information about the accesses executed by users for investigating who read/wrote confidential data.
- **Understandability:** As the property of data to understood by users, and is expressed in appropriate languages, symbols, and units in a specific context of use. Some information about data understandability can be provided by metadata. For example to represent a State (within a country), the standard acronym is more understandable than a numeric code.

The following five data quality characteristics are in place for the inherent point of view only:

- **Accuracy:** As the capability of data to correctly represent the true value of the intended attribute of concept or event in a specific context of use. Accuracy has two main aspects: Syntactic accuracy and semantic accuracy.
    - *Syntactical Accuracy* is defined as the closeness of the data values to a set of values defined in a domain considered syntactically correct.
    - *Semantic Accuracy* is defined as the closeness of the data values to a set of values defined in a domain considered semantically correct.
- **Completeness:** As the degree in which the subject data associated with an entity have values for all expected attributes and related entity instances in a specific context of

use. For example in an employee database, completeness is the extent to which each employee record contains the data required for the particular context.

- **Consistency:** As the property of data to be free from contradiction and coherent with other data in a specific context of use. Consistency can be either or both among data regarding one entity and across like data for comparable entities. A particular case of inconsistency is represented by synonyms: a dictionary of terms used to define data could be useful to avoid it. For example, an employee's birth date cannot be later than his "recruitment date."
- **Credibility:** As the property of data to be regarded as true and believable by users in a specific context of use. For credibility, certification from an independent and trusted organisation can be suggested.
- **Currentness:** As the property of data to be of the right age in a specific context of use. This data quality characteristic is violated for example in the case of the data not receiving a required update during a processing cycle or if the data receipt timestamp incorrectly identifies actual time of receipt.

The last three characteristics are in place only for the extended point of view:

- **Availability:** As the capability of data to be retrievable by authorised users in a specific context of use. A particular case of availability is concurrent access (either to read or to update data) by more than one user and/or application. Another case of availability is the capability of data to be available for a specific period of time or to be accessible also during managing operations such as backup.
- **Portability:** As a capability of data to be moved from one platform to another, preserving the existing quality in a specific context of use. This includes also the possibility to install and replace data in the destination platform.
- **Recoverability:** As a capability of data to be maintained and to preserve a specified level of operations and quality, even in the event of failure, in a specific context of use. Recoverability can be provided by features like commit/synch point, rollback (fault-tolerance capability) or by backup-recovery mechanisms. When a media device has a failure, data stored in that device should be recoverable.

MEASUREMENT THEORY FOUNDATIONS

All the models, investigated above tend to the necessity of quality affected attributes expressing numbers by means of numbers or more involved formal objects that can be processed by mathematical tools. The values obtained by that way, called measures, are during the evaluation, compared with indicators that describe the required level of quality for monitoring quality indicators.

Commonly speaking, the measurement can be characterised as a mapping of some segment of an empirical world into some numerical range or other mathematical structure, for example into a set of vectors. Such mapping has to reflect all properties and relationships that are a matter of our interest in the real world, as the properties and relationships between respective measures. Such a mapping can be not simple mapping. For the formalisation of that definition, the concept of structure must be defined. Structure is a set with a finite number of relationships and possibly a finite number of operations defined in this set. Next, the measurement is a mapping of the empirical structure into some formal structure that preserves all such relationships and operations. In mathematical terminology, such a mapping is called a structure homomorphism. Measurement is that homomorphism, including the method of its realisation. The function, which alone realises that mapping from the attributes' empirical structure into the formal structure, is called a measure. The value of that mapping for a fixed attribute is called a measure value.

Further, we shall confine ourselves only to the case of the integers or real numbers as a formal structure. For quality attributes in almost all situations, the priority relation from some view (for example "to be better," "to be more simple," or "to be more convenient") is in question. This relation shall be reflected as an inequality relation between respective measures, represented by numbers. Sometimes a concatenation operation on the empirical structure is interesting. In that case, it is required to reflect that operation by the respective concatenation operation between numbers, customarily by the addition of respective measures.

Measurement that is a structure homomorphism is customarily not determined uniquely. Usually some degree of freedom exists, for example on the selection of an etalon for the comparison, or in the choice of evolution scale. Such a freedom can be exactly characterised by a set of all admissible measure transformations. Admissible measure transformation is such a numerical function f, for which the following holds: If $\mu(x)$ is a measure, then the compose function $(f \circ \mu)(x) = f(\mu(x))$ is also a measure, therefore it is also a homomorphism of empirical structure into a formal one (preserve all monitoring properties of attributes and its relationships). The set of all admissible measure transformations creates, with respect to the operation of composing, a group. That group fully characterises a measurement scale type, abbr. scale type.

Formal sciences (mathematics in the first place) work with results of measurement. Using those tools creates consequences from measures. The results obtained are findings concerning measure values and relationships between those values. Such a finding has some meaning in the real world only if we can judge from those relations to respective relationships between respective attributes in an empirical world. Such transfer of findings from the formal into the empirical world is called an interpretation. Not everything correctly derived in the empirical structure, should be interpreted a real world finding. In the case when using some other equal eligible measurement leads to different result in the empirical structure, the interpretation of that finding is impossible. Such a finding in an empirical structure is not meaningful for an empirical world.

Most frequently used types of measure scale types were included in the Steven's hierarchy of scale types:

- **Absolute Scale Type:** The measure is determined uniquely. The group of all admissible measures transformation consists only from the identical transformation $f(x) = x$. Typical of such a measurement is the expressing of an attribute by a percentage of some ensemble, or the measure obtained as a ratio of two values, for example the ratio of the count failed test cases divided by the count of all test cases performed. We can perform each mathematical operation with measures of the absolute type. However, we lose the possibility to conclude about the measure of the attribute of a complex entity from the measures of its components attributes. For example if we know that in one program module there is 0.1% defect statements and 0.5% in the second module, we are not able without additional information to compute the percentage of defective statements in the complex program consisting from that two modules.
- **Ratio Scale Type:** Ratio scale type measures positive numbers defined as a ratio to some selected etalon. For example to a length of one meter, or mass of one kilogram, etc. A group of all admissible transformations creates all functions of the form

$f(x) = c \cdot x$, where $c > 0$ is some positive number. The value of the constant c depends on the ratio of two etalons in other equally eligible measurement (for example meters and yards or feet). This type of measurement is most favourable when we can also operate with the concatenation in the empirical structure. The measures preserve the priority order and also the concatenation operation. With measure values, all conventional mathematical and statistical operations are eligible and the results are always meaningful for interpretation, independent of the admissible transformation of measures. However, the ratio type measurement is not plausible every time. The possibility of ratio type measurement depends on the relationships between the priority relationship and concatenation operation on the empirical structure. A detailed discussion about the respective necessary and sufficient condition for the existence of ratio type measurement is out of the scope of this chapter. Those interested in this subject may refer to the monographs of Roberts (1979) or Krantz, Luce, Suppes, and Tversky (1971).

- **Interval Scale Type**: This measurement scale type has two degrees of freedom. The measures are real numbers, positive, negative, or zero. Zero and the margin of one unit is optional. A typical example of the interval type measurement is the temperature measurement in degrees of Celsius, Fahrenheit degrees, or Kelvin. A group of all admissible measure transformations is created by all functions of the form $f(x) = c \cdot x + d$, where $c > 0$ a d are real numbers. In our example, we cannot (with the exception of Kelvin) assign any empirical (physical) meaning to the operation of the temperature concatenation. Only the difference between two temperatures has a physical meaning, the energy consumed for a state change (in the case of the some necessary simplification of the problem). We have to be careful when interpreting the results obtained from an interval scale measurement. For example the statement that the temperature 20°C is twice as high as the temperature 10°C is not meaningful. The ratio of measures and percentage cannot be interpreted. From statistical moments only modus, median and arithmetical mean is meaningful, but geometrical mean and standard deviation (mean-root-square error) not. For the quality measurement, an interval scale type measurement is only rarely used. Sometimes only for the measurement of the both side aberrance from the expected value.
- **Ordinal Scale Type:** This type of measure scale is used in a case when only a preference relationship on an empirical structure is investigated and concatenation operation is not in question. Only a linear order of numbers is utilised. Any arithmetical operations are used. The attribute is described by a number on the principle "as better, as greater measure value" or "as better as lower measure value." The group of all admissible transformations is created by all strictly increasing: $x > y \Rightarrow f(x) > f(y)$, or strictly decreasing: $x > y \Rightarrow f(x) < f(y)$) functions. The example is whatever classification scale. Only such results obtained for measures that depend only on the order of numbers can be interpreted. The arithmetical operation results cannot be. From the statistical moments, only modus and median are meaningful. The arithmetical mean is not. This fact can lead to defective interpretations. The demonstration of that can be given by the "popular" calculation on the academic record as a mean mark, even though it is clear that for example in the case of the scale 1 = excellent, 2 = very good, 3 = good, 4 = fail; the differences be-

tween 1 and 2 and between 3 and 4 are dramatically poles apart. In contradistinction to a ratio scale type of measurement, the sufficient conditions for the existence of ordinal type measurement are fairly weak. The essential condition is only, that the preferences on the empirical structure shall create a weak order, not only a quasiorder or partial order with not comparable elements. The second necessary condition on the respective Birhoff-Milgram existence theorem is the existence of finite or countable infinite with respect to preferences orderdense subset of the empirical structure. This condition is obviously automatically fulfilled in an informatics application. The ordinal scale type measurement is widely used for the quality evaluation. Typically, these have to be used in the case of measures based on subjective judgment, where other more suitable possibilities are not available.

- **Nominal Scale Type:** This type is in fact not a measurement in the normal sense of the word. Only an appointment of an actual attribute into some subset of the set of all possible attributes is achieved. The numbers are used for the denotation of such subsets, but only as the label of such subset. Instead of numbers, any other distinguishable symbols, for example characters, can be used. The group of all admissible measure transformations is created by the set of all simple ($x \neq y \Rightarrow f(x) \neq f(y)$) functions. Of course, no results of any calculation or order comparison of measures can be interpreted. Only a modus which denotes the subset with the ultimate number of filed elements is meaningful. This type of measurement can be used only for the classification.

The discussed Steven's measure scale type hierarchy does not represent an exhaustive list of measure scale types. Sometimes a *logarithmic-interval type* of measures is used, for example for intensity measurement in decibels. The respective group of all admissible transformation is for this scale type formed be all functions of the form $f(x) = a \cdot x^b$, where $a > 0$, $b > 0$. Also the *difference type* of measurement, with the group of all admissible transformations created by all functions of the form $f(x) = x + c$ can be sometimes used.

The measurement scale type designation is the underlying factor for creating derived measures from the measures, which are already defined. It is essential for the decision as to which kind of operation and calculation can be applied to measure values to obtain meaningful information about the empirical world. In other words, to obtain a result that can be interpreted. Breaching this principle is the source of many failures in reasoning and judgment, while not only during the quality measurement and evaluation.

For each measure, its unique identification is necessary to indicate the measured attribute and measurement method, including sources of all input data. For derived measures, a measure from that measure shall also be derived, and an algorithm or formula for the derived measure calculation. Of course, for the given attribute, several alternative measures that require different measurement methods and different input data can exist for measures for software products, information systems and processes already mentioned. In addition, the stage of the system life cycle when the measure can be appointed, the quality characteristics, and subcharacteristics that are affected by the measure and for which personal participated in the system's life cycle need the information indicated by the measure value. Some of the specific problems concerning the measures for information products and quality processes will be discussed in the next paragraph.

INFORMATION PRODUCTS AND PROCESSES MEASURES

Information products and processes measures proposed in existing technical reports ISO/IEC 9126-2 (2001), ISO/IEC 9126-3 (2003), and ISO/IEC 9126-4 (2004) are mostly absolute scale type measures. These are based on the general principle of quality measure being the ratio of the number of favourable cases to the number of all investigated cases.

For example, to measure an external attribute "Functional implementation coverage" of the characteristic Functional suitability, and its subcharacteristic Appropriateness, "Functional implementation coverage" as $X = A / F$ or $X = 1 - B / F$, where A is the number of implemented functions detected in the evaluation and B is the number of missing functions detected in the evaluation, and F is the number of functions in the requirement specifications. Another example of the same kind can be the measure of the "Failure resolution" attribute for maintainability (analysability), defined as the ratio A / B, where A is the number of resolved failures and B is the number of actually detected failures, or the maturity attribute "Test overcome," defined as $X = A / T$, where A is the number of passed test cases during testing or operation and T is the Number of test cases to be performed.

The advantage of such kind of measures is that measure values are normalised into the interval [0, 1]. The closer the value is to 1, the better is the result gained. Such measures also have a lot of problems. First, as was mentioned above, the absolute measure value of components attributes says nothing about the measure of the overall system.

The other problems will be illustrated by the example of the Functional implementation coverage measure.

The first main problem is the question "What is one separate function." Functional requirements can be described in different levels. One complex function can be broken down into several less complex functions, and each of them into more and more detailed functions. In which level do functions have to be counted?

The next problem comes from the fact that not all functions have equal importance. The problem can be partially solved by adding some weighting w_j to each j-th function (or failure or test), $j = 1, 2, \ldots, N$, depending on the function necessity (failure severity or test importance), and normalise all N weights in such a way that $\Sigma_{all} w_j = 1$ and by using the more sophisticated formula $X = \Sigma_{i \in successful}\, w_i / \Sigma_{all} w_j$.

However such "weighted evaluation" does not cover all feasible quality requirements. Very often some functions are absolutely necessary and can not be overcome by any large number of functions with a lower but not zero importance. This situation cannot be solved using the importance weightings.

Therefore the more interesting way for the correction of measure evaluation formula probably lies in the determination of several levels of requirement importance. For example:

- LEVEL 1: Unconditional necessary.
- LEVEL 2: Important but not unconditional.
- LEVEL 3: Agreeable but not of high importance.

Let us, for example, suppose the following situation: We require N_1 functions of the level 1, N_2 of the level 2 and N_3 of the level 3. Our evaluation criteria is the following: All level 1 requirements must be fulfilled by the product; at least half of level 2 requirements have to be fulfilled; level 3 requirements are optional, its implementation is an advantage, but any large number of such advantages can cover one requirement of the level 2. In this situation, we can measure the respective attribute as the sum:

$$X = F_1 + F_2 / (N_1 + 1) + F_3 / ((N_1 + 1) \cdot (N_2 + 1)),$$

where F_1, F_2 and F_3 are the number of fulfilled requirements of the levels 1, 2, and 3, respectively. Such a measure whole-heartedly describes our three level preferences on the principle "the higher measure value – the better the attribute rating." The minimum acceptance criteria shall be $X \geq \square N_1 + H$, where H is the least natural number greater than or equal to $N_2 / 2$.

The "weighted" principle and "level principle" can be of course combined. To do it, it is necessary to choose the weights in such a way, that the sum of all the weights on each low level is smaller than the minimum weight on the level which is the next higher.

Unfortunately, in technical reports ISO/IEC 9126-2 (2001), ISO/IEC 9126-3 (2003), and ISO/IEC 9126-4 (2004), and also in the draft measures for SQuaRE, only primitive measures for each functional requirement, each failure etc. are equally counted and anticipated. The authors of this chapter are afraid of and the possible very limited use of such measures.

More perspective external measures can be based on the time duration measurement. For example, for time behaviour measure of performance efficiency is meant the mean or the maximum response time, defined as the user waiting time between the moment when he issues a request, and the application to finish the process. For some applications, the mean time is more interesting, for others the maximum response time can be critical. In addition, the mean or the minimum failure resolution time duration, the mean time duration between failures, or the ratio FT / OT, where FT is the time duration when the system is in a failure state and OT is the observation time, seems to be good indicator of external reliability and maintainability measures of the ratio scale type.

In many cases nothing else remains than to use only a subjective judgement. In this situation, we ask experts to choose an appropriate classification on a given discrete scale. Beginning with the binary answer YES/NO to a more detailed scale, for example 1, 2, …, 5 or 0, 1, …, 10. The odd number on the scale points seems to be more appropriate. The question should be how to aggregate several experts' individual evaluations. The problem of choosing an appropriate aggregation operator of individual ordinal measures of experts is far from trivial. As the measures are of the ordinal scale type only, there is no reason to prefer the arithmetic mean. Only the median is invariant with respect to admissible transformation of measures. A lot of possible aggregation operators exist, that differ depending on the importance of the options. For more details, see Vaníček, Vrana, and Aly (2008). In any case, during further usage of such a measure, it should be kept in mind that the measure is only of the ordinal scale type.

At the conclusion of the measure paragraph, it was stated that the problem of the selection of appropriate measures to cover most of characteristics in the ISO/IEC quality model is not easy to solve and is rather unclear. Nevertheless, the appropriate set of recommendable attributes and measures is probably the core of the ISO/IEC quality model. Without attribute and measure specifications, the whole quality model theory will have no foundation. Evaluations will be not objective and not be comparable with each other. There are many points of views to quality and generally, none should be underrated and omitted. On the other hand, the quality evaluation is not cost free and demands on the evaluation must be realistic. One possible way of solving this dilemma is to select a consistent overview of a less extensive set of base measures, or quality measure elements, that should be monitored during the system development, testing and exploitation stages. From such a set of elements, it should be relatively easy to compute various derived quality measures and also specify quality requirements indicators in such a way to be relevant for quality evaluation requirements in the current quality evaluation case.

It seems that each quality measure element can only be one of the following three types:

- Time duration of some event (ratio scale type measurement).
- The number of elements in some set of entities or events (ratio scale type measurement, not units but groups of units can be counted, for example a standardised pages of text).
- The ranking of the entity or event into some ordering category depending on subjective judgement (ordinal scale type measurement) or without preferences (nominal scale type measurement).

The authors of the chapter hope that on the basis of quality measure elements it could be possible to solve the attribute selection problem from the point of view of the acquirer and independent evaluator. There is the quality evaluation based on the external attributes of software product quality evaluation, and quality in use for quality evaluation of a system containing software evaluation utilisation process.

The authors are rather sceptical about the selection of appropriate internal quality attributes (quality predictors), essential for developers during the design and development stages of the system. The situation here is rather varied. Not only can the development systems and products be very different from the point of view of their targets, missions and addressed users groups, but also there are many different paradigms, methods, and tools used in design and development. Each paradigm, method, and tool calls for different types of documents and intermediate products during the design and development process as inputs and outputs of its subprocesses. The source data for internal quality evaluation should be gathered or derived from respective documents and intermediate products. Creating a reasonable summary set of attributes and measures that cover most of the possible situations is probably completely impossible. The only possible way for software developing organisation is to create, on the basis of general principles described in standards, its own system of internal attributes and measures for quality prediction, depending on its methods and tools used in addition to its own experiences. Of course, long term quality monitoring is a necessary pre-requisite.

THE PRESENT STATE OF THE SQUARE PROJECT AND ISO/IEC 250XX STANDARDS PREPARATION STAGE

In this paragraph, we shall describe the state of preparation of the ISO/IEC 250xx standard series as at November 2008 and some critical comments and potential risks and threats to its successful finalisation.

The SQuaRE project, on its opening in 2000, assumed the following square structure of standards: The general overview and umbrella documents in the centre; standards for the software product quality model; for quality measurement and measures; for quality requirements; and for quality evaluation processes on its sides. More detailed information can be found in the paper by Azuma (2001).

The full title of each SQuaRE document begins with words “Software engineering—Software product quality requirements and evaluation (SQuaRE).”

ISO/IEC 2500x – Product quality general division:

- **ISO/IEC 25000 Guide to SQuaRE**
- **ISO/IEC 25001 Planning and management**

ISO/IEC 2501x – Quality model division:

- *ISO/IEC 25011 Software quality model*
- *ISO/IEC 25012 Data quality model*

ISO/IEC 2502x – Quality model division:

- **ISO/IEC 25020 Measurement reference model and guide**
- **ISO/IEC 25021 Quality measure elements (technical report)**
- ISO/IEC 25022 Internal quality measures
- ISO/IEC 25023 External quality measures
- ISO/IEC 25024 Quality in use measures

ISO/IEC 2503x – Quality model division:

- **ISO/IEC 25030 Quality requirements**
- ISO/IEC 2504x Quality evaluation division
- *ISO/IEC 25040 Quality evaluation overview*
- ISO/IEC 25041 Evaluation modules
- ISO/IEC 25042 Developers' process
- ISO/IEC 25043 Acquires' process
- ISO/IEC 25043 Evaluators' process
- *ISO/IEC 25045 Evaluation Module for Recoverability*

Subsequently adopted norms without direct location on the initial square structure:

- **ISO/IEC 25051 Requirements for quality of Commercial Off-The-Shelf (COTS) software product and instructions for testing** (covers the old ISO/IEC 12119 [1994]: Information Technology – Software Packages – Quality requirements and testing)
- *ISO/IEC 25060 Common industry information for Usability – Overview*
- *ISO/IEC 25061 Common industry information for usability – Requirements for Context of use*
- **ISO/IEC 25062 Common Industry Format (CIF) for usability test reports**

Actual information about the state of standards publication can be obtained from the website of Joint Technical Committee-Subcommittee for Software and Systems Engineering (www.jtc1-sc7.org).

From the given list of standards and of its accomplishment stage, it is evident, that the top down approach was applied for standard series development. The general overview documents and guides are finished; the detailed documents are not yet ready. This approach proved to be effective for information systems and software design. However, doubts can arise about its use for standard series design. It is difficult to set definitions or write a guide for something, which is not ready and often not clear. Already it is evident that, after clarifying and finishing not only the managerial principles but also the professional technical problems, the published general standardisation documents need to be updated.

As has been mentioned above, apart from the time delay, the main threat to the SQuaRE project is the nonexistent agreement concerning the possible quality attributes and the choice of measures. We cannot be optimistic about the possibility of the SQuaRE research team to solve this problem in the foreseeable future.

It further seems that for a consistent problem description it will be useful to have at least a three level hierarchical structure of harmonised standard series structure and respective quality requirements and models. See Figure 3.

Several models, quality viewpoints, the large number of nominated characteristics and subcharacteristics, attributes and measures, and an extensive set of documents on the ISO/IEC 250xx series can discourage potential users from using quality standards.

EXPERT QUALITY ASSESSMENT

The requirement that software complies with quality standards is extremely important for software development teams and vendors. The general ISO 9000 series of standards for quality management apply mainly to development teams and for vendors of software systems. Specialised standards for software quality; data quality and

Figure 3. Model of system quality requirements

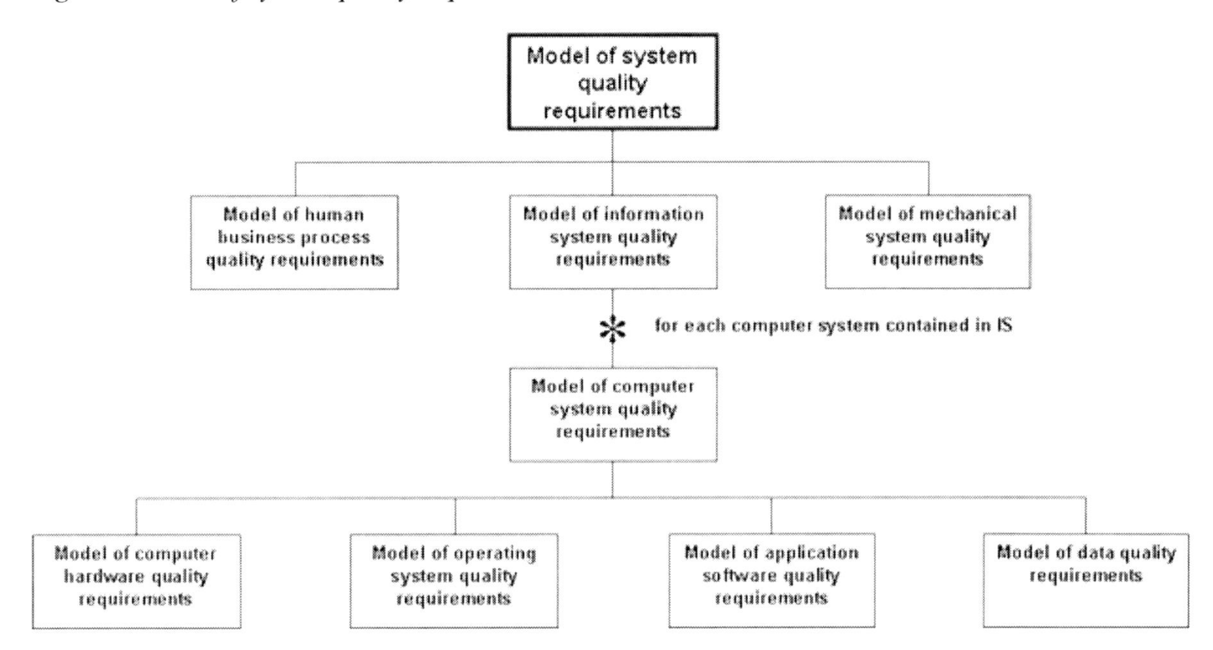

quality in use are dedicated mainly to system users and acquirers such as system integrators. They know their needs. From these needs, requirements can be derived and verified, to the extent to which individual products satisfy them. Indirectly, these specialised standards also have a great importance for developers and vendors, because they show where to focus to achieve assumed user requirements.

That is why, besides the international software quality standards, other alternative methods should also be used to help software end-users to assess quality of software from their users' perspective. A large group of methods has proven to offer to end-users a powerful instrument for quality assessment for both ready-made software, and also for emerging technologies. As a rule, these methods formulate and assess global quality features of certain types of software, and do not evaluate individual implementations. In this sense, this class of methods serves as a kind of "metarequirements" dealing with a "metaquality." These methods also enable one to assess the impact of an individual technology and its software representations in a certain business sector, and to estimate an expected time needed by this software to become mature. These methods can also evaluate and rate individual software vendors with respect to a number of factors.

The alternative approaches to the software quality assessment are based on expert evaluation, when experts from the given field conduct their assessment in accordance with a pre-specified well-proven methodology. It is clear that this kind of expert quality evaluation requires a number of independent and highly qualified experts in order that the final quality assessments and conclusions were trustworthy and reliable. These requirements are usually not affordable for the majority of end-users, including state or municipal governments and their administrations. Fortunately, some large, independent consultancy agencies exist, which are specialised in the above mentioned analyses. Software users, including smaller ones, can either buy their general research results or some special tailor-made analyses. Gartner Inc. can serve as an example of such a very prestigious agency, with a long tradition and high reputation (see e.g. www.gartner.com). They are specialised on assessment of a broad range of information technology aspects including software quality. As a source of data, they use 60,000 clients from 10,000 organisations, spread over the globe and across all industries. This helps Gartner to achieve unbiased results. Gartner research also has other important characteristics, like:

- A quantitative and qualitative blend, which also provides users with an explanation and interpretation of quantitative results.
- Peer review before any of their results are published.
- Proven, continually updated, and improved methodology guarantee an uncompromised quality of results.

That is why we select the Gartner methodology to explain how software users can utilise and take advantage of expert evaluation of the software quality. It can also explain how software users can gain descriptions of the situation important to them.

Gartner research is based on scenarios, key issues, and strategic planning assumptions in order to provide users with early advice, long before all the facts are publicly known and obvious. Gartner uses the following five-step research process in order to interpret trends accurately and to bring simplicity to complex situations:

- Building and refining scenarios concerning advances in IT over the next 5-10 years, and their impact.
- Conducting detailed surveys in order to collect information from IT users belonging to various categories and industries.
- Analysing emergence patterns to differentiate between temporary exposures and trends.
- Examining divergent viewpoints, testing and re-adjusting them in order to achieve the strongest conclusion for the client.
- Validating the findings of research against all available internal and external sources.

For the quality assessment of not only software but also the broader context like systems, technologies and trends, the following Gartner methodologies are particularly useful to software end users (e.g. within an e-Governance):

- Hype cycle.
- Magic quadrant.
- Market scope.
- Vendor rating.

We shall, therefore proceed to describe them in greater detail.

HYPE CYCLES

For simplicity, we shall further refer to various technical principles and their implementations (e.g. hardware devices, software packages, information systems, architectures or trends) as "technologies." New technologies or services (e.g. portals, Web services, object methodologies, open-source solutions, Wi-Fi, etc.) typically take various levels of visibility as they progress from the birth of an idea, though initial research, development, gradual exploitation, and utilisation until their replacement by another technology. Visibility of the technology does not depend only on its practical value and maturity. Typically, visibility of each technology goes through following five phases.

After a new principle, which can bring a new interesting usage, was discovered and published, research and development teams start to work in enhancing this principle. At this time the new idea is captured by media (both professional and lay ones) which start to speculate about the technology's potential and benefits. Consciousness about a new technology begins to rise, which becomes a technology trigger for further research and development. An excess of publicity creates inflated expectations in potential users. The visibility of technology gradually reaches a peak of inflated expectations. These expectations prove themselves as exaggerated and unrealistic. The real value of technology is below expectations, its products and services are mostly offered only as specimens or pilot solutions, they suffer from many early defects, etc. This causes embarrassment to the inventors (in research and development), suppliers, users and finally also the media which starts a negative promotion. That is why the visibility of technology quickly decreases until it drops into a trough of disillusionment and the technology is damned due to disappointment from unfulfilled expectations. Media again play an important role here in exaggerated underestimating of the real potential of the technology, in spite of the fact that in the meantime, producers have prepared the first generation products, which are already capable of useful operation. This trend usually causes many of the developers who were originally involved, to leave this technological segment. Taking advantage of collected experience, it is an opportunity for new or doggedly persistent companies to offer technology in a more advanced and usable form. The visibility gradually starts to rise and the technology recovers—it returns to the awareness of potential users. In this period, when the technology has been adopted by about 5% of potential users, that the second generation solutions become available and the number of implementations quickly grows. Finally, the technology is considered matured, its potential is well known; there is no longer a need to speculate about it. The visibility is consolidated and technology becomes a mainstream adopted when implemented by about 30% of potential users.

The time aspect is very important. Individual technologies pass through their life cycle at different speeds and it typically takes 2 – 10 years to reach maturity. It can sometimes take even longer, and some technologies never even reach maturity, become obsolete, and disappear before maturity. From the point of view of trend estimates, it is important to know:

- In which place of its lifecycle the technology is.
- How long it will take to reach the consolidated stage of maturity.

Both these parameters are presented in Gartner Hype Cycles reports. The Gartner Hype Cycles graph (as a part of the Hype Cycles report) graphically depicts the relationship between maturity of technologies and the level of their adoption/visibility. Additionally, the Hype Cycles reports show the time needed for technology to mature and its impact to end users belonging to a certain industry. They also present some hints concerning end-users' adoption strategy. The Hype Cycles report can answer question like: Which technology is viable right now and which need still time to mature? What is its impact to efficiently performing our tasks in our business? More than 800 analysts prepare and publish the Hype Cycles reports across 75 IT, business and consumer segments. Altogether, about 1,400 technologies are annually analysed in Hype Cycle reports. The Hype Cycles reports have gained a high and general appreciation among IT and business professionals for their entirety and clarity. Their results are generally accepted both by suppliers and users of technologies even in cases when the Hype Cycles report is critical with respect to a certain technology or when the Hype Cycles report users did not fully understand its semantics and interpretation of results.

In the Hype Cycles graph, the horizontal axis is for a degree of Maturity (and also for time) and the vertical axis is for the Visibility of a certain technology. A typical shape of the Hype Cycles graph in Figure 4 depicts how Visibility versus Maturity of an individual technology is perceived during its life cycle.

Every technology starts as a technology trigger and then gradually passes through stages: a peak of inflated expectations and a trough of disillusionment towards the plateau of productivity. Each considered technology is represented by its icon in the HC curve. The icons are labeled by a technology name. Because individual technologies move along the HC curve at different speeds, each

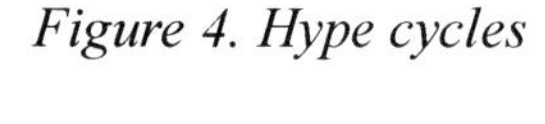

Figure 4. Hype cycles

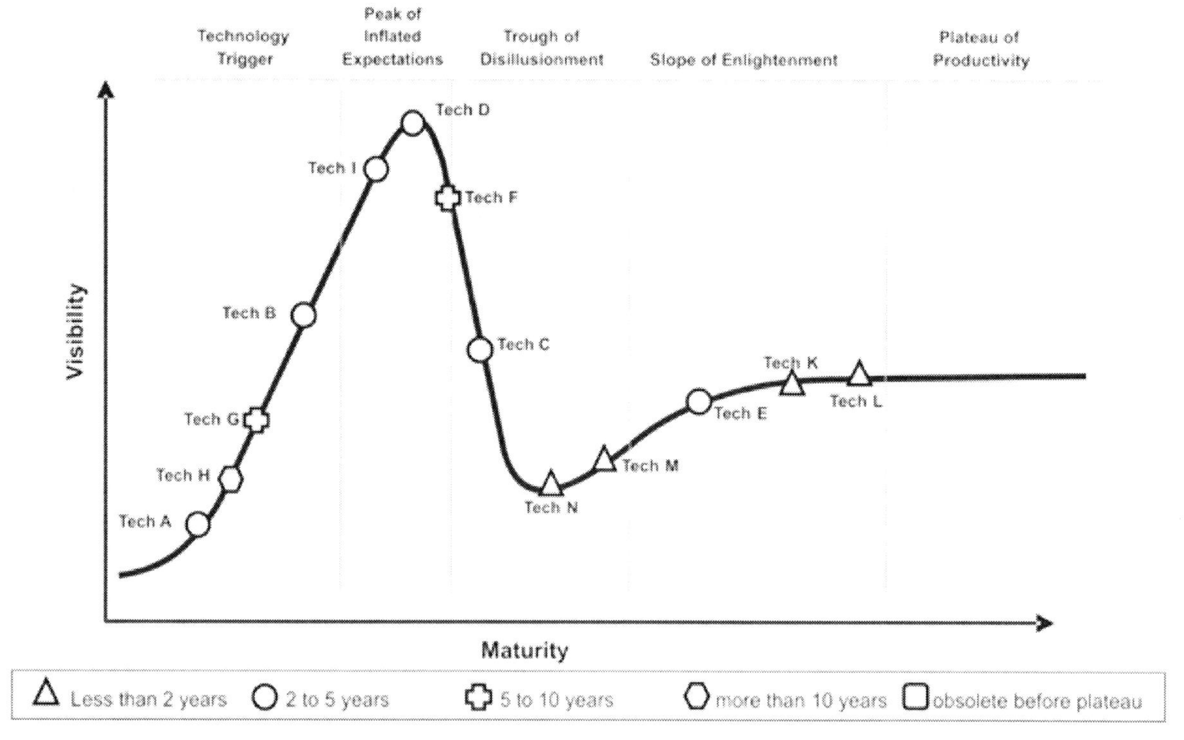

technology is assigned its specific attribute: a time to reach maturity. This is expressed by the shape of the icon. Typically, the following alternatives are considered for an estimated time to reach maturity:

- Less than 2 years.
- 2 to 5 years.
- 5 to 10 years.
- More than 10 years.
- Technology is obsolete before reaching maturity.

Not all of the 1,400 analysed technologies are needed or interesting for a single user. Instead, users need to analyse and compare performances only of those technologies, which are relevant to the users' industry or interests. That is why, Gartner publishes each year more than 60 specialized Hype Cycle reports which are thematically oriented and composed with respect to various technical aspects (e.g. content management, data management, communication applications, identity and access management, open source software, etc.), topics (e.g. analytic applications, business process management, system integration, human capital management, IT outsourcing, etc.), and industries (e.g. banking and investment services, government transformation, life insurance, healthcare, natural resources, transportation, higher education, etc.).

Hype Cycle reports, besides locating individual technologies into a Hype-Cycle graph, also contain a more detailed analysis of technologies. For each technology, information concerning the following aspects is published:

a. Definition.
b. Position and adoption speed justification.
c. User advice.
d. Business impact.
e. Benefit rating.
f. Market penetration.
g. Maturity.
h. Sample vendors.

We shall proceed to describe only the interpretation of benefit rating because names of the other aspects are self-explanatory. The potential benefits offered by the technology are sorted into four categories:

- **Transformational:** Where technology opens new approaches to conducting business across all industries, which cause an essential shift in the dynamics of the industry.
- **High:** Where technology offers new ways to perform both: horizontal and vertical processes, which can significantly increase revenue and bring important cost savings.
- **Moderate:** Where incremental improvements are possible to established processes, which can result in cost savings and increased revenue.
- **Low:** Where technology can slightly improve processes while increased revenue or cost saving are difficult to expect.

Comparing benefit rating with a time to maturity is expressed by a Priority matrix. This matrix can help users, technology managers, and planners to balance risks they are ready to bear with business benefits they would like to achieve. It helps to select an adequate investment strategy out of the following major options:

a. Invest aggressively if not already adopted.
b. Conservative investment profile.
c. Moderate investment profile.
d. Aggressive investment profile.
e. Invest with caution.
f. Invest with extreme caution.

The priority matrix can be expressed in the form of Table 1.

Hype Cycle reports and graphs help the end user to understand the evolution of software technologies and substantially improve the return of investments in emerging technologies.

MAGIC QUADRANT

An overall users' satisfaction is an important aspect of quality assessment. There are many factors contributing to users' satisfaction where selection of a proper vendor plays a dominant role. Users usually need to know what position vendors occupy in their market sector. For this purpose, Gartner offers an analytical instrument called the Magic quadrant. Two major aspects of vendors' properties are considered: Completeness of vision and Ability to execute (products)—these aspects are also coordinates in a Magic quadrant graph. Gartner define the following four categories of providers:

- **Niche Players**: Perform well in a certain market segment, but they have limited capability for innovation to outperform other providers. They are either new players in the market or they are usually oriented on a specific functionality or geographic region.

Table 1. Priority matrix

		Benefit rating			
		Transformational	**High**	**Moderate**	**Low**
Time to maturity	< 2 years	A	B	C	D
	2 – 5 years	B	C	D	E
	5 – 10 years	C	D	E	F
	> 10 years	D	E	F	F

- **Challengers**: Well positioned to execute but they are lacking in a sound strategy, innovation, and vision where the market is targeted. They usually have good human and financial resources.
- **Visionaries**: They know the direction of market evolvement. They usually introduce new technology, business models, and services but they still sometimes lack in the ability to execute.
- **Leaders**: Offer mature solutions, which meet demands of the present market. They also clearly demonstrate their vision, which is needed to sustain in leading position in evolving market.

The Magic quadrant graph is composed out of four quadrants, representing the above-described four categories of providers, see Figure 5.

Individual providers are assigned a certain position in this graph as labeled points so that users can compare their ability to execute, and their vision with respect to other providers. Criteria belonging to vision on the horizontal axis deal with the following aspects: understanding of market, marketing, and sales strategy, service and product strategy, vertical industry strategy, business model and innovation. On the other hand, criteria belonging to ability to execute on the vertical axis assess the following aspects: overall viability, product and service management, marketing, and sales execution, pricing, customer experience and operations. The magic quadrant provides users with an understanding of the position of potential technology suppliers, and describes their strength against users' specific expectation.

It compares all potential competitors in a stable market and it also answers the question about what a certain provider can presently deliver to meet immediate needs, and what can be expected from him in the future in innovating solutions to old problems.

Figure 5. Magic quadrant

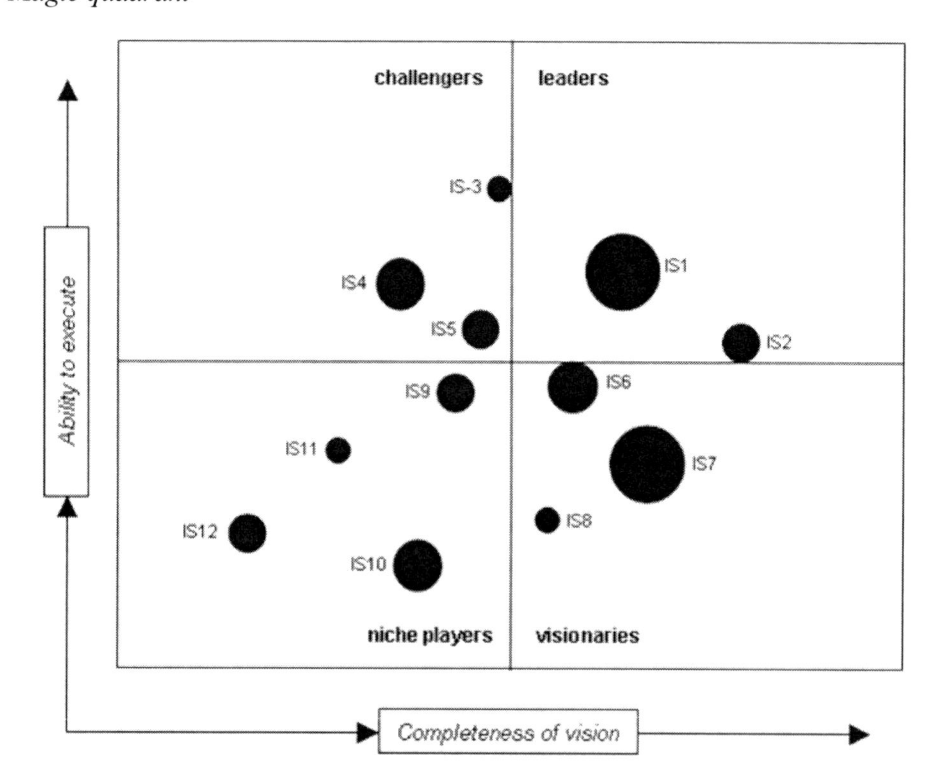

MARKET SCOPES

Gartner market scopes concentrate on distinctions between providers in emerging and mature markets and it provides an overall rating of the market itself. Market-scope rates and sorts every provider of technology into five categories: strong negative, caution, promising, positive, and strong positive. This rates both: the market itself as well as individual participating providers against visions and plans. It helps users to manage risks of investments in mature and emerging markets and allowing the emerging market to evolve prior to users' investments.

VENDOR RATING

If a user wants to engage a certain technology provider, he needs to get acquainted with both: technology products offered, and also with an evaluation of different aspects of the provider himself. The user should collect as many characteristics as possible for a full understanding of the potential of a provider to create a stable and sustainable partnership. Gartner rates providers into categories by a scale:

- **Strong Positive:** It is considered as a supplier of strategic solutions, products, and services. Customers can be comfortable in building a long-term relationship and partnership.
- **Positive:** Is strong in certain directions, but performance and execution in some areas are still developing. The customer can continue with all planned investments.
- **Promising:** Has a promising potential in some areas whilst execution is not consistent. The customer must consider possible changes in provider's status and their impact.
- **Caution:** Has problems in some areas. The customer should selectively survey its behaviour in relevant areas.
- **Strong Negative:** Has problems and difficulties in many areas. The customer should adopt a contingency plan.

This evaluation considers all important aspects, but particularly the following ones:

- **Strategy:** A long term capability and potential to formulate market opportunities.
- **Organisation:** It evaluates the strength of a provider's management and its capability to adapt to changes and to ensure the provider's growth.
- **Financial:** It assesses the revenue and profitability growth, history of R&D investments, how these aspects support stability, and its' cash position.
- **Products and Services:** It describes how the provider delivers and supports his offerings, how complete this offering is, and eventual gaps in his portfolio of offerings.
- **Technology:** It considers harmony with mainstream industry standards, whether complex solutions and suites are based on same common architecture, and how they enable extension and changes.
- **Sales and Distribution:** It evaluates how the provider develops partnerships which support and enable a mutually profitable long-term partnership.
- **Price Policy:** Focus is on fair and consistent prices and discounting policy.
- **Marketing:** It describes the ability of the provider to present a product and market vision in horizontal and vertical markets.

Vendor rating helps users to get oriented in how products, services and their providers are promising a long-term survival in the market, and also helps users to minimise the risk of orientation at individual providers while preserving the possibility to consider new alternate providers.

FUTURE TRENDS

One could expect that as in the past, the progress of information technologies will continue also in the future. The demands on functionality of this technology and also on its reliability and interface between the human and communication technology will increase. This all will cause increasing quality demands. These demands will focus particularly on functionality, reliability, user friendliness, and security. Users will become more and more demanding and they will require more thorough testing. Due to new automated activities, the quality characteristics improvement, and appropriate modifying quality standards, will remain a priority. Development of quality standards is thus an endless task.

Preparing, negotiating, and approving an international standard is a lengthy process, and thus published standards will always be one step behind the state-of-the-art. Therefore, it is important to use them with "good common sense" in order to avoid a situation when these standards would be an obstacle of progress. One should also give up illusions that standardisation can unify all possible approaches. Too many standards can make their utilisation difficult, when minor ones can hide more important ones.

The utilisation of standards during a quality assessment is definitely useful to avoid forgetting some substantial quality aspects and miss an opportunity to introduce maximum objectivity into an assessment process. The application of quality standards alone can never fully substitute for a general knowledge of problems, good common sense and personal experience, and experience arising from analogy.

CONCLUSION

This chapter outlines basic information about quality aspects of information systems and products, and it also provides information about international quality standardisation in this field. The principles contained in these standards are very important for the public and also for state administrations. Understanding the spirit of these standards can increase the effectiveness in use of information and communication technologies, and can reduce associated costs. The basic principles of quality management are contained in the standards of the ISO 9000 series. These standards are, however, too general to reflect specificity of individual information tools. The specific aspects of software and data quality will be addressed in the standards of the ISO/IEC 25000 series newly developed within a SQuaRE project. Besides an approach codified by quality standards, this chapter also provides information about selected supplementary methods which assess the quality of individual technologies and their vendors from the user's perspective. These methods can help in better orientation in an ICT market.

Public and state administration and governments are amongst the important customers of ICT vendors. Public and state administration and governments do not have facilities to develop ICT solutions in house; therefore, they are reliant on a market. Quality from the user's perspective is one of the important aspects when selecting a proper solution and its vendor. It is important to balance quality requirements with final cost. Too soft requirements could cause the acquired solution not to meet expectations. This may reduce the trust of citizens. Exaggerated requirements can lead to wasting public resources.

One of the common problems in public administration is caused by its de-centralisation. There is a certain tension between two principles: subsidiarity principle—when everything should be solved at the lowest possible level where all possible detailed information is available and ought to be handled by the lowest competent authority at the one hand; and a compatibility principle—when co-operation is a priority on the other hand. It also remains a question as to which level it is suitable to use outsourcing of solutions in a public administration. It can be more flexible but sometimes at the expense of reduced security and reliability. These, and also many other problems, should be carefully considered.

Even though a public and state administration does not develop ICT solutions in-house, responsibility for a requirements formulation and process of solution and vendor selection still remain on their shoulders. For this purpose, the public and state administration institutions need to have qualified independent experts to formulate and assess quality issues. To this aim, quality standards can become a valuable (but not exclusive) tool.

ACKNOWLEDGMENT

The authors wish to thank the Czech Ministry of Education for the support of this work within the framework of Grant No. MSM 6046070904 "Information and Knowledge Support of Strategic Control" and within the framework of Grant No. 2C06004 "Intelligent Tools for Content Assessment of Relevance of General and Specialised Data and Knowledge Resources."

REFERENCES

American Society for Quality. (2008). Basic concepts, glossary: Quality. *American Society for Quality.* Retrieved November15, 2008, from http://www.asq.org/glossary/q.html

Azuma, M. (2001). SQuaRE: The next generation of the ISO/IEC 9126 and 14598 international standards series on software product quality. In *Proceedings of the European Software Control and Metrics - Escom 2001*. London, UK: ESCOM.

Fenn, J. (2008). *Gartner's hype cycle special report for 2008*. Retrieved November 20, 2008, from http://www.gartner.com

ISO 9000. (2005). *Quality management systems – Fundamentals and vocabulary.* Geneva, Switzerland: International Organization for Standardization.

ISO 9001. (2000). *Quality management systems – Requirements.* Geneva, Switzerland: International Organization for Standardization.

ISO/IEC 12119. (1994). *Information technology – Software packages - Quality requirements and testing.* Geneva, Switzerland: International Organization for Standardization.

ISO/IEC 14598-2. (2000). *Software engineering – Product evaluation – Part 2: Planning and management.* Geneva, Switzerland: International Organization for Standardization.

ISO/IEC 14598-3. (2000). *Software engineering – Product evaluation – Part 3: Process for developers.* Geneva, Switzerland: International Organization for Standardization.

ISO/IEC 14598-4. (1998). *Software engineering – Product evaluation – Part 4: Process for acquirers.* Geneva, Switzerland: International Organization for Standardization.

ISO/IEC 14598-5. (1998). *Information technology – Software product evaluation – Part 5: Process for evaluators.* Geneva, Switzerland: International Organization for Standardization.

ISO/IEC 14598-6. (2001). *Software engineering – Product evaluation – Part 6: Documentation of evaluation modules.* Geneva, Switzerland: International Organization for Standardization.

ISO/IEC 15026. (1998). *Information technology – System and software integrity levels.* Geneva, Switzerland: International Organization for Standardization.

ISO/IEC 17598-1. (1999). *Information technology – Software product evaluation – Part 1: General overview.* Geneva, Switzerland: International Organization for Standardization.

ISO/IEC 25000. (2005). *Software engineering – Software product quality requirements and evaluation (SQuaRE) – Guide to SQuaRE.* Geneva, Switzerland: International Organization for Standardization.

ISO/IEC 25001. (2007). *Software engineering – Software product quality requirements and evaluation (SQuaRE) – Planning and management.* Geneva, Switzerland: International Organization for Standardization.

ISO/IEC 25020. (2007). *Software engineering – Software product quality requirements and evaluation (SQuaRE) – Measurement reference model and guide.* Geneva, Switzerland: International Organization for Standardization.

ISO/IEC 25030. (2007). *Software engineering – Software product quality requirements and evaluation (SQuaRE) – Quality requirements.* Geneva, Switzerland: International Organization for Standardization.

ISO/IEC 25051. (2006). *Software engineering – Software product quality requirements and evaluation (SQuaRE) – Requirements for quality of commercial off-the-shelf (COTS) software product and instructions for testing.* Geneva, Switzerland: International Organization for Standardization.

ISO/IEC 25062. (2006). *Software engineering – Software product quality requirements and evaluation (SQuaRE) – Common industry format (CIF) for usability test reports.* Geneva, Switzerland: International Organization for Standardization.

ISO/IEC 9126-1. (2001). *Software engineering – Product quality – Part 1: Quality model.* Geneva, Switzerland: International Organization for Standardization.

ISO/IEC 9126-2. (2003). *Software engineering – Product quality – Part 2: External metrics.* Geneva, Switzerland: International Organization for Standardization.

ISO/IEC 9126-3. (2003). *Software engineering – Product quality – Part 3: Internal metrics.* Geneva, Switzerland: International Organization for Standardization.

ISO/IEC 9126-4. (2004). *Software engineering – Product quality – Part 4: Quality in use metrics.* Geneva, Switzerland: International Organization for Standardization.

ISO/IEC TR 25021. (2007). *Software engineering – Software product quality requirements and evaluation (SQuaRE) – Quality measure elements.* Geneva, Switzerland: International Organization for Standardization.

Krantz, D. H., Luce, R. D., Suppes, P., & Tversky, A. (1971). Foundations of measurement: *Vol. 1. Additive and polynomial representations*. San Diego, CA: Academic Press Inc.

Roberts, F. S. (1979). *Measurement theory with applications to decision-making, utility and social sciences*. Reading, MA: Addison Wesley Publishing Company.

Vaníček, J., Vrana, I., & Aly, S. (2009). Fuzzy aggregation and averaging for group decision making: A generalization and survey. *Knowledge-Based Systems*, *22*(1), 79–84. doi:10.1016/j.knosys.2008.07.002

Chapter 11
From E-Government to E-Governance in Europe

Rebecca Levy Orelli
University of Bologna, Italy

Emanuele Padovani
University of Bologna, Italy

Carlotta del Sordo
University of Bologna, Italy

ABSTRACT

The influence of e-government on the modernization and growth of public sector initiatives in Europe has been deeply claimed. Little is known, however, about how the so-called shift from e-government to e-governance takes place in European governments. This chapter presents a view of both challenges and advantages of implementing e-governance strategies, by examining how closely and critically intertwined e-government and e-governance are in European countries.

INTRODUCTION

Over the past decades, there has been a significant evolution in public sector e-governments. With reference to Europe, there has been an overwhelming evolution of information systems and information technology over the last 20 years inspired by different strategies of implementation issued by the European Union over time. While the European Union has defined strategies and goals, which are valid for every member, each country has the freedom to adapt them to its particular social, administrative, and economic context. Therefore, differences arise in terms of span of services provided via e-government, quality of introduction of new information technology tools in governments, and level of e-governance achieved, in terms of e-participation, e-democracy, overcoming of the digital divide, and e-inclusion (European Commission, 2007a).

While previous studies have focused on the description of the introduction of e-government in European countries under different perspectives (Dunleavy, et al., 2006; EIPA, 2003; Hood, 1983), there seems to be a missing link: to what extent the recent so-called "shift from e-government to

DOI: 10.4018/978-1-4666-1909-8.ch011

e-governance systems" takes place in European governments? Research has delved into the factors that inhibit or promote the adoption of IT systems and e-government in general in the public sector. However, there is a lack of clarity in terms of the relationship between e-government and e-governance. E-government influences accountability and performance (Orelli, et al., 2010), which still remain the scope of any public management reform throughout the Western world (Bouckaert & Pollitt, 2005); thus there might be a strong connection between e-government and the ultimate goal of any e-governance initiative.

With the aim to investigate to what extent e-governance systems perform governance in the public sector, we judge as particular instructive to consider sixteen different national European cases. The cases were selected by different levels of e-government results achieved, i.e. high, medium and low ranking levels according to the European Commission indexes (European Commission, 2007b). The comparison among these countries will constitute the basis for the discussion about the influence of the shift from e-government and e-governance in European countries.

This chapter consists of four sections. The second section sets up the scene providing general information about e-government and e-governance in Europe and examines how closely and critically intertwined they are. The third section provides an overview of both European e-government and e-governance dimensions, and presents how typical they develop in the public administration, considering the experience of different EU countries. The main path and challenges of shifting to e-governance to support public governments are presented in the fourth section. Finally, in the conclusion we summarize the analysis highlighting the importance of shifting from e-government to e-governance in terms of advantages and challenges of such path.

E-GOVERNMENT AND E-GOVERNANCE IN THE EUROPEAN CONTEXT

To shed light on the shift from e-government to e-governance in the European Union (EU), an overview of definitions, actors, and main strategies for e-government and e-governance is presented.

E-government is the first form of extensive usage of IT in the public sector, and since mid 1990s it has represented a major concern of public innovation for European governments involved in the implementation of New Public Management-style reforms (Orelli, et al., 2010; Pina, et al., 2010). In 2000, the European Commission launched the *i2010 initiative* (European Commission, 2007a) aimed at creating a single European information space, strengthening investment and innovation in ICT research, and to support inclusion, better public services and quality of life through the use of ICT.

According to *EU's E-Government Main Policy Strategies* (E-government factsheet European Union; Nov 2008 Ed. 1.0), at present, the EU's strategies in the e-government field focus efforts on five priorities, namely: Inclusive e-government, Efficiency and effectiveness, High-impact key services for citizens and businesses, Key enablers, and E-participation. Considering these five EU's priorities - that will be better explained in the next section - the need arises for new connections that e-government requires and are supposed to match the declared strategies.

As a response to the increasingly complex and plural nature of public policy implementation and service delivery, New Public Governance (NPG) regime has emerged as a new trend, contemplating "the plural and pluralist complexities of the state in the twenty-first century" (Osborne, 2009, p. 7). While NPG retains several of the New Public Management (NPM) pillars, it emphasizes the relationship with the external environment and the inter-organizational relationships. Some authors have pointed out this feature using the idea of the governance of networks (e.g. Kickert, 1993; Considine, 1999; Salomon, 2002); therefore, technologies are asked to provide solutions which shift from the previous NPM regime to the NPG. Under this incremental perspective, e-governance can be defined as the use of information and communication technologies to support good governance (Heeks, 2001, p. 2). Consequently, three areas of intervention in the field of e-governance can be identified: e-Administration, e-Citizens and e-Services, and e-Society (Heeks, 2001, p. 2). E-Administration relates to that side of NPG more related to the traditional NPM basis, i.e. improvement of government processes by cutting costs, by managing performance, by making strategic connections within government, and by creating empowerment. E-Citizens and e-Services are intended to support the relationships between the public sector and the environment, by connecting citizens to government, talking to citizens and supporting accountability, listening to citizens and supporting democracy, and by improving public services. Finally, e-Society is devoted to inter-organizational relationships and comes from building interactions beyond the boundaries of government by working better with businesses, developing communities, building government partnerships, and building civil society.

Today there is a concern for cohesion, integration, and governance, both internal and external to EU member states. As the European Union keeps growing broader and embracing greater diversity, new needs and demands are arising such as for seamless public services across borders, essential to increase citizens' opportunities for mobility and for business in Europe. Even if not directly mentioned, it clearly appears the extent to which e-governance is embodied in EU's priorities, and the steam of pressure for e-governance enforcement in each of the 27 member states constituting the EU.

For e-governance to succeed a structured approach should be undertaken and a package of measures be delivered for those countries which wish to shift from e-government to e-governance (Heeks, 2001, p. 1; Salomon, 2002). In particular, there are six "drivers" that lead to a complete e-governance development (Heeks, 2001, p. 17), namely: Institutions [a], Laws [b], Leadership and strategic thinking [c], Human capacities [d], Technology [e], and Data systems [f].

First of all, e-governance can be progressed if there are *institutions* [a] in charge of coordinating, leading and driving e-governance initiatives. Such institutions act as means for facilitating e-governance, setting overall e-governance priorities, leading development and implementation of framework policies, standards and guidelines and acting as a focus for e-learning in the e-governance field. Institutions promote cross cutting e-governance infrastructure and applications and provide consultancy or facilitate inputs to individual e-governance projects. The EU is the first level institution that covers such responsibilities; the second level is represented by individual member states of the EU and by their second-tier government, depending on specific administrative structure in different countries.

The second driver consists of *laws* [b], which include laws and regulations required to allow for and to support the move to e-governance. E-governance requires a range of legislative changes including measures regarding electronic signatures, electronic archiving, data matching, freedom of information, data protection, computer crime, and intellectual property rights. Considering *leadership and strategic thinking* [c], a critical mass of strategic knowledge and skills within the government enough to support an e-governance strategy is required. Leaders who are able to insert e-governance into the agenda and make it happen represent a critical condition for a successful e-governance. This driver pertains to second-tier government and civil society leaders, and focuses on building awareness, confidence, and commitment to the e-governance process, which allow for high-level inputs and support for e-governance. Leadership training for current e-governance leaders is needed, focusing on their abilities such as leadership, interpersonal skills, strategic planning, and awareness of best practice. *Human capacities* [d] pertains to attitudes, knowledge and skills in place—especially within the public sector—that are required to initiate, implement and sustain e-governance initiatives. Key implementation capacities to be developed would likely include capacity to develop information systems, to manage projects and change, to be an 'intelligent customer'—who is able to raise project finance, specify needs, manage procurement, and manage vendors—to operate and maintain information systems. Human capacities can be addressed with training carried out within public institutions or with the adoption of innovative public-private partnership approaches, which are able to reduce reliance on the public sector. The fifth driver is *Technology* [e], which is represented by ICT infrastructure affected by e-governance. From an institutional perspective, it could well be valuable to explicitly separate out responsibility for the technical infrastructure underlying e-governance to a national ICT infrastructure body, which would have infrastructural responsibilities across all areas of ICT application, not just e-governance. Areas of interest under this driver might be e-procurement, e-payment, e-archiving; time stamping, e-mail policies and certificates, and official receipt.

The last driver is *Data systems* [f]. Management systems, records, and work processes in place should provide the quantity and quality of data to support the shift to e-governance. E-governance relies to a significant extent on existing data, systems, and processes. Alongside the introduction of ICTs for e-governance, there may therefore be the need for moves to reconstruct and renew the underlying data systems. Existing and planned good governance reforms ought to address this problem, paying attention to data issues whilst reconstructing work systems and work processes. Part of that integrated approach should be a recognition of the 'humanity of data'—the fact that data quality, data security and data sharing depend, at the root, on human motivations and values.

E-GOVERNANCE AND E-GOVERNMENT DIMENSIONS IN EUROPEAN COUNTRIES

This section provides an overview of the development of e-government and e-governance in the public sector, and describes how typically they are developed in public administrations, as experienced by different EU countries.

At present, EU's main policy strategies in the e-government field are explained in the *E-government Action Plan*. This was designed to help governments meet demands concerning the services they provide to citizens and businesses. Today, citizens demand better services, better security, and better democracy, while businesses demand less bureaucracy and more efficiency. Moreover, as the European Union keeps growing broader and embracing greater diversity, new needs, and demands are arising such as for seamless public services across borders, essential to increase citizens' opportunities for mobility and for business in Europe. The Action Plan focuses efforts on five priorities, according to *EU's E-Government Main Policy Strategies* (E-government factsheet European Union, Nov 2008 Ed. 1.0), namely: 1) Inclusive e-government: Governments, together with their agencies and other intermediaries, are increasingly integrating ICTs into their processes; 2) Efficiency and effectiveness: E-government significantly contributes to high user satisfaction, transparency and accountability, a lighter administrative burden and efficiency gains; 3) High-impact key services for citizens and businesses: To make high-impact services for citizens and businesses more widely available; 4) Key enablers: To put in place key enablers which provide the foundation for e-government systems to work together, and to build connections between ICT systems; 5) E-participation: To strengthen citizen participation and democratic decision-making by using new technologies. Each country adapts the European strategies to its particular context.

In order to assess the extent to which the shift from e-government to e-governance systems takes place in European governments, we now consider sixteen EU countries. For each of them we take into consideration two dimensions: the level of e-government and the level of e-governance. The following criteria have been taken into consideration for the selection of the analyzed countries: (a) countries ranked in the upper half section of the report (from 1 to 16 out of 31 countries considered); (b) representative countries from Napoleonic state/bureaucratic tradition, common low tradition, and federal tradition. As a result, sixteen countries have been selected, namely Austria [1], Slovenia [2], Malta [3], Portugal [4], The United Kingdom [5], France [6], Sweden [7], Estonia [8], Norway [9], Germany [10], Spain [11], The Netherlands [12], Finland [13], Belgium [14], Denmark [15], and Italy [16].

Starting from the e-government dimension, one can highlight the fact that the selected countries are positioned between Austria (leader country) and the mid-placed country, Italy, according to *The User Challenge Benchmarking the Supply of Online Public Services* (European Commission, 2007b). This report is the EC's benchmark study on electronic public services in Europe. It leverages the "Web-based" assessment of more than 5,000 public agencies and 14,000 Web pages providing the 20 public services in the 31 participating countries. The participating countries include the 27 Member States, plus Iceland, Norway, Switzerland, and Turkey (noted as EU27+).

The report shows outputs of public services delivered by governments across three main indicators calculated in a standardized manner in EU27+ countries. This provides a first base of comparison among e-government performances and development. In this 2007 benchmark, the three measured indicators are Full Online Availability (FOA), Overall Online Sophistication (OOS), and User Centricity (UC).

FOA measures the number of public services fully available online, while OOS provides an indication of the extent to which the online provision of services is based on new models of front and back-office integration, the reuse of available data and to what degree the idea of pro-active service delivery is embedded. Lastly, UC provides a measure of user's confidence in e-services in terms of security and convenience, of the possibility to choose a multi-channel access, and of site's compliance with international standards of accessibility. Within EU27+, FOA, and OOS have a high correlation. The average scores on sophistication (76%) are generally higher than the fully-availability online (58%). As to UC, only three countries score more than 30%, i.e. Bulgaria, Norway, and Austria. This high score for Bulgaria results from a policy for which call centers are linked as an alternative delivery channel to most of the public services, whilst Austria is also country leader as to FOA.

Considering the e-governance dimension, documents provided by the European Union on its official website are the primary source of data (http://epractice.eu). Each country provides a document called "E-government factsheet," which is regularly updated and represents the only official and systematic analysis that can be used to assess e-governance implementation. In the cases where data presented in the factsheets were neither adequate nor complete, we have investigated specific official documents cited in the factsheet.

As opposed to the e-government dimension, which can be measured through two indicators (FOA and OOS), for the e-governance dimension there are no available indicators (United Nations, 2002; 2010). In order to measure e-governance in the selected sixteen countries, we adopted Heeks' framework based on six drivers that can be made operative as follows (Heeks, 2001, p. 17). In order to assess "*Institutions*" [a], we investigated on the presence of building strategic institutions for e-governance, because e-governance can only be progressed if institutions exist and act as a focus for awareness and as a mean for e-governance facilitation. The presence of a subject [a1] and a strategy [a2], at least one, aimed at coordinating, leading, and driving e-governance is required. As for Austria, on one hand 'Digital Platform' is a strategic subject that acts with the aim of ensuring an active participation of all government levels, by strengthening the culture of cooperation and coordination between all stakeholders, and by fostering a sustainable development of e-government through large-scale implementation of interoperable and secure solutions. On the other hand, an Austrian strategy in this field is stated as "Cooperation: existing applications and infrastructures will have to work together in order to reach the desired level of efficiency" (Source: eGovernment Factsheet - Austria: pp. 12-13). "*Laws*" [b] has been analyzed in order to understand if there is an adequate legal framework aimed at permitting and supporting the move to e-governance. These laws and regulations should concern electronic signatures [b1], electronic archiving [b2], data matching [b3], freedom of information [b4], data protection [b5], computer crime [b6], and intellectual property rights legislation [b7]. "*Leadership and strategic thinking*" [c] is developed through the assessment of the critical mass of strategic knowledge and skills within government, which must be sufficient to support an e-governance strategy, in terms of second-tier government [c1] or civil society leaders [c2], which foster and coordinate a successful e-governance strategy. Indeed a critical pre-condition in successful e-governance for development is a person or small group of people, leaders with visions who insert e-governance in the agenda and make it happen. With reference to "*Human capacity*" [d], we analyzed whether the required attitudes, knowledge and skills were in place to initiate, implement and sustain e-governance initiatives. We particularly refer to training activities [d1], and partnership [d2]. As to "*Technology*" [e], we investigated the span of services that enable the provision of traditional government services through new IT technologies, namely E-procurement [e1], E-payment [e2], E-archiving [e3], Time Stamping [e4], E-certificates [e5], and Official digital receipt [e6]. In order to measure "*Data Systems*" [f], we analyzed whether the management systems, records, and work processes in place were able to provide the quality of data in terms of authorization and identification processes [f1], and the data sharing to support the move to e-governance [f2].

We now preliminarily analyze the variables associated whit the six drivers that measure e-governance adoption (Table 1). As shown, country by country we indicate the presence (°) or the absence () of all the considered variables that measure the six defined drivers.

As can be seen, there is a strong presence of e-governance variables in the sixteen EU countries considered, though not homogeneous. The most advanced drivers in terms of implementation are Institutions and Data Systems. Concerning Institutions, the presence of such driver can be related to the pressure stemmed by the European Union to foster definition of appropriate subjects and strategies for e-governance. The presence of e-governance Institutions is seen as a pre-requisite for the introduction of a path that leads from e-government to e-governance.

On the other side, Data Systems is linked to the presence of data with good quality, which can be shared among networks. Again, a system aimed at implementing and monitoring the provision of good shared data is crucial for any country moving towards e-governance. Leadership and strategic thinking and Human capacity are two drivers to which European countries should pay more attention, because human capabilities are crucial to a complete transition to e-governance. Finally, the provision of Laws and Technologies is not as advanced as it might be expected in EU countries, considering the lack of several variables in this area.

Table 1. Adoption of e-governance in EU countries

	Institutions		Laws							Leadership and strategic thinking		Human capacity		Technology						Data Systems	
	[a]		[b]							[c]		[d]		[e]						[f]	
	a1	a2	b1	b2	b3	b4	b5	b6	b7	c1	c2	d1	d2	e1	e2	e3	e4	e5	e6	f1	f2
Austria	°	°	°	°		°	°			°	°	°	°	°	°	°	°	°	°	°	°
Slovenia	°		°	°		°	°			°		°		°	°		°	°		°	°
Malta	°		°				°	°	°	°	°	°	°	°	°			°	°	°	°
Portugal	°	°	°	°		°	°	°		°	°	°	°	°	°			°	°	°	°
UK	°	°	°		°	°	°			°	°	°	°	°	°	°		°		°	°
France	°	°	°			°	°			°	°			°	°			°		°	°
Sweden	°	°	°			°	°			°					°			°		°	
Estonia	°		°			°	°	°		°		°		°		°		°		°	
Norway	°	°	°			°	°			°			°	°	°	°		°			°
Germany	°	°	°			°	°			°				°	°			°		°	°
Spain	°	°	°	°		°	°			°		°		°				°		°	
The Netherlands	°	°	°			°	°			°		°		°	°	°		°		°	°
Finland	°		°			°	°					°		°	°	°		°		°	°
Belgium	°		°			°	°			°		°		°	°	°		°		°	
Denmark	°		°			°	°							°	°	°		°		°	
Italy	°		°			°	°			°		°		°	°	°		°		°	

Note: a1 = subject; a2 = strategy; b1 = electronic signatures; b2 = electronic archiving; b3 = data matching; b4 = freedom of information; b5 = data protection; b6 = computer crime; b7 = intellectual property rights legislation; c1 = second-tier government; c2 = civil society leaders; d1 = training activities; d2 = partnership; e1 = e-procurement; e2 = e-payment; e3 = e-archiving; e4 = time stamping; e5 = e-certificates; e6 = official digital receipt; f1 = quality of data; f2 = sharing of data; ° = presence of the variable, () = lack of the considered variable.

THE SHIFT FROM E-GOVERNMENT TO E-GOVERNANCE: PATH AND CHALLENGES

In order to understand the extent and the direction of the shift from e-government to e-governance in European governments, we start considering e-government and e-governance measures in the selected EU countries.

The first column of Table 2 shows the e-government level both in terms of Full Online Availability (FOA) and Overall Online Sophistication (OOS), while the second column shows the e-governance level measured by the synthetic indicator calculated through the analysis proposed in the previous section. As can be seen, the best in class e-government adopters in the European Union are not the best e-governance implementers as well. Particularly, we can point out a correspondence between levels of e-government and e-governance in countries like Austria and UK, as they take the first positions in both ranking. On the other hand, other countries like Slovenia or Sweden, for example, are very well placed in terms of e-government (ranked 2nd and 4th in OOS), but do not maintain the same level in terms of e-governance, with the 7th and the 11th position respectively. In other words, a discrepancy between e-government and e-governance implementation among advanced European countries

Table 2. Overall e-government and e-governance in EU countries

	E-government				E-governance	
	OOS		FOA			
	Value	Rank	Value	Rank	Value	Rank
Austria	0.99	[1]	1.00	[1]	0.87	[1]
Slovenia	0.96	[2]	0.90	[3]	0.57	[7]
Malta	0.96	[2]	0.95	[2]	0.69	[4]
Portugal	0.90	[3]	0.90	[3]	0.84	[2]
Uk	0.90	[3]	0.89	[4]	0.82	[3]
France	0.87	[4]	0.70	[7]	0.60	[6]
Sweden	0.87	[4]	0.75	[6]	0.43	[11]
Estonia	0.87	[4]	0.70	[7]	0.48	[9]
Norway	0.86	[5]	0.78	[5]	0.63	[5]
Germany	0.84	[6]	0.75	[6]	0.46	[10]
Spain	0.84	[6]	0.70	[7]	0.60	[6]
The Netherlands	0.83	[7]	0.63	[9]	0.63	[5]
Finland	0.82	[8]	0.67	[8]	0.43	[11]
Belgium	0.80	[9]	0.60	[10]	0.49	[8]
Denmark	0.80	[9]	0.63	[9]	0.38	[12]
Italy	0.79	[10]	0.70	[7]	0.49	[8]

emerges, and it would be interesting to try to analyze such phenomenon more deeply.

Under this perspective, two figures can help clarify the situation. Starting from the first one (Figure 1), e-government (in terms of FOA) and e-governance dimensions are presented. By the distribution of the considered countries, we can identify two main situations.

A set of countries have a considerably high level of e-government in terms of full on line availability, but a relatively low level of governance. Sweden, Germany, Estonia, Italy, Finland, Denmark, and Belgium are 'following' stronger implementers, namely Norway, Spain, France, and The Netherlands. Here the difference may arise in terms of attention to e-governance issues. In fact, both groups have the same level of attention to e-government issues (FOA between 0.6 and 0.8), but a different level of governance that is below 0.6 in the former and between 0.6 and 0.8 in the latter. This situation seems not to be linked with different administrative traditions. Hence, for example, Napoleonic tradition countries, such as France, Spain, Belgium, and Italy, do not perform in the same way, as it may be expected, in terms of e-governance.

On the top of the figure, there are two different groups. 'Followers' again, represented by Malta and Slovenia, and 'pioneers,' including Austria Portugal, and the UK. We witness the same level of FOA in terms of e-government again—between 0.9 and 1—and again different levels of e-governance—between 0.6 and 0.8 for the first group and more than 0.8 for the second one.

To complete the analysis, a second figure is presented (Figure 2), where e-government (in terms of OOS) and e-governance dimensions are compared. This second dimension is included in order to assess both dimensions of e-government usually presented by official EU reports. Even if

Figure 1. Specific e-government and e-governance dimension in EU countries (FOA)

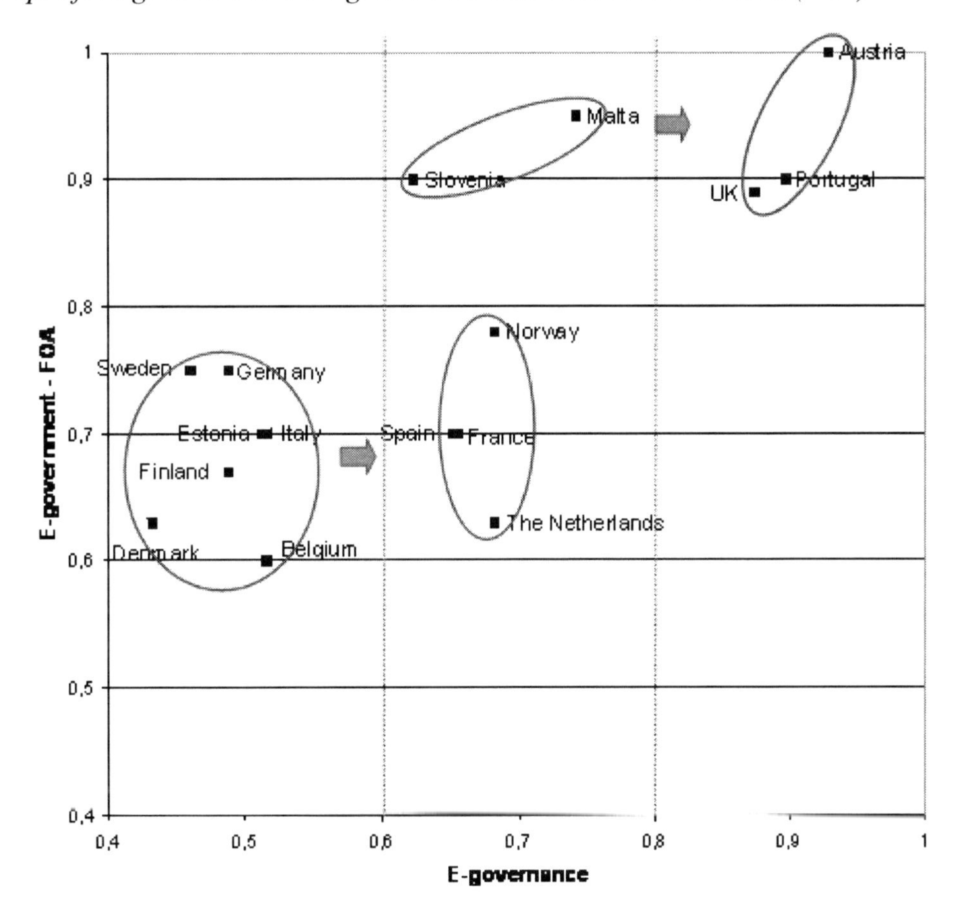

OOS dimension presents a reduced variance than FOA in the sixteen countries analyzed, Figure 2 shows a situation that is similar to the previous presented.

According to Heeks' framework applied to EU countries, we can point out that there is no automatic correlation between e-government and e-governance implementation in EU countries. Implementing an e-government strategy and succeeding in such practice is not enough to guarantee a coherent development of e-governance practices as well. In fact, there are some actions or strategies that tend to both e-government and e-governance directions, whilst there are other e-government actions that have to be selected according to the orientation towards e-governance that each country decides to have.

CONCLUSION

This chapter attempts to answer the question, *"To what extent and in which direction does the recent so-called 'shift from e-government to e-governance systems' take place in European governments?"* Given the growing importance of e-governance in the public sector, discussion has not been as robust as it should be regarding the role of e-governance in understanding new challenges. With this purpose, we combined the areas of e-government and e-governance in order to analyze how closely and critically intertwined they are in European Union countries. After having explained how typical e-government and e-governance look in the public administration among different EU countries, we can draw some conclusions about the shift from e-government to e-governance.

Figure 2. Specific e-government and e-governance dimension in EU countries (OOS)

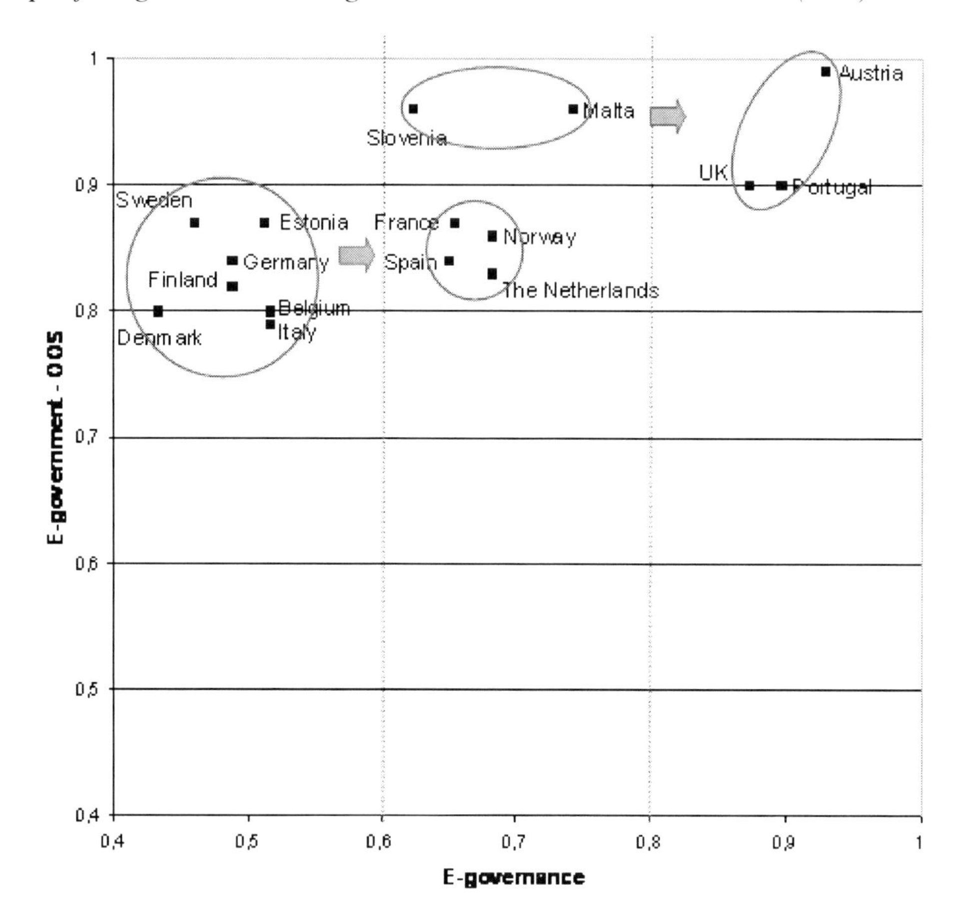

At this stage, there are two main areas of evidences that we can point out. First of all, not all the e-government strategies and actions undertaken by EU governments lead to e-governance development. As a consequence, countries should carefully verify which aim they introduce new technologic tools for and if such introduction is aligned with their expectations in terms of shift from e-government to e-governance.

Secondly, there are some key areas that must be strengthened in order to achieve a complete shift from e-government to e-governance. The first lack seems to depend on 'Laws,' and is connected to a juridical environment that is at the basis of any kind of action in the field of e-governance. 'Leadership and strategic thinking' and 'Human capacity' are other two crucial and underdeveloped areas, connected to a more general lack of attention to human capacities in the development of e-governance.

The view and the findings of e-governance proposed in this chapter invites policy makers to a more careful consideration of e-governance pressures and implementation that different e-governance strategies bring with them.

REFERENCES

Bouckaert, G., & Pollitt, C. (2005). *Public management reform: A comparative analysis*. Oxford, UK: Oxford University Press.

Considine, M., & Lewis, J. (1999). Governance at ground level: The frontline bureaucrat in the age of markets and networks. *Public Administration Review*, *59*(6), 467–480. doi:10.2307/3110295

Dunleavy, P., Margetts, H., Bastow, S., & Tinkler, J. (2006). *Digital era governance: IT corporations, the state, and e-government*. Oxford, UK: Oxford University Press.

EIPA. (2003). *eGovernment in Europe: The state of affairs*. Brussels, Belgium: Publishing Atlanta.

European Commission. (2000). *eEurope action plan*. Retrieved from http://www.europa.eu.int

European Commission. (2007a). *i2010 - Annual information society report 2007*. Brussels, Belgium: European Commission.

European Commission. (2007b). *The user challenge benchmarking: The supply of online public services*. Brussels, Belgium: Capgemini.

Heeks, R. (2001). Understanding e-governance for development. *IDPM i-Government Working Paper, 11*.

Hood, C. (1983). *The tools of government in the digital age*. New York, NY: Palgrave Macmillan.

Kickert, W. J. M. (1993). Complexity, governance and dynamics: Conceptual explorations of public network management. In Kooiman, J. (Ed.), *Modern Governance: New Government-Society Interactions*. London, UK: Sage.

Orelli, R. L., Padovani, E., & Scorsone, E. (2010). E-government, accountability, and performance: Best-in-class governments in European union countries. In Reddick, C. G. (Ed.), *Comparative E-Government* (pp. 561–586). London, UK: Springer. doi:10.1007/978-1-4419-6536-3_29

Osborne, S. (2009). *The new public governance*. London, UK: Routledge.

Pina, V., Torres, L., & Royo, S. (2010). Is e-government leading to more accountable and transparent local governments? An overall view. *Financial Accountability & Management*, *26*(1), 3–20. doi:10.1111/j.1468-0408.2009.00488.x

Salomon, L. M. (2002). *The tools of government*. Oxford, UK: Oxford University Press.

United Nations. (2002). *Benchmarking e-government: A global perspective*. New York, NY: UN Publishing.

United Nations. (2010). *E-government survey*. New York, NY: UN Publishing.

ADDITIONAL READING

Garson, G. D. (2003). *Public information technology: Policy and management issues*. Hershey, PA: IGI Global.

Garson, G. D. (2006). *Public information technology and e-governance: Managing the virtual state*. Sudbury, MA: Jones and Bartlett.

Hacking, I. (1983). *Representing and intervening*. Cambridge, UK: Cambridge University Press.

Heeks, R. (1999). *Reinventing government in the information age*. London, UK: Routledge. doi:10.4324/9780203204962

Heeks, R. (2006). *Implementing and managing egovernment*. London, UK: Sage.

Heeks, R., & Bailur, S. (2007). Analyzing e-government research: Perspectives, philosophies, theories, methods and practice. *Government Information Quarterly*, *24*(2), 243–265. doi:10.1016/j.giq.2006.06.005

Kudo, H. (2010). E-governance as strategy of public sector reform: Peculiarity of Japanese policy and its institutional origin. *Financial Accountability & Management*, *26*(1), 65–84. doi:10.1111/j.1468-0408.2009.00491.x

Lapsley, I., & Miller, P. (2010). The e-government project. *Financial Accountability & Management*, *26*(1), 1–2. doi:10.1111/j.1468-0408.2009.00495.x

Miller, P., & Rose, N. (1990). Governing economic life. *Economy and Society*, *19*(1), 1–31. doi:10.1080/03085149000000001

OECD. (1997). *Managing across levels of government: Part one: Overview*. Paris, France: OECD.

OECD. (2003). *The e-government imperative*. Paris, France: OECD.

Osborne, D., & Plastrik, P. (2000). *The reinventor's handbook: Tools for transforming your government*. San Francisco, CA: Josey-Bass.

Reddick, C. G., & Frank, H. A. (2007). E-government and its influence on managerial effectiveness: A survey of Florida and Texas city managers. *Financial Accountability & Management*, *23*(1), 1–26. doi:10.1111/j.1468-0408.2007.00417.x

Rosslyn, E., & Cheryl, H. (1996). Visions of excellence in Australian and Canadian human services organizations. *International Journal of Public Sector Management*, *9*(5), 109–124. doi:10.1108/09513559610146384

Santos, R., & Heeks, R. (2003). *ICTs and intra-governmental structures at local, regional and central levels: Updating conventional ideas. E-Government Short Papers No. 7*. Manchester, UK: University of Manchester.

Saxena, K. B. C. (2005). Towards excellence in e-governance. *International Journal of Public Sector Management*, *18*(6), 498–513. doi:10.1108/09513550510616733

Walker, R. M. (2006). Innovation type and diffusion: An empirical analysis of local government. *Public Administration*, *84*(2), 311–335. doi:10.1111/j.1467-9299.2006.00004.x

Chapter 12
Core Values:
e-Government Implementation and Its Progress in Brunei

Kim Cheng Patrick Low
Universiti Brunei Darussalam, Brunei

Mohammad Habibur Rahman
Universiti Brunei Darussalam, Brunei

Mohammad Nabil Almunawar
Universiti Brunei Darussalam, Brunei

Fadzliwati Mohiddin
Universiti Brunei Darussalam, Brunei

Sik Liong Ang
Universiti Brunei Darussalam, Brunei

ABSTRACT

In this chapter, e-Government and national cultures of the island republic of Singapore and the Sultanate of Negara Brunei Darussalam (henceforth Brunei), both small countries, are examined. The authors discuss the salient core values in the two national cultures that enable e-Government to be successfully implemented or at least have the right ingredients to be successful.

INTRODUCTION

Singapore and Brunei are both considered to be comparatively small countries situated in South East Asia. The aims of the chapter are to examine e-Government and the role of national cultures and its core values in Singapore and Brunei that enable e-Government to be successfully implemented, and thus make e-progress and assist economic growth and development (see Figure 1).

To put it simpler terms, e-Government can be defined as the administrative processes of the government as well as the latter's facilitative interaction with the public or the citizenry. In addition, e-Government is used to serve citizens, support businesses, and strengthen societies (Lee, 2007).

DOI: 10.4018/978-1-4666-1909-8.ch012

Figure 1. The respective positions of Singapore and Brunei in South East Asia

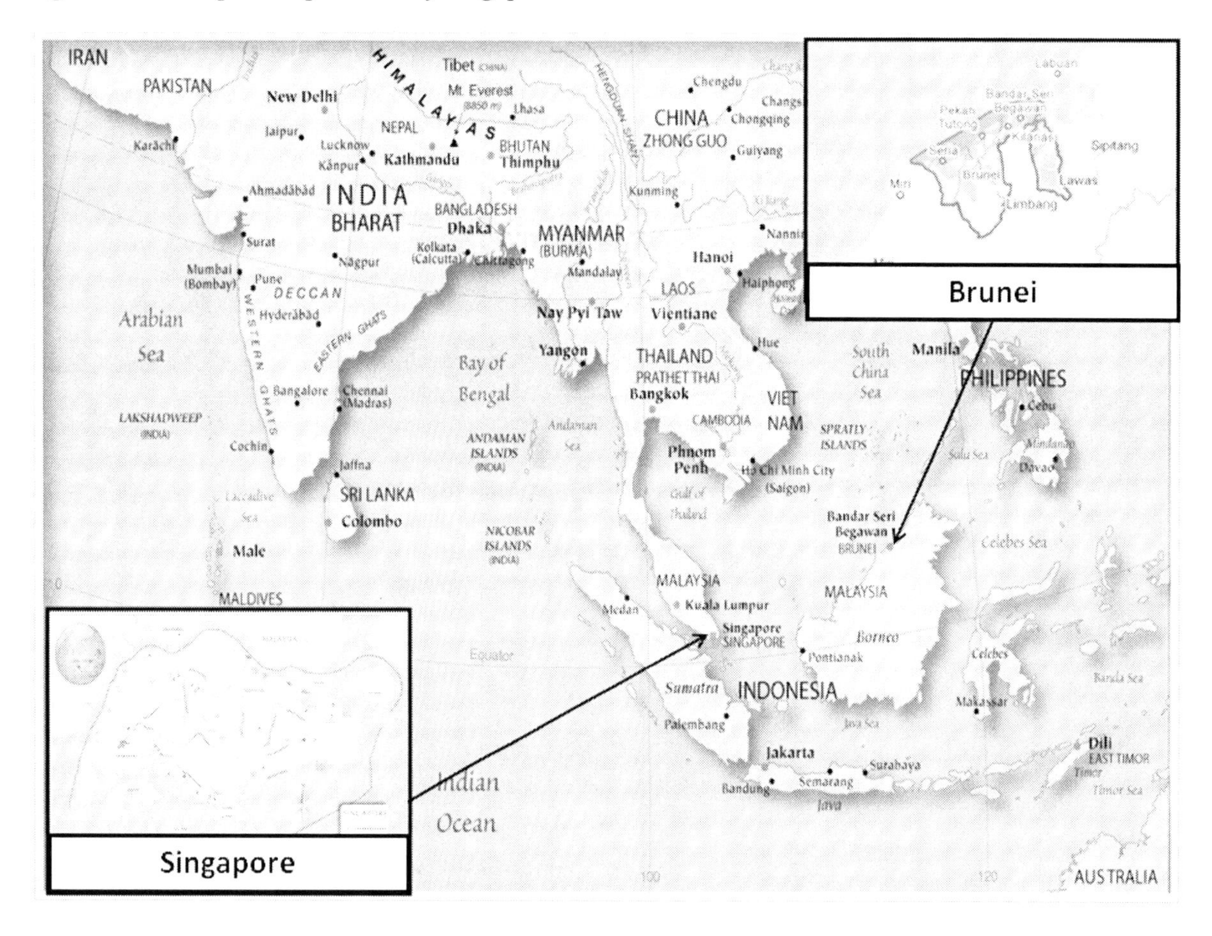

A culture is usually taken as "the way we do things around here," and values about "how things ought to be" are shared amongst the people (Mead, 1994; cited in Hill, 2009). In other words, core values establish the foundation of a culture. Hence, national culture can be defined as a set of core values held by its people, and these core values are the people's key beliefs or convictions, something very close to their hearts (Low, 2009, 2005). Interestingly, values are people's or organization's (nation's) priorities; they also provide purpose and a sense of direction, setting the standards and giving us a sense of right and wrong (Low, 2011).

In e-Government, culture is one of the critical factors in enabling economic growth by streamlining government processes, providing better access to information, and promoting a suitable enterprise environment to further facilitate greater business growth.

THE CRITICAL SUCCESS FACTORS OF E-GOVERNMENT IMPLEMENTATION

The researchers felt that in order to make e-government a success, it is very important that a nation should have a top down approach in which the government with a dynamic and robust management team (including a strong ICT support), strategically planned, installed, and made available ICT infrastructure at the critical places of the country for current and future networking purposes. Hence, good government provides good leadership with the necessary infrastructure and strategic maintenance (Low & Ang, 2011; Low, 2011, 2009a, 2009b, 2005); and this is then followed by implementation of good e-government policy for the public and the citizenry at large (Rahman, et al., 2012).

On the other hand, a nation also requires a bottom-up approach in which the government enables the people to participate and own the e-government process. This meant that the people need to be enlightened, engaged, empowered, and entertained (The 4Es). Firstly, by enlightening the people, one makes people aware of the e-government system, and when people understand the system well, they can obtain facts and data supply by the government when they need them. Secondly, from time to time, one can also engage the people via education and training sessions and this would help them to improve their confidence and trust in using the system. Thirdly, one can also motivate the people by enabling them to involve in the e-government process and to see the progressive changes that they have made for the betterment of e-government. And lastly, when entertainment goes electronic, and paying cashless, the people would accept the e-process as a way of life and that it would be part and parcel of their daily living. When e-ticketing or purchases of cinema/drama and opera tickets go online and there would be much ease and convenience for the people with easy adoption of the e-process (Low & Ang, 2011). The top-down and bottom-up approaches of e-government implementation can be illustrated in Figure 2.

Figure 2. The key success factors of e-government implementation

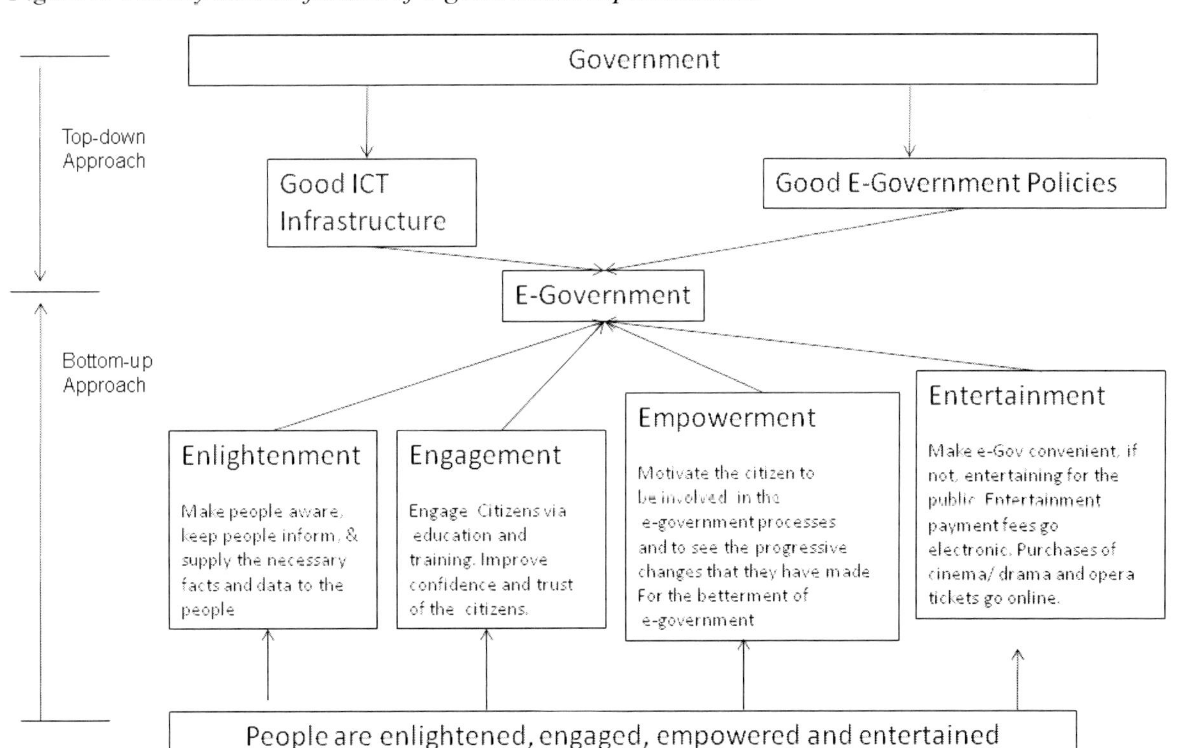

Extract from Low & Ang, 2011

E-GOVERNMENT, NATIONAL CULTURE, AND SUPPORTING CORE VALUES FOR E-GOVERNMENT

At the outset, it is appropriate to appreciate this point—English is the *lingua franca* of ICT, and what at the first level helps and reduces the challenges in e-Government is that the use of the English language is commonly used in both Singapore and Brunei Darussalam.

E-Government modernises business processes by enabling more accurate, 24/7 responses to citizen requests, and linking transaction accounts in different agencies (Dutton, 1996). It, in fact, fortifies good governance practices, which include managerialism, accountability, transparency, and freedom of information, rule of law, and combating corruption. These may be stimulated by commitments under international agreements, and by competitive pressures.

Some countries adopt e-Government in ways that normally reinforce traditional bureaucratic structures, cultures, and links from administration to citizens and politics; in Singapore case, e-Government makes its civil bureaucracy more responsive. In this aspect, the core values of efficiency, effectiveness (Low, 2002; cited in Low, 2005; Low, 2009, 2008) and reliability (Mahizhnan & Andiappan, 2002) of Singapore are an asset when implementing e-Government. These values have been said to assist the city-state Singapore to be successful in implementing its e-Government (Mahizhnan & Andiappan, 2002). Singapore, as a city-state, has been known for its efficiency and effectiveness in its administration.

Besides, in achievement-oriented and pro-active Singapore (Low, 2009, 2005), e-Government enables service integration while this has yet to be developed in Brunei Darussalam. The island-Republic has jurisdictions that enable comprehensive systems where a Web-portal or smart card integrates information and services from various government agencies to help citizens and other stakeholders get seamless service without needing to know about the responsible government agency. Thus, users can obtain services across different geographic levels of government within the same functional area, and across different functions. As an example of the latter, a citizen can submit a change of address on her driving license, and the change is automatically registered with the health, elections, and tax departments, thus avoiding the need for multiple or duplicate filings. Citizens can also use these portals to make payments and other transactions, obtain a checklist of things to bring when applying for services in-person, find answers to Frequently Asked Questions (FAQs) and engage the services of relevant commercial enterprises.

Singapore's success is further augmented by its Government's continuous efforts in laying down the required support, infrastructure, and amenities (Low, 2011, 2009, 2007, 2005). In Brunei, His Majesty Sultan Haji Hassanal Bolkiah Mu'izzaddin Waddaulah has been spearheading economic development and progress, bringing the country to higher heights of economic growth. In fact, Low and Yusob (2008) has argued that the national philosophy of MIB (*Melayu, Islam dan Beraja*: Malay Islamic Monarchy) has led the Sultanate to achieve much growth with good direction from the top as well as care and concern for its people's welfare. Much unity also exists, with many Bruneians seeing the Sultan as the provider, caring and showing concern for their welfare and wellbeing. Bruneians too enjoyed many benefits and these, among other things, include no income tax, free education, and health services.

Brunei Darussalam is a small country with less than 400,000 people. Small can be beautiful. Like small City-State Singapore, Brunei can also tap its smallness and turning its smallness into a competitive advantage. Being small, Brunei can easily set up its IT/ICT infrastructure.

Wanting to be less dependent on petrol and gas, the Majesty's Government is also looking at ways to diversify the Sultanate's economy. It has been laying the necessary infrastructure such as the building of highways, flyovers and more roads and improving the ports to facilitate business and economic growth. Bandar's size will also be expanded (Ibrahim, 2007). The Sultanate's efficiency and administrative effectiveness can be enhanced and, like Singapore, 'make compact' through the e-government process, with its people attuned to e-government acceptance and practices.

Education and investments on skills growth, training, and development can bring nations to the threshold of economic success. It is generally believed that better and higher education can contribute to human capital growth and a higher level of productivity. Here, the national cultures of both Singapore and Brunei can be said to stress on education (Low, 2011, 2008; Low, 2002; cited in Low, 2005; Said Ya'akub, 2007). In Singapore, the Government invests in university education, seeking to make local universities world-class and best in the region. The same applies to Brunei Darussalam, with various education schemes—including the Sultan's Scholars—being implemented to boost education among the Bruneians. According to UNESCO 2007 statistics on education, 94.9 percent of adults and 99.6 percent of youths are literate in Brunei Darussalam (UNESCO 2007 statistics). "Many years ago, Ministry of Education imposed all government and private schools to have computing lessons in their curriculum starting with primary classes," said most school teachers when interviewed.

An educated workforce and citizenry makes e-Government readily accepted with its implementation smoother. A good Bruneian example of this has been cited by Low, Almunawar and Mohiddin (2008) was the introduction of the Hariis system within the Ministry of Health. Developed purely by the Human Resource (HR) specialists of the ministry, the information system which provides portal and Information Communication Technology (ICT) support for the Ministry is greatly accepted and widely used by the management and staff of the ministry. This reflects a good start, as it moves away from "mere data collection but also a mindset acceptance of the change to come" (input from several Ministry of Health's staff). Besides, it also critically serves as a solid action step to move forward in e-Government implementation (see Figure 3).

Figure 3. Similarities of values between Singapore and Brunei and conducive factors for the implementation of e-government

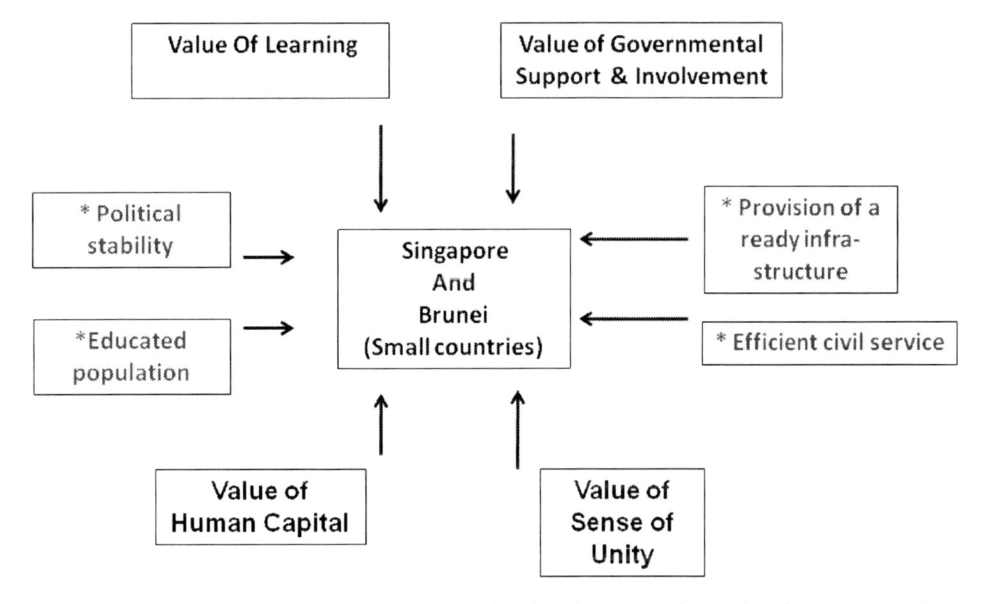

Nonetheless, unlike Singapore, which has a good IT professional pool, one of Brunei's key challenges in the progress of e-Government is that of lack of human resources skilled in Information and Communications Technology (ICT). The issue is expected to pose an even greater challenge in the future when the number of e-Government projects increase (e-Government Media, 2009, p. 67). Singapore has various IT training bodies both in the private and public sectors, and in this regard, Brunei's institutions such as Universiti Brunei Darussalam and Institute of Technology Brunei need to train and produce a pool of IT/ICT professionals as well as developing the IT/ICT capabilities and capacities.

KEY GUIDING PRINCIPLES FOR GOOD E-GOVERNMENT

Of critical importance too, the political will must be present; this is to ensure that goals and aims of e-Government is being implemented and well carried out. Through e-government, businesses are facilitated and people served—"Every service that can be delivered electronically shall be." Overall, reflecting action-oriented efficiency and effectiveness, Singapore's e-Government guiding principles, where deem fit, ought to be emulated by Brunei, and they are:

- Every service that can be delivered electronically shall be.
- The customer shall supply or update personal information only once.
- Those without home access shall have access to public delivery network.
- Staff must assist those who need special help like the elderly.
- All services shall be "customer-centric" and not "agency-centric."
- Physical visits shall be kept as low as possible (Mahizhnan & Andiappan, 2002).

The Government of Brunei has as its vision "an e-smart Government in the 21st century," its mission "to establish electronic governance and services to best serve the nation" (e-Government Media, 2009, p. 67), and its guiding principle, providing "Citizen-centric services."

POTENTIAL PROBLEMS, CAVEATS, AND THE PROGRESS FOR E-GOVERNMENT INITIATIVES AND IMPLEMENTATION IN BRUNEI

The success of e-Government depends to a large extent on the public's adoption and acceptance of a new way of doing things—an e-lifestyle. Singapore has been successful with its e-lifestyle (Mahizhnan & Andiappan, 2002). Kifle and Low (2009) has affirmed the view that "[strong] leadership is defined in terms of directing and completing the whole implementation of e-government, getting the results as well as winning the people over in the cause and actions," and this is tantamount to building a supporting IT culture. And what was and is true of Singapore is also relevant to Brunei. There was thus a strong need to build up "a national IT culture that involved massive public education campaign. The plan called for the involvement of schools, the mass media and other grassroots organizations to create awareness and promote IT literacy" (IDA, 2006, p. 5, commenting on the Singapore situation).

In Singapore, various campaigns were held annually to raise public awareness of the e-lifestyle. These campaigns in Singapore stress on the 4Es—e-Learning, e-Entertainment, e-Communications, and e-Transactions—to provide Singaporeans five strong reasons why they have got to be connected to information communication technology. To move citizens from awareness to adoption of an e-lifestyle, a series of thematic online fairs are also planned to boost consumer confidence in online services such as shopping for groceries, purchasing travel packages, and banking online.

However, in Brunei, more of such an e-lifestyle has yet to be part and parcel of the prevailing culture, and it needs to be encouraged. One Institut Technologi Brunei (ITB) survey shows only 31 percent of Bruneians used e-banking (Hazair, 2007). It also reported that "the popularity of personal computers and the advent of Internet banking presented both an opportunity and a challenge for the banking industry," meaning also to e-lifestyle in Brunei. The survey also reported that that 14 percent reported that "they still worried about security in online banking." Online banking, somewhat new to Brunei, was first introduced by Hong Kong and Shanghai Banking Corporation (HSBC) in 2000.

Nonetheless, a recent newspaper article has indicated a "good response to HSBC online services" (Hosni, 2009, p. 4). It is said that Internet banking services have made life easier and are gaining its appeal among "highly sophisticated people who are always on the move." It is also gaining attraction among bank customers since it affords customers "the luxury of saving a lot of time and energy by providing a new service where electricity and water bills can now be paid through Internet banking."

As such, positively speaking, some evidence of e-lifestyle is, more or less, making a toehold, and that is very encouraging. Besides, the younger generation of school children is, in fact, pretty much exposed to ICT and e-ways. Based on interviews with four hundred and twenty-three schoolteachers and citizens, the feedback was that: "all youths in Brunei are computer literate and they are capable in surfing the net.," "In most ways, our school children are IT-savvy," and "they are more open and oriented towards new technology."

Evident in the researchers' interviewees with several Singaporeans (six from the Infocomm Development Authority: IDA and Institute of System Science: ISS and four businesspersons and e-Government users during the period 21 to 26 September 2009) on the Singapore e-Government implementation and the ways in which it manages or leads the change, the prevailing views were that in Singapore:

Government initiated, felt like top down but explanations were made. [But more critically it] managed change with the benefits highlighted to the citizenry. Newspapers gave key information and highlighted benefits too.

The public buy-in was vital to ensure the smooth implementation, support, and acceptance. In Singapore, another "core belief that helps the e-Government implementation process is that change needs to be well handled or managed. It is more or less about change management issues and problems" "Change is well accepted and must be managed" (several interviewees' input; mentioned six times).

Senior citizens who were illiterate or not well educated found it difficult to cope with technology; besides, they grew in a setting where education then was not readily available. In addition, how these were then overcome in Singapore? (These could be lessons for Brunei Darussalam in its e-Government initiatives and implementation, which may be a topic for another article.)

In Singapore, the Government set up a number of Citizens Connect Kiosks in Community Clubs (CCs), and these are strategically located in places where there are high human traffic. There are a total number of 27 such kiosks. In addition, these include in places such as the CPF building and the Income Tax building at Novena. The Citizens Connect Kiosks are manned by IT (Information Technology) ambassadors who trained IT personnel and are paid IT employees to assist citizens to use the IT. A critical factor in managing the change is that these IT ambassadors lent a hand to up the high-touch factor that facilitated the change: IT acceptance and e-Government process and habit.

Similarly, in Brunei Darussalam, greater public awareness on e-government and its benefits would be useful; this is indeed much needed and it can bring about more support to the e-government process and implementation. The youth organizations can also assist to disseminate the information of e-Government efforts to its members. TelBru, several months ago, has also lowered its e-speed/ Internet fees, and this means taking the Internet and things e to the people, further making easy the public's increasing e-acceptance and growing of the e-habit and ways. The e-Government media's publications in the Sultanate's major newspapers such as Borneo Bulletin are also another positive step that serves to educate the public as well as showing the determination to ensure the progress of e-Government initiatives and implementation in Brunei Darussalam.

Like Singapore (Low, 2009, 2005), Brunei indeed enjoys much peace and political stability, and this is advantageous (Mohiddin & Low, 2008). After all, since the more stable is the organisation (nation), the more stable would be the information systems providing better and effective service to the end-users. In this regard, Mohiddin's (2007) study supports the point that peace and stability indeed helps in e-implementation.

Stability in Islam is in the faith of following the religion, and what is critical and helpful in e-Government is the fact that Islam is not against modernisation as long as it is not against it. The Islamic teachings in Brunei continued to be reinforced in every possible ways without neglecting the progress and development of the nation globally (Mohiddin & Low, 2008). The influence of Islam is more visible in the public sector compared to private; nevertheless, the private sectors are continuously trying to adapt the Islamic way of life as much as possible. Non-Muslims are expected to give their due respect.

In Brunei, the need arises to reduce the amount of paperwork that is currently said to exist in many government departments ("slow processing within the government," an unnamed Universiti Brunei Darussalam lecturer, cited in Han, 2007, p. 1), and that must be done either as a prelude or alongside the implementation of e-Government. To tap synergies and generate faster processing, greater coordination among ministries is also needed (Kifle & Low, 2009), and this is vital for the smooth implementation of e-Government in Brunei.

Anecdotal evidence and talking to several expatriates and locals appear to suggest that little innovation or risk taking exists in Brunei. Although much paperwork, bureaucracy, and red tapes abound, things are still rather conservative, with many regulations, laws, and orders to be followed, still the positive aspects appear to exist. Bruneians are said to hold the value of risk innovation. Besides, innovation/risk taking does have positive impact on System Quality in Brunei. This might be because organisations in Brunei have to respond effectively and appropriately to present and future needs and challenges (Mohiddin, 2007). Kifle and Low's (2009, p. 274) (*italics, authors' emphasis*) study appears to confirm this when they indicate the findings that, "innovation in Brunei Public Sector (also) comes from the lower and middle level. This is *a positive indication* that they can be creative and there is strong leadership potential from within (the ministries)."

Interestingly, just as in Singapore's case—"the strong political will of the government leaders has led to success of the island's Republic e-Government initiatives and implementation" (Singaporean respondents' input; mentioned seven times), Brunei is also blessed with this; the Government of Brunei is indeed determined to have its own "e-smart Government in the 21st century" (e-Government Media, 2009, p. 67).

The authors here would also argue that yet another important cultural ingredient in Brunei when it comes to the implementation of e-government is its overall cultural value of people-orientation. Bruneians are also said to be feeling-oriented, and according to Mohiddin (2007), this value of people orientation has positive impact on Information Quality, System Quality, and Service Quality. Bruneians are people-orientated and hence a

Collectivist society. Bruneians' strength relies on unity in whatever circumstances, and this helps as a boon when comes to implementing systems and e-government. In Brunei and in organisations within the Sultanate, a strong preference for group decisions (face-to-face meetings are still favoured) and consultative management exists, and employees act in the interest of their in-group, not necessarily of themselves. Employees usually do best in in-groups as in, for example, the introduction of the HARIIS system within the Ministry of Health mentioned earlier.

Interestingly, in implementing the e-government process, it is good to have a combination of high-tech and high-touch so that the people feel involved or engaged so that they own the process. Because of this feeling-orientation of Bruneians, generally speaking, they would ensure or at least take steps to ensure that the people factors are attended to and/ or at least be sensitive to the people issues such as job displacements/ alignments or cuts if these would occur. The people can be assured, at least, not feeling alienated.

CONCLUSION

To sum up, very much like Singaporeans, Bruneians should be achievement-oriented, continuously improve, and forge ahead in e-government implementation. In addition, like any other nations, to succeed, Bruneians need to close any gaps that may exist. Besides, as pointed out by His Majesty, Bruneians ought to grow intellectually, benchmark and "not take things for granted by living extravagantly, but rather be mindful of the fortunate circumstances they (the Bruneians) find themselves in" (Said Ya'akub, 2007, p. 1).

True, e-Government is a challenge to Bruneians yet opportunities abound. Although at this point in time, the Brunei Government's development of e-Government may just be starting or budding, the Sultanate's national culture has the necessary DNA and template for success, and from here, the efforts to e-Government should be an ongoing process with continuous improvement (embracing the Malay saying of "*sedikit-sedikit lama-lama menjadi bukit*," translated, bit by bit, it eventually becomes a hill). It is a never-ending journey where destinations are only temporary stopovers; and that one-mindedness plus resilient attitude should also be adopted and built upon.

ACKNOWLEDGMENT

The authors of this chapter would like to acknowledge the sponsorship and support of Universiti Brunei Darussalam (research grant no. ubd/pnc2/2/rg1[122]). This has enabled them to prepare this chapter from the sponsored research.

REFERENCES

Dutton, W. H. (Ed.). (1996). *Information and communication technologies: Visions and realities*. Oxford, UK: Oxford University Press.

e-Government Media. (2009, November 7). The progress of e-government initiatives in Brunei Darussalam. *Borneo Bulletin*, p. 67.

Han, S. (2007, July 26). Overcome red tape to be more productive. *The Brunei Times*, p. 1.

Hazair, H. (2007, September 20). Speed, security stifling online banking growth. *The Brunei Times*, p. 1.

Hill, C. W. L. (2009). *International business: Competing with the global marketplace* (7th ed.). New York, NY: McGraw-Hill Irwin.

Hosni, A. (2009, November 27). Good response to HSBC online services. *Borneo Bulletin*, p. 4.

Ibrahim, K. (2007, July 23). Size of BSB to increase 10-fold. *The Brunei Times*, p. 1.

IDA. (2006). *The great campaign 1986 – 1991*. Singapore, Singapore: Infocomm Development Authority.

Kifle, H., & Low, K. C. P. (2009). e-Government implementation and leadership – The Brunei case study. *Electronic . Journal of E-Government*, *7*(3), 271–282.

Lee, B. Y. (2007). *Singapore e-government*. Paper presented at the 4th Ministerial e-Government Conference. Lisbon, Portugal.

Low, K. C. P. (2005). Towards a framework & typologies of Singapore corporate cultures. *Management Development Journal of Singapore*, *13*(1), 46–75.

Low, K. C. P. (2007). The cultural value of resilience – The Singapore case study. *Cross-Cultural Management: An International Journal*, *14*(2), 136–149. doi:10.1108/13527600710745741

Low, K. C. P. (2008). Core values that can propel e-government to be successfully implemented in Negara Brunei Darussalam. *Insights to a Changing World Journal*, *4*, 106–120.

Low, K. C. P. (2009). *Corporate culture and values: The perceptions of corporate leaders of cooperatives in Singapore*. Saarbrucken, Germany: VDM-Verlag.

Low, K. C. P. (2009a). Strategic Maintenance & leadership excellence in place marketing – The Singapore perspective. *Business Journal for Entrepreneurs*, *3*, 125–143.

Low, K. C. P. (2011). *Typologies of the Singapore corporate culture*. Retrieved from http://www.GlobalTrade.net

Low, K. C. P., Almunawar, M. N., & Mohiddin, F. (2008). *E-government in Brunei Darussalam – Facilitating businesses, creating the competitive edge*. Paper presented at the International Conference on Business & Management - Creating Competitive Advantage in the Global Economy. Bandar Sewi Begawan, Brunei.

Low, K. C. P., & Ang, S. L. (2011). e-Government should be for the people and owned by the people – The Brunei perspective: Lessons to be learnt from others. *Business Journal for Entrepreneurs*, *1*, 117–135.

Low, K. C. P., & Mohd Zain, A. Y. (2008). *Creating the competitive edge, the father leadership way*. Paper presented at the International Conference on Business & Management - Creating Competitive Advantage in The Global Economy. Bandar Sewi Begawan, Brunei.

Mahizhnan, A., & Andiappan, A. (2002). *e-Government – The Singapore case study*. Retrieved from http://www.infitt.org/ti2002/papers/60ARUN.PDF

Mead, R. (1994). *International management: Cross cultural dimension*. Oxford, UK: Blackwell Business.

Mohiddin, F. (2007). *Information systems success in Brunei: The impact of organisational structure and culture*. (Unpublished PhD Thesis). Curtin University of Technology. Bentley, Australia.

Mohiddin, F., & Low, K. C. P. (2008). *e-Government & national culture – The way forward*. Paper presented at the Human Resources Compendium: Current Global Trends in HR Best Practices. Bandar Sewi Begawan, Brunei.

Rahman, M. H., Low, K. C. P., Almunawar, M. N., Mohiddin, F., & Ang, S. L. (2012). E-government policy implementation in Brunei: Lessons learnt from Singapore . In Manoharan, A., & Holzer, M. (Eds.), *Active Citizen Participation in e-Government: A Global Perspective*. Hershey, PA: IGI Global. doi:10.4018/978-1-4666-0116-1.ch018

Said Ya'akub, I. (2007, December 4). Be single-minded to succeed. *The Brunei Times*, p. 1.

UNSCO. (2007). *Education in Brunei Darussalam statistics*. Retrieved from http://stats.uis.unesco.org/unesco/TableViewer/document.aspx?ReportId=121&IF_Language=eng&BR_Country=960

Chapter 13
E-Government in Bangladesh: Prospects and Challenges

Noore Alam Siddiquee
Flinders University, Australia

Md. Gofran Faroqi
Ministry of Public Administration, Bangladesh

ABSTRACT

In this chapter, the authors delineate the overall policy and institutional framework of e-government from the perspective of Bangladesh. Recognizing the current government's attempt at branding the country as "Digital Bangladesh," the authors explore major e-government programs and initiatives in operation. Most importantly, they eloquently elaborate on the constraints and challenges facing Bangladesh in its pursuit of electronic governance and also shed light on the way forward.

INTRODUCTION

Since its rise in the 1990s, e-government has quickly become a key element of public sector reform drives around the world. Currently it is widely regarded as an effective means to promote efficiency in government operations and to improve the delivery of public services. Generally defined as the application of Information and Communication Technology (ICT) in the public service, e-government is, in fact, an extension of e-commerce in the public sector where modern technology, especially the Web-based Internet, is used to deliver public services in a much more convenient, customer-oriented, and cost-effective manner. Most scholars, however, see e-government as more than mere application of ICT for it involves the process whereby the citizens and businesses are able to interact and execute transactions with government electronically (Haque, 2002; West, 2008). Thus, it requires the government to improve its internal operations by transforming the process in which public services are delivered. It also requires a modification of the entire range of relationships that the government has externally with citizens and businesses and other governments.

It is believed that this new approach to governance will not only facilitate easy and hassle-free access to services, it will also ensure efficiency,

DOI: 10.4018/978-1-4666-1909-8.ch013

transparency, and public participation in the process of governance. As such, it has come to be seen as a solution to many of the perennial problems of the public sector. It promises to make governmental operations more efficient, cut red tape, and simplify bureaucratic procedure, reduce transaction costs, and make governmental operations transparent. In addition, e-government is believed to promote innovations in the delivery of public services, offer increased flexibility in the service use and foster people's participation and empowerment (see UN, 2008). In developing countries, e-government has assumed an additional significance: it is expected to help eradicate poverty, boost national economic growth, reduce bureaucratic complexity and corruption, and establish good governance (Hoque, 2006).

While e-government offers enormous benefits and enjoys massive popularity across the globe, its implementation has proved to be far from an easy and straightforward task for it requires transformation of organizational structures and business processes, mobilisation of significant human, technological and financial resources, among others. Resource constraints, lack of political support, lack of technology, inadequate infrastructure, illiteracy and the paucity of skilled human resources often pose a formidable challenge. The task gets compounded in the context of developing countries where such problems are acute. At the same time, digital divide between rural and urban people, the poor and rich, the literate and illiterate makes service accessibility to all next to impossible. As such, the vast majority of the population remains unaffected by the new mode of governance and service delivery notwithstanding huge promises offered by variety of governmental initiatives in the field. This chapter illustrates some of these points by a detailed examination of the state of e-government in Bangladesh—a developing country in South Asia. It provides an overview of Bangladesh's journey to e-government to date and some of the constraints and challenges the country is facing in this regard. Mainly based on secondary literature, the chapter is organised into three sections. The first section outlines the key policy issues and institutional framework for e-government in Bangladesh. Section two focuses on the current status of e-government in Bangladesh highlighting some of the major programs and initiatives. The final section analyses some of the key challenges confronting Bangladesh in terms of e-initiatives and online service delivery.

E-GOVERNMENT IN BANGLADESH: THE POLICY AND INSTITUTIONAL FRAMEWORK

As elsewhere, in Bangladesh e-government has lately become a key element of the drives for improving governance and service delivery. Faced with increasing demands from citizens especially from the business community to have access to online services and the concomitant development in the private sector, e.g. e-commerce and online banking the government in Bangladesh has made it a priority. Although emphasis to modernize public administration with the help of science and technology was put as early as in 1972, the actual process in this regard began in the 1990s. Having assumed power in 1996, the *Awami League* government took major steps towards adoption of ICT in the public sector. It agreed, to connect the country with Information Super Highway, designated the ICT as a thrust sector, waived taxes on computers and computer accessories, and formed a Task Force with the Prime Minister as its head. All this has paved the way for rapid computerisation and ICT penetration in the country. The process of e-government implementation gained a fresh momentum following the Public Administration Reform Commission Report, 2000, that emphasised on the speedy implementation of the electronic government (Hoque, 2006). Subsequently, the broader framework of strategies and actions for e-government has been outlined in different policy documents.

The formulation of the National ICT Policy, 2002, is one of the most significant developments in e-government. The policy has outlined the visions of e-government with the objective of improving the quality of services to the citizens. It declares that 'the government shall use ICT systems within the public administration to improve efficiency, reduce wastage of resources, enhance planning, and raise the quality of services.' It further provides that 'the government shall implement ICT systems to provide nationwide coverage and access by any citizen to the government databases and administrative systems which can be used to extend public services to the remote corner' (GOB, 2002). Among others, the policy envisaged the establishment of telecommunication infrastructure nation-wide, connecting to submarine fibre optic cable networks, extending Internet facility to rural areas, establishing cyber kiosks in post offices, union, and upazila (sub-district) complexes, setting up an Internet exchange and increasing the bandwidth capacity and availability. It has also emphasised the spread of ICT in governmental agencies across the country including ministries, divisions, departments, autonomous bodies, and all district and Upazila headquarters and Union Parishad offices. Under the policy, each Ministry/ Division must have an ICT unit comprised of ICT professionals and it must set up websites with policy documents and updated information.

Although the policy falls short of identifying specific strategies or action plans for implementing these visions, it does outline the broader framework for e-government. Subsequent changes and modifications to the ICT policy identified action plans, the actors, the expected deliverables and benefits and goals for short, medium, and long-term development. The revised ICT Policy announced in 2009 starts from a singular national vision of enhancing social equity through the use of ICT; it outlines specific objectives, strategic themes and action plans for realizing the visions into reality. The policy envisages expanding the use of ICT with the ultimate aim of instituting an accountable, transparent, and responsive government. ICT is seen as a driving force supportive of the national goal of becoming a middle-income country within 10 years and joining the ranks of developed countries within 30 years (GOB, 2009).

Since the implementation of ICT policy required legal framework as well as management structures, the government introduced relevant legislation and institutional mechanisms. The ICT Act passed in 2006 aims to prevent crimes related to computer fraud, hacking, damaging programmes and data, and launching computer viruses. The ICT (Electronic Transactions) Act, 2006 was designed to facilitate e-commerce by safeguarding the online transactions. It is also intended to foster electronic filing in government agencies and ensure efficient delivery of electronic records (BCC, 2009). The Official Internet Connection and Usage Policy 2004 outlines the framework for the use of Internet facilities in government offices and determines its management and financial aspects, such as 'entitlement,' 'approval procedures,' and 'cost limits' (Hoque, 2006. p. 350). Meanwhile the current *Awami League* government has articulated the vision of 'Digital Bangladesh' making it a part of the nation's development strategy. Visions of Digital Bangladesh include developing human resources ready for the 21st century, connecting citizens in ways most meaningful to them, bringing services to the doorsteps of the citizens' and making the private sector and market more productive and competitive through the use of ICTs. This was followed by a number of policy measures. The ICT Policy 2009, ICT Act 2009, Right to Information Act 2009, various local government acts promulgated since 2009 have laid the foundation for making Digital Bangladesh a reality.

Efforts have also been made to create and strengthen the management structure for the implementation of e-government. The structure involves a number of organisations with the Ministry of Science, Information and Communication Technology in the leading position. It has been assigned the responsibility for the development and promotion of the ICT sector. Accordingly, its programmes cover expanding the IT education, bringing changes to laws, involving the private sector and NGOs, standardising the public websites and so on. The Bangladesh Computer Council (BCC), an autonomous body established by Act No IX of 1990 for the promotion of all kinds of ICT activities in the country, works under this ministry. The BCC has been made responsible for computerisation, providing advisory services to different online initiatives, conducting ICT training courses for government officials and running development projects related to the expansion of ICT (BCC, 2009).

The Prime Minister's Office (PMO) plays the central role on policy matters related to e-government. E-government Cell established within the PMO is responsible for providing guidance as well as oversight and coordination of the implementation of e-government programs. In particular, it is responsible for preparation and implementation of e-government vision, strategic goals, and flagship projects. It is also responsible for the development of infrastructure strategy and conducive legal and regulatory framework. To support E-Government Cell, a separate program known as Access to Information (A2I) was introduced in 2006. It seeks to ensure the suitability of e-initiatives and programs to national priorities and support the development of new projects on ICT application. With its extensive policy advice and project intervention supports A2I has become the *de facto* hub of Digital Bangladesh especially in matters of strategy development, planning and technical assistance (PMO, 2010).

The National ICT Task Force (NTF) is headed by the Prime Minister and represented by public, private, and civil society organisations. It advises the government and oversees the policy implementation. Immediately below it is the Executive Committee, headed by the Principal Secretary to the Prime Minister, with the mandate to execute the e-government directives of NTF. The telecommunication infrastructure and regulatory managements are the responsibility of the Ministry of Post and Telecommunications. Bangladesh Telecommunications Regulatory Commission (BTRC), an autonomous body under the ministry, regulates the telecom providers—both fixed and mobile lines. It also oversees the licensing for the VSAT operators and Internet Service Providers (ISPs), the development of digitization schemes, regulation of tariffs and setting of standards (BTRC, 2009).

The Ministry of Finance plays an important role by incorporating the development and strategies of e-government in annual budgets and allocating resources for the purpose. In the 2009-2010 budgets, for example, the vision of 'Digital Bangladesh' is outlined with priority attached to the building of ICT infrastructure, mainly the Internet connectivity and telephone coverage for those who are least advantaged and living in rural areas. The budget envisages to bring all upazilas within Internet coverage in the next five years and to have fibre optical lines across the country by the year 2010. For increasing ICT literacy and developing a trained work force, the government plans to introduce compulsory computer and technical education at the secondary level by 2013 and at the primary level by 2021. Apart from the government agencies a number of private organisations, e.g. Bangladesh Computer Samity (BCS), the Bangladesh Internet Service Providers (ISP) Association, and the Bangladesh Association for Software and Information Services (BASIS) have important roles in e-government. They advocate for broadening of e-government, render expert services on ICT, lobby for ICT policy decision and enforcement of existing ICT laws, and provide training for human resources development.

TOWARDS DIGITAL BANGLADESH: MAJOR E-GOVERNMENT PROGRAMS AND INITIATIVES

The success of e-government depends to a large extent on development of necessary infrastructure, Web initiatives and the development of human capital. This would allow the government to offer services electronically, enhance its operational efficiency and achieve other objectives. With this in view and especially the vision of Digital Bangladesh, the government has initiated and implemented a variety of e-government programs. Although Bangladesh lagged behind other countries of the region in making serious drives for e-government and started late, yet some of the initiatives have marked a major shift in the service provision demonstrating massive potentials. This section focuses on the state of e-government in Bangladesh, major programs currently underway and the types of services that are available online under such programs.

Web Initiatives

The most obvious frontline presence of e-government is ensured through the Web initiatives. The UN survey 2008 shows that Bangladesh's Web measure index 0.3512 has improved compared to the previous situation with .092 in 2003 (UN, 2008). Progress has been made in interactive and enhanced stages as well as the number of government websites has also increased significantly. Websites are now available in a number of important offices including the Prime Minister's Office, President's Office, key ministries, divisions, departments, higher courts, audit and accounts office, board of revenues, central bank and others (NWPB, 2009). A host of projects for providing online services such as Hajj management, electricity and water management, birth registration, railway ticketing, and so forth are also presented through Web initiatives. A recent study on 50 ministries and divisions has found that in 2006 38.9% of them were involved in at least one project on e-government although 16.7% were at the initiation stage (Hoque, 2006).

Those with e- government projects are reported to be making progress in terms of providing user-friendly and customised services. For example, the Finance Division has developed a customized software for budget planning, sensitivity and impact analysis, financial projections, and preparation of various reports. It has also created a software to facilitate interface between the development and revenue budgets. The Bangladesh Bureau of Educational Information and Statistics (BANBEIS) has launched a GIS map-based software that provides information on the density of academic institutions in particular regions and relevant information related to education (Sobhan, et al., 2004). The Planning Commission has a website for useful searchable databases of all projects under Annual Development Programme (ADP) undertaken in the last three years (Planning Commission, 2009).

Under the Support to ICT Task Force, a number of e-government projects are underway in several ministries and departments. The Ministry of Education, the Ministry of Labour and Employment, the Ministry of Expatriate Welfare and Overseas Employment, the Dhaka Passport Office, the General Manager (Postal) North's Office, the Department of Agricultural Marketing, the Land Record Office in Manikganj, are just a few. Many other government offices including the Bangladesh Bank, the Ministry of Law, Justice, and Parliamentary Affairs, and the Dhaka City Corporation are also in the process of implementing important e-government initiatives (Sobhan, et al., 2004). Major government agencies now can be accessed using *Bangladesh.Gov.bd*—the national Web portal of Bangladesh—which has links to all important public institutions. This provides the Web addresses of all major government agencies including the President's Office, the Prime Minister's Office, and the Council of Ministers, Parliament, the Judiciary, statutory bodies, 41 ministries and divisions. It also contains important information related to education, disaster management, passport, and income tax matters. Furthermore, it includes information on how to start a business, foreign direct investment, weather and climate, current events, tourism and national statistics, government forms, postal services, stock exchanges, and currency rates (NWPB, 2009).

Online Services

While the ultimate objective of e-government is to make services accessible to people without constraints of time and distance and to integrate the service delivery of various agencies to enable the citizens to access them via one portal, currently Bangladesh is at the early stage of the e-government implementation. It would be fair to say that Bangladesh has moved from emerging presence to enhanced stage and to some extent to the interactive stage. As such, members of the public are able to get some jobs done, besides accessing useful information, online. Forms of different services have been made available online. Besides the national portal of Bangladesh, a separate website contains major digitized forms called 'Bangladesh Government Digitised Forms—e-citizen services application.' Sixty-five different kinds of government forms including forms for birth registration, driving license, citizenship, family pension, passport, immigration, No Objection Certificate (NOC), Visa Application Form, utility bills form, telephone connection forms are available in the website (PMO, 2009).

One of the important online services available around the world is e-taxation. Bangladesh has made inroads in this area although currently scopes for online transactions are limited. Forms for Tax Identification Number (TIN) and Tax Returns are available online and can be downloaded for filling in, but they cannot be submitted online. However, since July 2009 e-taxation has been in place, which allows Large Tax-Payer Units (LTUs) to submit the declaration and make online payment. Plans are underway for broadening the scope of e-taxation by covering VAT and individual income taxes (NBR, 2009). Payment of utility bills is another widely used service of e-government. Citizens are now able to carry out electronic payments for utility bills. Service providers such as Power Development Board (PDB), Dhaka Electricity Supply Authority (DESA), Water and Sewerage Authority (WASA), Titas Gas, and BTTB have all made several contracts with different banks and mobile operators for their bills to be paid electronically using Internet banking, ATMs, Ready Cash card, Q-Cash card, POS, SMS, and other channels. However, currently these services are only limited to cities and the urban population.

Besides those above, various ministries and departments of the government have introduced innovations in their respective domains of service delivery. Some of the successful websites to this end are the Online Hajj Information Management website of the Ministry of Religious Affairs, the Electronic Birth Registration Information System of Rajshahi City Corporation, the integrated website and MIS in the Department of Roads and Highways, the Bangladesh Bank website and so on. Here we present two successful cases of such innovations and a few ongoing projects:

Online Hajj Information Management System

To provide innovative online services to the increasing number of pilgrims every year to perform the Hajj, the Ministry of Religious Affairs (MRA) launched a website in 2002, with its own funding and initiatives of the then Minister. This website provides the list of all registered pilgrims along with their pilgrim pass, photos, and other personal details. By putting one's passport number down a user can know the most up-to-date information of any individual Hajj performer. Its interactive feature entails the exchange of message options that facilitate easy and reliable communication between pilgrims and their relatives. It also enables access to policies and issues such as health instructions, baggage rules, authorised agents and their contact information, necessary forms, news on death bulletins, flight delays, lost and found items, and weather conditions (MRA, 2009). The disclosure of the above information, which was kept secret in files earlier, has increased transparency and reduced the scope of corruption. The list of agents available on the website also helps users to verify whether or not a particular agent is authorized and to obtain contact information for individual agents. Furthermore, the website sets an example of a tripartite partnership business model consisting of the government, the private sector, and the public. This model works well and helps reduce governmental burden as this is managed by charging small fees from individual pilgrims. It also makes the common people aware of ICT use (Sobhan, et al., 2004).

Electronic Birth Registration Information System

The Rajshahi City Corporation (RCC) has taken a landmark step in developing an Electronic Birth Registration Information System (EBRIS) that provides citizens with a unique identity card required for various social services, including education and health care. The Birth Registration Information System introduced in 1998 (done manually at that time) with funding from the UNICEF was subsequently transformed to computerised database and integrated with the immunization programs of the Public Health Department. It enables the citizens to register births from a ward office in a few seconds. The database generates birth certificate and immunization card for each newborn baby. It also generates list of babies due for immunization on a particular day (Sobhan, et al., 2004). Because of its successful back office integration in births and immunisation schedules and linkage with different offices, it has been cited by the United Nations e-report 2008 as one of the connected stages of e-governance in Bangladesh, where major e-government initiatives are at a rudimentary stage.

Back office integration has helped in the retrieval and enquiry of data in seconds, avoiding duplication of entries, by multiple agencies. Tracing the records of registered births, starting from immunization requirements to school enrolment status, helps the RCC to monitor the needs of a community and take necessary remedies (Akther, et al., 2007, pp. 43-45). Perhaps the most significant impact of the EBRIS project is that it has contributed to a significant increase in the rate of birth registration and immunization within RCC areas where currently birth registration rate is close to 100% and immunization rate is 84% (Akther, et al., 2007). Besides, the EBRIS system has increased the transparency of RCC as users can easily access the information regarding the status of all registered children with regard to immunization, schooling, and other social services (Sobhan, et al., 2004).

Digitisation of Land Records

An important e-government initiative currently underway is the digitisation of land records system. Some of the projects already completed in this area include the Modernisation of Land Administration Project (1995-98), Computerisation of Land Management Systems of Dhaka City Project (2004-06), Dhaka, Manikganj and Gazipur Collectorate Land Record Room Computerization Project (2005-07), Digital *Khatian* Preparation Program (2005), and the Coastal Land Zoning Project (2009). These projects represent partial attempts at digitization of land records and their results have been encouraging. The government has remained committed to spread the benefits of such experiments throughout the country and is considering public-private partnerships as an option for their wider replication. The government expects this to be modeled on the automation project of the Chittagong Customs House in which a private provider—Data Soft—runs the project sharing port revenues for a specific period of time. The IT firms will invest to develop integrated software to provide services from one-stop centres, currently being provided by more than 10 government offices. Documents will be uploaded to a database to which the government officials and people, in some cases, will have access. It is expected that the project would not only reduce costs and help generate significant revenues for the government but also ensure the delivery of land management services to clients within minutes in a sharp contrast to days or weeks required under the existing system (see, *The Daily Star*, 1st March, 2010).

Online Banking

Bangladesh has made strides in the application of ICT in the banking and financial sector. Bangladesh Bank (BB)—the central bank of the country—has made an extensive use of ICT in all of its operational spheres including internal management, monetary strategy, banking supervision and e-banking with a goal to be a paperless organisation. Digitisation of its internal management covers steps like automation of export receipts; import payments; invisible receipts; invisible payments; scheduled bank advances, deposits, bills, and debits; co-operative bank advances, bank assets/ liabilities; summary statements. This also covers central accounts of the country; loans and grants; exchange rates; monetary survey; broad money survey; salary bill and management of provident fund of its own staffs; assets/ liabilities; export form matching; wage earners' remittance; among others (Rahman, 2010).

While there are a number of initiatives, perhaps e-banking overrides others as this service touches most directly the needs of the citizens. The bank's steps for electronic banking include the introduction of automated clearing cheques, which facilitates inter-banking clearing payments within hours. E-payment gateway, which is likely to begin soon, will facilitate electronic transactions and quick payment connecting all banks for interbank transactions (Sobhan, et al., 2004; Rahman, 2010). For facilitating e-commerce, it has allowed other banks to online money transactions, payment of utility bills through Internet, intra-bank transfer of money, credit card payment of trading goods and services and mobile banking. Using the mobile phone operator's facility, many banks are allowed to extend banking services to the doorsteps of the mass people. An account holder can now check the transaction history of his account; download the statement; know about the status of deposited securities and so forth. Furthermore, three commercial banks are allowed to facilitate inward remittance transfer with assistance from mobile companies with their outlets at different places of the country. More recent development is the use of mobile technology by BB for monitoring and supervising status on agricultural and SME loan. The Credit Information Bureau (CIB) online, expected to be introduced in October 2010, is likely to enable banks and financial institutions to access the database containing credit details of individual borrowers, owners and guarantors (Rahman, 2010).

Union Information Service Centres

To spread the online services at the grass root levels the local government division of the Ministry of Local Government, Rural Development, and Cooperatives has made a significant move for the development of information kiosks at union level—the lowest unit of local government. Known as the Union Information Service Centres (UICS) will be connected to information super highway by the year 2012 to provide information and online services. By the end of 2012 a total of 4501 UISCs have been established across all Union Parishads in the Country (A2I, 2010).

Table 1. Telecommunications infrastructures in South Asia

Country	Internet Index	PC Index	Cellular Index	Main Telephone Line Index	Broad band Index	Infrastructure Index
Bangladesh	0.003	0.027	0.085	0.008	0.000	0.0246
Bhutan	0.035	0.017	0.028	0.042	....	0.0244
India	0.061	0.017	0.095	0.038	0.007	0.0435
Maldives	0.075	0.164	0.578	0.133	0.049	0.1959
Nepal	0.010	0.005	0.022	0.022	0.000	0.0119
Pakistan	0.086	0.006	0.1430	0.034	0.001	0.0540
Sri Lanka	0.023	0.039	0.168	0.093	0.004	0.0656

Source: Compiled by authors based on UN (2008)

Development of IT Infrastructure

The Telecommunications Infrastructure Index is a composite index of five primary indices related to Internet users, PCs, main telephone lines, cellular telephone, and broad banding. Table 1 shows Bangladesh's relative position in South Asia in terms of telecommunications infrastructure development.

It is clear that Bangladesh's position in infrastructure is lower than that of India, Maldives, Pakistan, and Sri Lanka. In fact, it has the lowest index in Internet usage in the region, available to only 0.31% people. It appears that the overall infrastructure index is relatively better than countries like Bhutan and Nepal due to its improved weight in cellular phones.

As of 2008, Bangladesh's position in land phones and broad banding has been very fragile and weak. Only 0.72 per cent of people have land phones and broadband users are negligible (UN, 2008). One major initiative in this regard is the liberalisation of telecommunication industry in the 1990s, thus allowing the private operators to operate in fixed-line and mobile sectors. This has resulted in increased telecommunication networks and subscribers especially mobile phone subscribers. To speed up the Internet bandwidth, fibre optic cables have also been laid through many parts of the country. Bangladesh joined the Global Information Superhighway (submarine cable connectivity) in 2006 through the SEA-MEA-WE 4 consortium. As a result, BTTB has begun to reduce the charges for both the fibre optic and existing VSAT connections. Internet Service Providers (ISPs) have also increased to 200. Many of them are running or patronising approximately 600 cyber cafes around the country, 250 of them are in Dhaka City (DANIDA, 2006). Since broadband connections are mainly concentrated in cities, the establishment of a national bandwidth backbone to carry domestic Internet traffic is being discussed at the highest levels (Hoque, 2006).

Though with 2.42 PCs per 100 persons Personal Computer Index in Bangladesh is relatively better than that of its neighbours, it is still inadequate to address the requirements of e-government. Since the 1990s, computer accessories are either tax free or with very limited tax. Consequently, the PC/server market in Bangladesh is growing fast. The growth is further consolidated by the PC demands of large enterprises, multinational corporations and more importantly, government agencies that collectively account for almost half of total PC/servers that are imported (DANIDA, 2006). However, along with the low rate of computer penetration, the concentration of these consumers in cities and urban areas poses a challenge to the development of e-government across the country. A study by the Bangladesh Bureau of Statistics (BBS) and BCS showed that IT concentration was the highest in Dhaka—the capital city (72.76%), followed by Chittagong, Rajshahi, Khulna, Sylhet, and Barisal with small percentages for each (Hoque, 2006, p. 358).

As noted earlier, the success of e-government hinges largely on the establishment of an affordable and broadly accessible ICT infrastructure to deliver and use online services. With the recent steps as outlined above, clearly Bangladesh is making slow but important progress in terms of infrastructure to be able to provide e-services throughout the country. However, even if infrastructure is adequate and robust, technology itself is unlikely to make huge difference unless it is managed and used by people with sufficient knowledge in the field.

Development of Human Capital

Competent human resources are very important for e-government since they indicate both the citizens' preparedness to engage in transactions electronically as well as the capacity and skills of those who are involved in service provision. In the Human Capital Index measured by UN, Bangladesh's position is better than in the two other indices (i.e. the Web measure index and the infrastructure index) with an overall value of 0.5033. This value is lower than that of India (0.6195), Sri-Lanka (0.8137), Maldives (0.8617), and Nepal (0.5176), but slightly better than those of Pakistan (0.4659) and Bhutan (0.4867) (UN, 2008, pp. 279-281).

In view of its importance for the success of e-government the government in Bangladesh has initiated variety of measures seeking to develop human capital in the ICT sector. To ensure basic computer and Internet literacy the government's initiatives range from introducing 'Computer Science' subjects at school level, facilitating graduate courses at tertiary level to arranging computer-aided education for the general masses. To cater for the IT learning demands of younger generation, the government has designed the course curricula for both secondary and higher secondary levels with 'Computer Science' as an optional subject. It has also introduced computer equipments in many schools and colleges. The government is now planning to make Computer Science a compulsory subject. The government aside, the private sector and non-governmental organisations are launching projects to introduce computers and train the students in schools and other institutions. However, such efforts have had limited impact especially in rural areas due mainly to lack of electricity, poor location, unavailability of trained personnel, and low motivation and awareness (Hoque, 2006, p. 360). As the IT orientation is not quite widespread at the secondary level, the industry relies more on the tertiary level IT professionals. For developing leadership in change management and process reengineering of e-government, government officials are being trained. The BCC runs various programmes to indoctrinate government officials on IT (BCC, 2009). Recently, 800 government officials have received such training (Barkatullah, 2008). Other governmental ministries and agencies also provide trainings to government officials (Hoque, 2006).

At the tertiary level there are 73 universities in the country; these include some 21 public universities, most of which offer some form of IT courses and relevant degrees. Besides, there are 20 government and 87 private polytechnic institutes that provide diploma or vocational certificates in IT. There are another 1500 private ICT training institutions that provide IT training to the general public. Some agencies such as the Department of Youth, Department of Women's Affairs, and BCC also have their own ICT training programmes for citizens with facilities at the divisional and district levels (DANIDA, 2006, p. 23). These efforts have helped increase IT personnel in the country in recent times. According to the BCS reports, the total number of IT professionals in 2006 was more than 25,000. This represents an increase of 12.5% from the previous year (DANIDA, 2006, p. 16). These trends of increasing IT literacy will help to provide back office support for e-government, notwithstanding the fact that overall improvement in human capital index is essential for making the ordinary citizens aware and capable of using e-services.

ASSESSING E-GOVERNMENT IN BANGLADESH: CONSTRAINTS AND CHALLENGES

The foregoing discussions show that Bangladesh has made slow but steady progress towards the implementation of e-government. The government has undertaken and implemented wide variety of programs in areas like Web measures, infrastructure development, and even in terms of developing human capital—the ultimate operators as well as recipients of e-government services. The 2008 ranking for e-readiness of Bangladesh, as measured by the UN, is 142. This places it in the 6th position in the South Asian region, which remains far below the world average and is the lowest ranking position region in Asia. However, it must be noted that recently Bangladesh's position has improved considerably from 162 in 2005 to 142 in 2008. This improvement in the ranking is attributed to Bangladesh's progress in Web measurement in the enhanced and interactive stages and the growth of mobile telephones. Yet, compared to other nations of the world and many other countries in South Asia this ranking is abysmally low and far from the level required for providing online and integrated services to the citizens. It is nothing surprising that in a poor country like Bangladesh the pursuit of e-government and online service delivery has been hampered by a range of challenges and constraints. This section focuses on some of these challenges and limitations.

Web Related Constraints

Notwithstanding various governmental initiatives as outlined above presently e-government in Bangladesh is beset with a number of Web-related problems and challenges. First, Web initiatives in most government agencies can only be considered to be at the *'emerging presence level'* with very slow advancement towards other progressive levels such as interactive or connected stages. The UN Survey 2008 shows that 88% websites in Bangladesh are utilised for emerging presence, 42% for enhanced and interactive stages respectively, only 6% for transactional purposes, and 7% for connected stages. The overall utilisation is 31% across all five stages. The low percentages in transactional and connected stages indicate that most Web attempts are not for online services and transactions; rather they mostly present some static information. Second, the content of various websites has improved very little over time in providing services to the citizens. Hoque (2006) has found that out of 50 ministries and divisions only 28 have Web initiatives. Most of them contain very rudimentary forms of information such as 'About Us,' 'Contact Us,' 'Email Address,' etc. Features such as 'Forms,' 'Notices,' 'Publications,' 'Search,' and 'Privacy Policy' are very rare. Some websites display outdated contents and some sites even sometimes pop up with blank pages. Websites rarely have the options for interactive features that contain complaints, comments, suggestions, and compliments, statistical queries with a pull-down menu. Instead, the sites mostly provide detailed information on traditional bureaucracy such as 'organizational structure,' 'hierarchy,' 'activities,' 'achievements,' 'top personnel,' etc. Most ministries do not have the customised software; the purpose of the website is to serve only the internal operation of the ministry or division, the user groups being only the staff (Sobhan, et al., 2004). Thus, they cannot attract a wide range of users since very little attention is given to citizens' demands and choices. Besides, websites that provide services to citizens and businesses are not widely known to the public (Hoque, 2006, p. 360-361).

Third, online initiatives that provide important services such as e-procurement, e-taxation, registration and so forth are limited. Though Public Procurement Rules (PPR) 2006 require the government to implement e-procurement, it is only at the initial stage since the Central Procurement Technical Unit (CPTU) of the Ministry of Planning is expected to launch e-tendering system on a pilot basis in 2010 (MoP, 2009). Similarly, e-taxation is only made accessible for large taxpayers and corporate bodies leaving the vast majority of individual income tax payers out of such service. Other services such as vehicle registration, licensing, permits, and passports do not offer scope for online transactions. Fourth, integrated portals connecting and sharing various departments of the government and then delivering one-stop service to the citizens are hardly present. There are very little exchanges between different ministries and departments in the centre of administration - the secretariat. In fact, the current e-government drives in the Bangladesh Secretariat are confined mainly to 'planning and strategy formulation, connectivity, and infrastructure, procurement of technology, and website creation' (Hoque, 2006, p. 360).

Inadequate Infrastructure and Digital Divide

One of the most important hurdles to providing e-services in Bangladesh is inadequate infrastructure namely PCs, telephone lines (both fixed and mobile) and Internet especially broadband facilities. Although telephone connections—both fixed and mobile—have increased considerably compared to the past, it is still not sufficient to serve the whole country and the huge population. This is especially true for rural population and the vast majority of poor people who cannot afford the phones. The number of PCs has not increased much as currently only 2.42% of the population have PCs (UN, 2008). This low rate is further exacerbated by the concentration of PC users in cities and urban areas. Thus, it comes as no surprise that Bangladesh has performed poorly in the e-readiness index. In a country where per capita income is roughly US $500 (Gunter, 2008), it is very unlikely that the vast majority of people will be able to afford PCs and hence Bangladesh will see any dramatic improvement in this index in the near future.

Indeed, the record of South Asia as a region in general and Bangladesh in particular is hardly satisfactory. This is due to the fact that South Asia has the lowest ranking in Asia and even the country with the highest e-readiness score places itself below the world average. Bangladesh finds itself in the 6th position among seven countries in the region. Although Bangladesh's Web measure index has improved lately, infrastructure index and human capital index have remained poor compared to that of countries like India, Maldives, and Sri Lanka. It is clear from Table 2 that Bangladesh's position has remained not only far below the world average the nation has also performed poorly among most of the South Asian countries. Nepal is the only country in the region that is behind Bangladesh in terms of e-readiness index and rank.

More frustrating is that available infrastructure and technology are used for purposes other than providing online services. Although the Government of Bangladesh is a big consumer of PCs, they are mainly used for word processing purposes and clerical needs, rather than for providing on-line services to citizens and for improving governmental efficiency and responsiveness. Many officials seem content with their offices being equipped with computers and other accessories and rarely do they make effective use of those wares. In fact, it has been observed that computers in government offices are mainly used for word processing, personal e-mail correspondence, accounting, and Internet browsing (Hasan, 2003; Hoque, 2006). Similarly, although the number of mobile phones has increased significantly, they are not widely used for providing or accessing online services. Even more serious

Table 2. E-readiness index for South Asia

Country	Web Measure Index	Telecom. Infrastructure Index	Human Capital Index	E-government Readiness Index	Overall Ranking
Bangladesh	0.3512	0.0246	0.5033	0.2936	142
Bhutan	0.4012	0.0244	0.4867	0.3074	134
India	0.4783	0.0435	0.6195	0.3814	113
Maldives	0.2943	0.1959	0.8617	0.4491	96
Nepal	0.2876	0.0119	0.5176	0.2725	150
Pakistan	0..4247	0.0540	0.4659	0.3160	131
Sri Lanka	0.3946	0.0656	0.8137	0.4244	101

Source: Compiled by authors based on UN (2008)

problem is the massive digital divide that keeps the vast majority of the population virtually unaffected by the change. As noted earlier, there are only 2.42 PCs per 100 people and most of them are in large cities and towns. Tele-density remains low with 13.25 telephone subscribers (fixed and mobile phones combined) per 100 persons. The Internet diffusion rate remains extremely low, largely concentrated in major cities and urban areas often with very slow and technically defective VSAT line. Currently only 0.03 per 100 people have access fixed broadband connections (UN, 2008, 2010). The spread of broadband network countrywide through fibre optic cables has remained a goal. What all this suggests is that that the vast majority of the people are unable to reap the benefits of e-government, which is compounded further by problems of poverty, illiteracy, and widespread corruption in the society. Broadening and deepening public access to online services has remained a formidable challenge for the government.

Inadequate Human Capital

Far from being seen as a device with wider application and relevance, ICT in Bangladesh is still seen as a hardware and software industry operable only by technical experts. Lack of /inadequate literacy among the general masses and lack of efficiency among the government servants who are responsible for implementing e-government initiatives undermine governmental drives. It is ironical that despite being one of the most populous countries in the world, inadequate human capital has been among major problems of e-government in Bangladesh. First, lack of literacy poses a formidable challenge to the development of a competent user group in the country. Currently Bangladesh's Human Capital Index is only 0.5182 with adult literacy rate of 53.5% and the combined gross enrolment of primary, secondary, and tertiary schools being 48.46% (UN, 2010). This low level of literacy makes it difficult for nearly half of the population to understand any meaning of e-government. Second, since 'Computer Science' has been made optional at the secondary level many students do not choose the subject. In addition, though many projects are underway to provide IT in schools and colleges and other institutions for facilitating IT literacy among the students the use of those equipment, especially at rural areas, is far from optimal because of such setbacks as interrupted power supply, poor location, unavailability of trained personnel, low motivation and awareness.

Third, although the gross enrolment for IT education at the tertiary level is increasing, and a huge number of ICT graduates from different institutions are coming out every year, they cannot be absorbed into the employment market and this leads to mismatch between market demands and IT education (Hoque, 2006). Finally, lack of capacity among the policy makers and managers of e-government poses a significant challenge for effective implementation of e-initiatives. A study (Imran, et al., 2008) found that the most significant barrier to the adoption of e-government in Bangladesh is the lack of knowledge among the government decision makers (26%). This is followed by attitudinal problems (15%), lack of political will (13%), lack of planning and strategy (12%), and infrastructure (11%). This is corroborated by the fact that only a few ministries and government agencies are capable enough to handle the ICT in their work processes (Sobhan, et al., 2004). Finally, surpassing all this, resource constraints present the biggest problem for the management in designing and implementing significant e-government projects. As indicated, some of the successful e-government projects were financed by external donors. Due to termination of donor funding many e-government projects have been discontinued in the midway (Sobhan, et al., 2004). To tackle such constraints and problems the need for sufficient resource allocation through budgetary means is underscored. As such, an ICT Development Fund has been created through block grants in the annual budget to be accessed by public sector entities through a competitive process (GoB, 2009). However, this has yet to produce a sizeable amount that could fund major e-government programs.

CONCLUSION AND THE WAY FORWARD

It is obvious that e-government in Bangladesh is still in its infancy in many respects and despite some innovative schemes, there is a long way to go before the vision of e-government is translated into reality. It is also clear that Bangladesh's position is far below the world average and one of the lowest in South Asia. In the Web measure index, though the number of websites has increased in recent times, many of them are yet to advance beyond presence levels consequently failing to offer substantial scope for interaction, engagement, and transactions. Most websites contain some static information, outdated contents, and fewer options for citizens to involve in dealings. There are limited initiatives for the provision of online services; integration of websites to provide one-stop services has remained a vision. Bangladesh also lacks a robust infrastructure necessary for e-government. In the telephone sector, though the number of mobile subscribers has increased in recent times, it only covers approximately 30% of the population; mobile phones are not widely used to provide online services. Only 2.42% of the population can afford PCs with heavy concentration in urban areas. Internet accessibility is even much lower and, alarmingly, it is mostly concentrated in urban areas and, in many cases, with the slow and technically defective VSAT connection. Given this and other limitations, Bangladesh is lagging behind many developing countries especially those in the region in terms of e-readiness and online service provision, transaction and integration.

For a developing country like Bangladesh, challenges of e-government are manifold and there is hardly any quick fix available. Since e-government is an evolutionary process, one important imperative is the sustained drive to bolster e-initiatives and innovative programs building on the successes of and lessons from the existing programs. However, such attempts would make little difference if the vast majority of the population continues to remain untouched by such programs. Therefore, it is vital to bridge the digital divide. Since PCs and Internet connections are not affordable to most citizens, the alternative would be establishing Internet kiosks in small towns and rural areas for the provision of e-services. Establishing Internet kiosks for community access has been an effective model in low

Internet penetration countries such as Cambodia, India, Pakistan, and Sri Lanka. Bangladesh has a lot to learn from the experiences elsewhere. The kiosks can provide most traditional services of government currently being provided manually by the upazila administration such as those in agriculture, land records, social welfare benefits, public auctions, education, and health. NGOs that have a wider base in rural areas can help in the running and establishing such kiosks. Another alternative for PCs can be mobile phones where the SMS can be a viable method to render some services. Mobile network covers almost all parts of the country and it emerged as a potential solution to the fixed line telephone deficit in remote areas. So, it can be used to make information available regarding, weather forecast, disaster warning, payment of utility bills, and taxes of different types. Recent steps for making available the SSC and HSC results and payment of utility bills through mobile phones have earned wide appreciation. Mobile phones can also be used as modem for Internet transmission on a wider scale as many operators have that facility in rural areas.

Substantial improvement in the current situation would require massive investment and concerted efforts on part of the government to develop infrastructure and skills needed for successful e-government. However, one of the problems the government faces in this regard is the lack of financial resources. On the one hand, the government is unable to make massive investment required for the purpose and on the other hand, donor support is unpredictable and often for a limited in scale and period. Given this, one option that appears promising in scaling up e-government is the public-private partnership (PPP). Though this is by no means a panacea, PPP model appears to have worked well in similar contexts and seemingly, there is market appetite for such ventures. Obviously, it requires strong public service leadership to be able to forge effective partnerships to take the advantage of the information age. While there is a range of other imperatives, to our mind the most important requirement is the strategic leadership supported by unfettered commitment of the political and administrative elites to the tasks that lie ahead. Without such leadership and commitment, the vision of Digital Bangladesh is unlikely to be translated into reality.

REFERENCES

Akther, M. S., Onishi, T., & Kidokoro, T. (2007). E-government in a developing country: Citizen-centric approach for success. *International Journal of Electronic Governance*, *1*(1), 38–51. doi:10.1504/IJEG.2007.014342

Barkatullah, T. M. (2008). ICT human capacity building. In *Country Report: Bangladesh*. Dhaka, Bangladesh: Bangladesh Computer Council.

BCC. (2009). *About us*. Retrieved from http://www.bcc.net.bd/

BTRC. (2009). *About us*. Retrieved from http://www.btrc.gov.bd/

DANIDA. (2006). *Business opportunity study within the IT and telecommunication industry in Bangladesh*. Copenhagen, Denmark: Ministry of Foreign Affairs.

Government of Bangladesh. (2002). *National information and communication technology (ICT), policy 2002*. Dhaka, Bangladesh: Ministry of Science and Information and Communication Technology.

Government of Bangladesh. (2009). *National information and communication technology (ICT), policy 2009*. Dhaka, Bangladesh: Ministry of Science and Information and Communication Technology.

Gunter, B. G. (2008). Aid, debt and development in Bangladesh. *UNDP Policy Dialogue Series, 38*. Dhaka, Bangladesh: UNDP Bangladesh. Retrieved from http://www.bangladeshstudies.org

Haque, M. S. (2002). E-governance in India: Its impacts on relations among citizens, politicians and public servants. *International Review of Administrative Sciences*, *68*, 231–250. doi:10.1177/0020852302682005

Hasan, S. (2003). Introducing e-government in Bangladesh: Problems and prospects. *International Social Science Review*, *78*. Retrieved from www.questia.com/PM.qst?a=o&se=gglsc &d=5002081994

Hoque, S. M. S. (2006). E-governance in Bangladesh: A scrutiny from the citizens' perspective. In R. Ahmad (Ed.), *The Role of Public Administration in Building a Harmonious Society,* (pp. 346-365). Napsipag, Manila: Asian Development Bank.

Ministry of Planning. (2009). *e-Government procurement*. Retrieved from http://www.cptu.gov.bd/EProcurement.aspx

Ministry of Religious Affairs. (2009). *Hajj management information system*. Retrieved from http://www.bangladeshdir.com/webs/catalog/hajj_management_information_system__ministry_of_religious_affairs__bangladesh.html

NWPB. (2009). *Portal home*. Retrieved from http://www.bangladesh.gov.bd/

Planning Commission. (2009). *About planning commission.* Retrieved from http://www.plancomm.gov.bd/about.asp

PMO. (2009). *Government digitized forms online*. Retrieved from http://www.forms.gov.bd/

PMO. (2010). *Access to information (A2I) programme, quarterly progress report*. Retrieved from http://www.gov.bd

Rahman, A. (2010). *Digital Bangladesh bank*. Paper presented at the National Seminar on Vision 2021: Challenges for Engineering Profession. Chittagong, Bangladesh.

Sobhan, F., Shafiullah, M., Hossain, Z., & Chowdhury, M. (2004). *Study of e-government in Bangladesh*. Dhaka, Bangladesh: Bangladesh Enterprise Institute.

UN. (2008). *E-government survey- From e-government to connected governance*. New York, NY: Division for Public Administration and Development Management.

UN. (2010). *United Nations e-government survey 2010*. New York, NY: Department of Economic & Social Affairs.

West, D. M. (2008). *Improving technology utilisation in electronic government around the world, 2008*. Retrieved from http://www.brookings.edu/.../2008/0817_egovernment_west.aspx

Chapter 14
Combating Corruption through e-Governance in India

Durga Shanker Mishra
State Government of Uttar Pradesh, India

ABSTRACT

Studies have shown a prevalence of high level of corruption in the Indian Administrative System, which adversely affects the day-to-day lives of common citizens. This chapter examines the role of e-governance in combating corruption in delivering public services. Through a literature review assessing the outcomes of a few e-governance initiatives related to improving service delivery in different parts of India, this chapter argues that even though technology assists in instituting a transparent, accountable, consistent, reliable, and efficient system for delivery services, it cannot overcome corruption by itself. It will require political will, focused administrative strategy, business process reengineering for simplifying and opening up the system, and persistent efforts to ensure that corruption entrepreneurs do not subvert the gains of the technology.

INTRODUCTION

The poor and disadvantaged populations in developing countries are not able to make the best use of the public services effectively on account of lack of access due to social, economic, physical, informational, and other barriers; and inadequate mechanisms to provide their feedback on complaints/ views/ requests to the service providers and policy makers (Shadrach & Ekeanyawu, 2003, p. 2). Quoting a study in Bangalore city, they indicate a prevalence of widespread corruption in delivery of services by public agencies. Srinivas and Nayar (2007, p. 20) cite a survey of the Bolangir district of Orissa, stating that 63% of respondents expressed that government welfare schemes are riddled with corruption. They refer Harish C. Saxena, Member, National Advisory Council, saying that weak governance resulting in poor service delivery, excessive regulation, and wasteful expenditure are factors impinging on social indicators. Based on a detailed field study

DOI: 10.4018/978-1-4666-1909-8.ch014

of petty corruption in India, a Centre for Media Studies Report (2006, p. 1) states, "*Despite a reduction in reporting of corruption in 2005, a large cross section of households had to pay bribes to avail public services in 2005. In case of five public services (Police, Land Administration, Judiciary, Electricity, and Government Hospitals) covered in the CMS corruption study, more than 10 million households had paid bribes during the year for availing services.*"

Addressing an official ceremony, Dr. Manmohan Singh, Prime Minister of India, pointed out that corruption is a social cancer eating away vitals of institutions of governance and the society and is a threat to national well-being. Reiterating national resolve for providing corruption-free, transparent, accountable, responsible, and responsive governance system, he urged for zero tolerance to corruption and a multi-pronged approach to stem the rot (CBI Bulletin, 2006, p. 11). Following the recommendations in World Bank (1997, pp. 157-158) and World Bank (1998, pp. 85-93), the Bank and other donors insist on good governance practices in the reforms agenda. However, Demmers et al. (2004, p. 10) quote World Development Reports 2000/2001 and 2002 in emphasizing that there is no blueprint for good governance. Considering the discourse on good governance, Dwivedi and Mishra (2007, p. 705) summarize transparency, rule of law, accountability, incorruptibility, sensitivity, and ethical behaviour as key factors that determine good governance. Critical among these is designing the systems for corruption-free delivery that requires all other factors to be automatically entwined as necessary elements.

CMS Report (2006, pp. 25-28) suggests e-governance as a possible strategy to safeguard citizens from corruption in day-to-day dealings with public agencies and emphasizes that combating corruption is vital to sustaining good governance. This survey corroborates World Bank (2003, pp. 195-196) findings that the poor are the worst affected by petty corruption in delivery of services. Various initiatives (including e-governance) have been attempted in different parts of India (and elsewhere) for improving the delivery system, especially making it pro-poor, who have the least voice and virtually no options to exit in view of financial implications. World Bank (2006) examines 25 such innovative cases in service delivery across India. In this chapter, the author scrutinizes relevant cases from this report, other published materials, and his own experience in working over two and half decades as a member of the Indian Administrative Service (IAS), to analyze whether adopting the strategy of e-governance in delivery of public services would bring good governance? Eradicating corruption is the crux of good governance and therefore, the chapter focuses on this specific issue in depth.

Most of the studies on public services reforms involving e-governance address the issue of improvement in efficiency and public convenience and are generally silent on the eradication of corruption because it is invisible in collusive sharing of benefits by the giver and the taker; its disclosure by anyone would attract legal action and generally there is no base-line for comparison. Appreciating these limitations, the author triangulates available information regarding specific cases with his own experience, in analyzing the role of e-governance in improving transparency, accountability, consistency, and such other factors, which limit the opportunities for corruption. He argues that e-governance may reduce corruption provided there is political support, focused administrative leadership, business process reengineering, sustained effort, and other conditions, as have been discussed in the chapter.

The chapter has five sections. After the introduction, the second section discusses the theoretical perspective and prepares a framework for analysis. Section three reviews the literature on the matter with specific focus on relevant cases and personal experiences of the author to draw correlation between e-governance strategy and combating corruption from public sector delivery system. The available information is analytically assessed in the fourth section for drawing inferences. The final section summarizes the findings.

THEORETICAL PERSPECTIVE

The role of state (federal, state, or local governments) in providing public services is still quite significant even after two decades of liberalization, deregulation, and privatization, which commenced in the early 1990s in India. State has major responsibility to protect the interest of the poor and vulnerable sections of the society, who may not have means to receive services at competitive rates from the market. The author feels that the advantages of privatization in terms of cost reduction and innovations put forward by Shleifer (1998, p. 141) do not hold good for this category of citizens because of their low capacity to pay, geographical scatter, and unorganized voice. Such reforms may heighten the gap between poor and rich (or elites) leading to alienation of this population. Thus, such attempts need to be given due care to ensure equitable distribution of services at affordable rates, especially to the lower economic strata.

CMS Report (2006, pp. 1-4) points out high prevalence of corruption in providing basic services to the urban/ rural population. 62% of respondents had first-hand experiences of paying bribes or using contacts to get their jobs done in public offices. This study on tracking corruption in eleven public services extrapolates that 145.4 million households paid bribes to the tune of Rs 210.68 billion during the year 2005 (CMS, 2006, pp. 13-14). This means that an average of nearly Rs 1,500/year was paid as an extra undue money by the households in receiving such services through the year. N. Vittal (1999, p. 2), a former Central Vigilance Commissioner of India, states that in the post-independence era, initially individuals corrupted the institutions but later the corruption has become institutionalized and inevitable culture in public life.

In a study of metropolitan water supply and sanitation of India, Davis (2004, pp. 56-61) mentions five major areas of corruption: speed money for seeking new connections/ getting repairs, collusive action for illegal connection or falsifying bills, commissions in award of contracts for works and services, kickbacks in execution of works/ procurement of equipments, and transfer of personnel to plump positions. More or less, this is true in delivery of other public services. CMS Report (2006, p. 25) envisages three broad situations for bribery/ extortion in delivery of public services; they are when the customer is unaware of supply position or his right and so there is scope to create artificial scarcity and demand price for the provision of service, there is actual scarcity and official has discretion in allocation and so he/ she can claim premium, and for speeding of the delivery by facilitating queue jumping. Dwivedi and Mishra (2007, p. 724) give a mathematical construct of corruption, which is Corruption = Discretion + Mystification – Accountability; i.e. corruption increases with increase in discretion/ mystification or decrease in accountability and vice versa. Illustrating from delivery of health services, Savedoff and Hussmann (2006, p. 13) infer that asymmetric information (i.e. lack of transparency), high discretion to service deliverers and lack of accountability make the system vulnerable to corruption.

E-government means usage of Information and Communication Technology (ICT), i.e. Local Area Networks (LAN), Wide Area Networks (WAN), Internet, mobile, and computing technology to transform government by making it more accessible, effective, and accountable by providing access to requisite information, enabling public to interact with officials, making operations transparent, and offering public services online. UNDP (2004, p. 2) refers e-governance as multi-faceted use of ICT for improving collective governance that includes making delivery of services more accessible, efficient and responsive. Prabhu (2006, p. 1) conceives it as a form of e-business in governance comprising of processes and structures involved in deliverance of services and allowing interactions electronically. Finger and Pecoud (2003, p. 5) envisions role of e-governance in public services in delivering, monitoring and policy making by three steps, viz. *e-government,* i.e. providing better quality and efficient delivery, may be in partnership with private sector or civil society organization, *e-regulation,* i.e. regulating price, quality, and accessibility, and *e-democracy,* i.e. making policy by taking feedback of public and other stakeholders.

From the above, one may infer that implementation of e-governance in delivery of public services would automatically curb corruption in view of the following:

- There is *no direct contact* between providers and receivers, thereby breaking the chain that offers the scope for negotiating 'I (receiver) lose less and you (provider) gain some (bribe)' or extortion by public officials;
- *Discretion is significantly curtailed* by standardization of decision-making processes, which are normally performed by computer itself;
- *Improved transparency and demystification* of processes and procedures by providing access to all relevant information online;
- *Increased accountability* by having possibility of detecting wrong-doing (queue jumping/ favouring/ disfavouring) or non-doing through computerized trail;
- *Citizens' empowerment* by allowing them to get services at any kiosks or at home (*choice*), checking status on Web through allotted reference number (*public monitoring*) and giving feedback on concerns/ complaints to higher authorities (*voice*); and
- *Efficient computerized monitoring* by higher authorities regarding specific action undertaken by cutting-edge level service providers.

THE EXPERIENCE

In a real life situation, any transformation creates winners and losers, and therefore, a participatory strategy taking care of various reactions, nuisances, implications, and road bumps rather than straight linear approach could be more effective in achieving the objectives (Grindle & Thomas, 1991, pp. 149-150). Initially, there were big celebrations and laudable claims on e-governance initiatives in India and public thought that this would usher a new era of corruption-free good governance. Dey (2000, p. 305) writes:

"Common man felt that he would benefit immensely from good governance via e-governance (which is nothing if not citizen-friendly) and opts for it; the government itself is faltering- walking and talking haltingly. People have long been a harassed lot in their relationship with the government- proverbial 'babudam' with endless forms, regulations, by-laws, paper- work, delays, secrecy, authoritarianism and negativism- and they would like to be out of all these nightmarish experiences. They would not take this anymore and therefore, demand for 'good governance,' slogan for 'paperless office' and cry for technology as if e-governance provides panacea."

While eradicating corruption had been on the reform agenda of developing countries since early 1990s, Davis (2004, p. 54) asserts that very few e-governance initiatives directly confronted the issues of transparency and corruption with stated goal of reducing rent-seeking behavior of civil servants in delivery of services. World Bank (2006, p. 17) states that even though e-Seva was initiated on instance of the then Chief Minister (CM) of Andhra Pradesh Chandrababu Naidu and became a successful model in delivery of public services with his active support, tactically neither CM nor civil servants claimed e-Seva as an attempt to curb corruption, instead it was projected as a programme to enhance citizens' convenience and reduce drudgery of public servants. Even Administrative Staff College of India (ASCI), Hyderabad, which surveyed impact of e-Seva in 2002, was silent on this issue. Assessment of few other popular e-governance initiatives in India from the published literature is summarized below:

- *Bhoomi: Computerization of land records in Karnataka:* The project concerning computerization of nearly 20 million land records of 6.7 million farmers in the state for offering online services for issuing record of Rights, Tenancy, and Crops (RTC) and getting mutation of land done through 177 public kiosks, took nearly 7 years to stabilize by setting up a state-wide networked hardware system, developing software, entering data, and getting those verified by revenue staff and concerned land holders, making necessary statutory amendments, organizing manpower and preparing stakeholders for the change. Survey conducted by Public Affair Centre, Bangalore in July 2002, quoted by World Bank (2006, pp. 21-22), Bhatnagar and Chawla (2004, pp. 12-13), and UNDP (2004, p. 5) disclose that 79% users found availing services under Bhoomi through kiosks operators to be simple, 74% could get error-free records, and 93% of detected errors could be rectified instantly; 72% got records in very first visit (hardly 5% were so fortunate in manual system), only 3% had to pay bribes as against 66% earlier, and average bribe per user reduced from Rs 152.46 earlier to Rs 3.09. Bhoomi empowered farmers, who can deal with kiosks operators without any fear as customers by paying nominal fee of Rs 15. First-cum-first-serve principle in mutation curbed opportunity for queue-jumping and computerized monitoring for time-bound delivery eliminated scope for arbitrary corrupt practices.
- *e-Governance of Stamps and Registration in Maharashtra and Andhra Pradesh:* Stamps and registration departments, which were considered as one of the most corrupt delivery systems in these states, were computerized. Literature shows that while sustained efforts were made in Maharashtra involving reengineering of the processes and systems followed by e-governance initiatives with laid down service standards for reward or penalty associated with compliance of norms, in Andhra Pradesh only processes were *ITized*[2] leaving scope for touts to exploit the system. Quoting report of Caseley's survey in 2005, World Bank (2006, pp. 30-31) confirms significant reduction in corruption in registration of documents in Maharashtra. 93.5% of respondents mentioned that they did not pay any '*extra*' fees, only 6% said that they had to pay bribes. 100% claimed that registration was done within 30 minutes and most of them did not take help of any touts, showing a significant turnaround from corrupt practices. However, Caseley (2004, pp. 1152-1155) and World Bank (2006, pp. 25-26) mention that Andhra Pradesh initiative did not lead to reduction of corruption in registration of documents. Empirically Caseley estimates annual payment of bribes in two computerized sub-registration offices to be far more than non-computerized offices. The data show that speedier services made possible through application of IT increased corruption. Caseley explains that this was due to ICT reform was top-down with no involvement of stakeholders, lack of political will to eliminate the role of Deed-Writers (DWs), who were the *corruption*

entrepreneurs[3], as has been conceptualized by Granovetter (2004, p. 15), short tenure of reform champion, and possible benefits of corruption being shared by higher ups in the department.

- *Gujarat Computerized Inter-State Check Posts:* System for collection of taxes in ten inter-state border check posts was computerized. Referring to earlier survey of Indian Institute of Management, Ahmadabad (IIMA) and its own study in February 2005, World Bank (2006, pp. 23-25) points out that this project experienced bumpy road-map. Initially introduction of new procedures and technology led to increased revenue but with the transfer of reform champion and quick successive changes of four transport commissioners within 18 months, check post inspectors, i.e. *corruption entrepreneurs*, got opportunity to subvert the system on the pretext of machines not working properly and old practices of corruption and harassment unraveled. However, the state could ill-afford the huge loss in revenue the reform had shown and therefore, the new stable leadership subsequently rectified the subversion. Report quotes that revenue increased from Rs 560 million in 1998-99 to Rs 2,910 million in 2003-04.
- *e-Governance in Hyderabad Metropolitan Water Supply and Sewage Board (HMWSSB):* With active support of the government and media, reform champion took various bold steps in improving service delivery. This included setting up Single Window Cell (SWC) to receive applications for new water/ sewage connections and process them in time-bound manner (against earlier system wherein customer had to chase 'no objection certificates' with 14 offices, each being potential corruption points), providing 24x7 telephone helpline called Metro Customer Care (MCC) for receiving complaints and coordinating expeditious actions, launching Customer Redressal Efficiency System (CRES) software for monitoring performance data against laid down standards under citizens charter and therefore, holding frontline and middle level accountable for service quality and compliance. Davis (2004, p. 263) estimates that these measures decreased payment of average bribe of US$ 22 earlier to US$ 1 per case, which is not a bribe but a token of appreciation. Caseley (2006, p. 531) concludes that the initiative succeeded in curbing corruption by triangulating accountability relationship between citizens, senior managers, and frontline workers. Studies have shown that significant improvement in transparency and accountability, media participation and substantial reduction in discretionary powers of employees, led to reduction in corruption.
- *Jhansi Jan Suvidha Kendra (JJSK):* This is a recent e-Governance initiative undertaken by the district administration in Jhansi to help bring the government closer to the citizen. It is completely new approach in the arena of public grievance redressal through effective application of computer and Web technology along with mobile, and landline phones. As soon as the complainant calls on the toll free number, the grievances are recorded automatically on an audio file and stored in a software program. The complainant gets a Unique Grievance Number (UGN) and the concerned officers are intimated on their mobile phones via SMS instantly. Grievances are categorized as one of A/B/C, according to the severity and the time within which they should be resolved. Various reports are generated by the software for weekly monitoring and the quality of redressal/ disposal, which is monitored by the District Magistrate, the

administrative head of the district. Reports are continuously uploaded on the website and the concerned person can check the status of the complaint anytime by just calling JJSK or checking the status online. The redressal of grievance by public authority is intimated to the complainant in due time and is checked through SMS/phone call. This initiative is yet to be assessed independently and reviewed in the published literature. However, the author has interacted with the stakeholders and got quite positive response in terms of time-bound and transparent resolution of public grievances and thereby reduction in the scope for corruption.

- *Overseas Citizenship of India (OCI):* The author was involved in planning, developing, testing, and operationalizing this e-Governance initiative in the Ministry of Home affairs, Government of India, while he was posted as Joint Secretary (Foreigners). With the statutory amendment for creating a new class of Oversees Citizens of India from among the Persons of Indian Origin (PIOs), who had migrated after independence to other countries in search of educational, commercial, business, employment, or any other pursuits and taken up the citizenship of those countries, a Web based secured system was developed to enable such persons to apply for OCI online from any part of the world at any time 24x7x365. Applications are processed electronically giving access to the latest position to the applicant through his unique computer generated reference number. Applicant is delivered OCI Visa and Certificate of Registration through courier service within a month from the date of receipt of relevant documents and fees by the concerned offices, which is electronically acknowledged. So far, over 600,000 PIOs have received OCI visa and certificates with high level of satisfaction of the customers and no complaints regarding corruption or harassment.

ANALYSIS AND INFERENCES

Cases discussed in the previous section, literature review, and personal experience of the author point out a positive correlation between implementation of e-governance in delivery of public services and curbing corruption. However, effectiveness of this relationship has been found to be dependent on the following factors:

1. **Leadership Support:** This is the most crucial element, which is evident from the different experiences of registration and stamps departments of Maharashtra and Andhra Pradesh.
2. **Continuity of Reform Champion:** Successes of e-registration in Maharashtra, HMWSSB, Bhoomi, JJSK, and OCI initiatives, uneven journey of e-check-post in Gujarat, and failure of e-registration in Andhra Pradesh make this inference quite clear.
3. **Process Reform:** Konana (2007) says 'don't automate, obliterate,' i.e. eliminate the complexity in the processes. Dey (2000, p. 304) argues that hardware and software constitute only 10-15% of the problem, remaining over 85% is organizational management problem, which is internal to the government. These viewpoints covey the importance of Business Process Reengineering (BPR) in making systems simple and clear before any digitization. One needs to understand delivery system design thoroughly to eliminate opportunities for bribe or extortion. Exercises undertaken by HMWSSB initiatives, e-registration in Maharashtra, Bhoomi, and OCI initiatives corroborate this assertion.

4. **Careful Systems Design:** HMWSSB initiatives, e-registration in Maharashtra, e-Seva, e-check-post in Gujarat, JJSK, and OCI initiatives confirm that design of delivery points should take care that the transactions are in public view, citizens have choice to avail the facility from any centre, they can check the status anywhere at any time, and can voice their complaints without fear of being reprimanded on the face by the authorities. Briefly, system design should improve transparency and empower citizens.
5. **Citizen-centric Approach:** Saxena (2005, p. 512) concludes that citizen-centric rather than techno-centric approach in e-governance initiatives is required for achieving desired outcomes. A background paper prepared by Department of Administrative Reforms, Pensions, and Grievances (DARPG) of Government of India for Chief Secretaries Conference in 2006 advocates this approach (DARPG, 2007, pp. 3-4). This is evident from the successful outcomes of Bhoomi, HMWSSB and e-registration in Maharashtra initiatives.
6. **Clear Service Standards:** Bhatnagar (2003, p. 28) contends that corruption reflects the power distance between civil servants and the public and would remain even after introduction of e-governance. Therefore, service norms and delivery standards should be very clearly spelled out to ascertain accountability. This can be done through the citizens' charter as was done in HMWSSB case, or by fixing up standards for reward or punishment to the provider like e-registration in Maharashtra.
7. **Centralizing and decentralizing together for different purposes:** Technology enables provision of services in a decentralized way with various networked counters providing wider choice for customers and at the same time monitoring delivery in centralized manner by higher-level authority (Brown, 2005, p. 251). HMWSSB initiatives, Bhoomi, and e-registration in Maharashtra evolve this point noticeably.
8. **Convergence of Services:** For making the initiative sustainable, it should offer basket of services relevant to the citizens as one-stop-shop. The choice of services can be ascertained through Participatory Rapid Appraisals (PRA) or any other survey instruments. Heeks (2001, p. 71) confirms that PRA was used in ICT design of Gyandoot[4]. e-Seva is another successful initiative with host of services. Convergence improves viability of the model in the long term.
9. **Sustained Effort:** Initiatives in e-governance require sustained effort or else the losers of the e-governance initiative, especially *corruption entrepreneurs*, as in case in e-check-post in Gujarat or CARD[5] thwart the system. Such persons are looking for opportunities to use excuses to fail initiatives in order to revert to traditional system to allow their corrupt practices.
10. **Public-Private Partnership:** Public-private partnership helps in mobilizing additional resources, i.e. finance, manpower, and technology, and thereby reducing initial burden on the government, shielding civil servants from public and therefore, offering more flat organization for delivery wherein citizens can claim services on equal-footing and, inculcating business culture for getting value for money (user fees). Initiatives of e-registration in Maharashtra, e-Seva, Gyandoot, e-check-post in Gujarat confirm this.
11. **Change in Mind Set:** DARPG (2006, p. 11) argues in favour of paradigm shift from 'lack of trust,' the colonial legacy to 'trust in citizens' in reengineering of public services. Climate of trust and confidence changes governance style and ICT provides adequate powers to identify the rule-breakers. Further, civil servants need to be sensitized with reali-

ties of poverty and ground-level problems to develop 'sense of calling,' as were done in HMWSSB case.

12. **Support of Media and Civil Society:** This is quite critical in anti-corruption e-Governance initiatives. Harris and Rajora (2006) feel the need for improving awareness and participation of civil society in e-governance initiatives for increasing the benefits. Davis (2004, pp. 66-67) and Caseley (2006, pp. 538-542) spell out the active role of media and NGOs in successful anti-corruption and efficient service delivery by HMWSSB.

From the above analysis, viability of e-governance strategy in eradicating corruption from delivery of public services may be questioned due to conditions mentioned above for the initiatives to get desired success. The foregoing discussion infers that e-Governance does not provide a panacea but is a vital and powerful tool which can multiply the coverage (decentralized outreach) and control (centralized monitoring), increase efficiency, enhance transparency, augment reliability and consistency, ensure accountability, slash discretion and therefore arbitrariness in decision-making, and make services available at affordable prices. However, there are many organizational roadblocks, the most important is the complexity of delivery systems, which needs to be reengineered for simplification with citizen-centric approach and various others as have been mentioned in the previous paragraph. *Corruption entrepreneurs* are looking for opportunities for subverting the initiative to serve their personal interests and that needs to be strictly guarded by the IT initiative champions.

Haque (2002, p. 246) argues about the socio-economic 'digital divide' in India, by which he means that a large portion of population does not have access to computers and Internet. Added to this, there are issues of illiteracy, poor or no knowledge of English, and intermittent power availability in rural areas, where large number of users live. He, therefore, concludes that this instrument will serve the interest of only elites and add to disparity. Mehdi (2005, p. 484) mentions negative implications of 'digital divide' and not being 'e-ready.' Prabhu (2006, p. 4) states that public policy should address this significant constraint in implementation of e-governance initiatives. All these arguments are partially right. Quoting Civil Service Chronicle, Pani, Misra, and Sahu (2004, p. 207) mention that there are only 1.65% Internet users in India as against 8.96% global average, 2.65% in China and 58.5% in USA. There were only 11 computers per 1000 population in India in 2004 as compared to global average of 27 per 1000 and 500 per 1000 in USA. However, Vision for Eleventh Plan (2007-2012) envisages achieving 65 computers and 40 Internet subscribers per 1000 population by end of the Plan (Working Group on IT for Eleventh Plan, 2007).

These statistics show the large gap in availability of computer and Internet facilities to the common people, especially in rural areas. However, the author considers the 'digital divide' argument to be partially right because it amounts to fundamental chicken and egg dilemma—that is which one came first? Should the country wait for large-scale per capita computer and Internet availability for initiating e-governance drives or use appropriate strategy to harness the strength of technology for the benefit of the citizens? Author favours second approach and analysis in the chapter supports this claim. Public kiosks can multiply the outreach and various other reforms may help connecting people to new services. Bhoomi or Gyandoot is fully rural based projects, whose successes have been acclaimed internationally. Computerization of railways reservation system in India is a vibrant example of ICT initiative for driving out corruption from countrywide network. Therefore, on the basis of his own first-hand experiences in governments (Federal and State) and foregoing discussions, author endorses Backus (2001, p. 4) model for implementation of ICT initiatives with 'start small (*tactics*), scale fast (*strategy*), and think big (*vision*)' for knocking down the ghost of corruption from public services delivery systems. The country could then gradually move from e-operations to e-regulations and e-democracy, as per Finger and Pecoud (2003, p. 6) model discussed earlier, in providing good governance through e-governance.

CONCLUSION

This chapter addressed the issue of combating corruption from systems involved in delivering public services. Though e-Governance offers a viable strategy to reduce corruption from delivery system, yet it is not a panacea. It assists change and creates efficient, consistent, transparent, and accountable system, with no scope for arbitrary discretion, and empowers people by facilitating them to check the status online from anywhere and give feedbacks/ complaints without fearing wrath of public servants (Misra, 2005, pp. 290-292). However, it cannot solve all problems of corruption and overcome barriers for civic engagement. Analysis shows certain conditions to be fulfilled for making the initiatives succeed in curbing corruption. It has been observed that e-governance is a process that requires planning, sustained dedication of resources, and strong political will to eradicate corruption. This does not happen just by purchasing computers or other hardware and launching an informative static website. Continued interactive effort with clear long term vision to provide corruption-free good governance in delivery of services and strategy for scaling up successes, with citizen-centric business process reengineering and changed paradigm of trust can make the difference. Fourth Report (2007, p. 141) of Second Administrative Reforms Commission mentions that if the storage, retrieval, processing and transmitting power of IT is properly harnessed, it can make governmental processes more transparent and objective and so reduce the scope for corruption. However, before IT is invoked, it is necessary that existing procedures are properly reengineered and made computer adaptable.

The author thinks that there is a research gap in establishing a direct link between e-governance strategies and anti-corruption outcomes. With large scale investments in e-governance on so many initiatives, more empirical studies need to be undertaken on the subject to find out reaction and behaviour of possible losers, i.e. *corruption entrepreneurs* and those who are happy paying the bribes for getting the services swiftly, and to understand the impact of the initiative in curbing corruption from delivery of public services to give better insight into this issue.

REFERENCES

Backus, M. (2001). *E-governance in developing countries*. Research Brief No. 1. New Delhi, India: International Institute for Communication and Development (IICD).

Bhatnagar, S. (2003). E-governance and access to information . In Hodess, R., Inowlocki, T., & Wolfe, T. (Eds.), *Transparency International Global Corruption Report 2003: Special Focus Access to Information* (p. 28). London, UK: Pluto Press.

Bhatnagar, S., & Chawla, R. (2004). *Online delivery of land titles to rural farmers in Karnataka, India*. Paper presented in 'Scaling up Poverty reduction: A Global Learning Process and Conference'. Shanghai, China.

Brown, D. (2005). Electronic government and public administration. *International Review of Administrative Sciences*, *71*(2), 251. doi:10.1177/0020852305053883

Bulletin, C. B. I. (2006). *Prime minister Dr Manmohan Singh delivering speech on the occasion of foundation stone laying ceremony*. New Delhi, India: Government Publication.

Caseley, J. (2004, March 13). Public sector reform and corruption: CARD façade in Andhra Pradesh. *Economic and Political Weekly*, 1152–1155.

Caseley, J. (2006). Multiple accountability relationships and improved service delivery performance in Hyderabad City, South India. *International Review of Administrative Sciences*, *72*(4), 531. doi:10.1177/0020852306070082

DARPG. (2006). *Background note: Improving public service delivery – Management aspects and technology applications*, (pp. 3-4). Retrieved June 8, 2007 from http://darpg.nic.in/arpg-website/conference/chiefsecyconf/csc.doc

Davis, J. (2004). Corruption in public service delivery: Experience from South Asia water and sanitation sector. *World Development*, *32*(1), 56–61. doi:10.1016/j.worlddev.2003.07.003

Demmers, J., Jilberto, A. E. F., & Hogenboom, B. (2004). Good governance and democracy in a world of neoliberal regimes . In Demmers, J., Jilberto, A. E. F., & Hogenboom, B. (Eds.), *Good Governance in the Era of Global Neoliberalism: Conflict and De-Politicization in Latin America, Eastern Europe and Africa* (pp. 1–19). London, UK: Routledge. doi:10.4324/9780203478691_chapter_1

Dey, B. K. (2000). Challenges and opportunities-A future vision. *The Indian Journal of Public Administration*, *46*(3), 305.

Dwivedi, O. P., & Mishra, D. S. (2007). Good governance: A model for India . In Farazmand, A., & Pinkowski, J. (Eds.), *Handbook of Globalization, Governance and Public Administration* (pp. 701–741). New York, NY: Taylor & Francis.

Finger, M., & Pecoud, G. (2003). From e-government to e-governance? Towards a model of e-governance. *Electronic . Journal of E-Government*, *1*(1), 5.

Government of India. (2007). Fourth report: Second administrative reforms commission. In *Ethics in Governance*, (p. 141). Retrieved August 9, 2007 from http://arc.gov.in

Granovetter, M. (2004). *The social construction of corruption*. Paper presented at the Conference 'The Norms, Beliefs, and Institutions of the 21st Century Capitalism: Celebrating the 100th Anniversary of Max Weber's the Protestant Ethic and Spirit of Capitalism'. Ithaca, NY. Retrieved May 25, 2007 from http://www.stanford.edu/dept/soc/people/faculty/granovetter/documents/The%20Social%20Construction%20of%20Corruption%20June%2005.doc

Grindle, M. S., & Thomas, J. W. (1991). Implementing reform: Arenas, stakes and resources . In *Public Choice and Policy Change: The Political Economy of Reform in Developing Countries* (pp. 149–150). Baltimore, MD: Johns Hopkins University Press.

Haque, M. S. (2002). E-governance in India: Impact on relations among citizens, politicians and public servants. *International Review of Administrative Sciences*, *68*(2), 246. doi:10.1177/0020852302682005

Harris, R., & Rajora, R. (2006). *Empowering the poor: Information and communication technology for governance and poverty reduction- Study of rural development projects in India, a UNDP-APDIP publication*. New Delhi, India: Elsevier Press. Retrieved May 22, 2007 from http://www.apdip.net/publications/ict4d/EmpoweringThePoor.pdf

Heeks, R. (2001). *Understanding e governance for development*. Manchester, UK: University of Manchester. Retrieved June 4, 2007 from http://unpan1.un.org/intradoc/groups/public/documents/NISPAcee/UNPAN015484.pdf

Konana, P. (2007, January 29). Can IT-enabled services lower corruption?. *The Hindu, Thiruvanathapuram Edition*.

Mehdi, A. (2005). Digital government and its effectiveness in public management reform. *Public Management Review*, *7*(3), 484.

Misra, S. (2005). E-governance: Responsive and transparent service delivery mechanism . In Singh, A. (Ed.), *Administrative Reforms: Towards Sustainable Practices* (pp. 290–292). New Delhi, India: Sage Publications.

Pani, N., Misra, S. S., & Sahu, B. S. (2004). *Modern system of governance: Good governance v/s e-governance*. New Delhi, India: Anmol Publications.

Prabhu, C. S. R. (2006). *E-governance: Concepts & case studies*. New Delhi, India: Prentice Hall of India.

Report, C. M. S. (2006). *Tracking corruption in India-2005: Towards sustaining good governance*. New Delhi, India: Centre for Media Studies. Retrieved from http://www.cmsindia.org

Savedoff, W. D., & Hussmann, K. (2006). Why are health systems prone to corruption? In *Transparency International Global Corruption Report 2006: Special Focus Corruption and Health* (p. 13). London, UK: Pluto Press.

Saxena, K. B. C. (2005). Towards excellence in e-governance. *International Journal of Public Sector Management*, *18*(6), 512. doi:10.1108/09513550510616733

Shadrach, B., & Ekeanyawu, L. (2003). *Improving the transparency, quality and effectiveness of pro-poor public services using ICTs: An attempt by transparency international*. Paper presented in the 11th International Anti-corruption Conference. Seoul, South Korea.

Shleifer, A. (1998). State versus private ownership. *Journal o f Economic Perspectives, 12*(4), 141.

Srinivas, A., & Nayar, L. (2007). Elephant must remember: Growth story, yes, but there are over 300 million yet to see it. *Outlook, the Weekly News Magazine, 47*(15), 20-22.

UNDP. (2004). E-governance. *Essentials*, *15*, 2.

Vittal, N. (1999). *Applying zero tolerance to corruption*. Retrieved June 28, 2007 from http://www.cvc.nic.in/vscvc/note.html

Working Group on IT for Eleventh Plan Report. (2007). *Recommendations of working group on eleventh five year plan 2007 - 12: Information technology sector*. Retrieved June 7, 2007 from http://planningcommission.nic.in/aboutus/committee/wrkgrp11/wg11_IT.pdf

World Bank. (1997). *World development report: The state in a changing world*. Oxford, UK: Oxford University Press.

World Bank. (1998). *Assessing aid: What works, what doesn't, and why*. Oxford, UK: Oxford University Press.

World Bank. (2003). *World development report 2004: Making services work for poor people*. Washington, DC: The World Bank.

World Bank. (2006). *Reforming public services in India: Drawing lessons from success*. New Delhi, India: Sage Publications India.

ENDNOTES

1. The opinion expressed in the chapter is that of the author and does not reflect the views of the Government of Uttar Pradesh, where the author is currently posted as Principal Secretary to the Hon'ble Chief Minister.
2. Expression used for application of information technology.
3. Master manipulators of social networks and influences who make personal gains at the cost of the state.
4. Internet connected kiosks for providing multiple services to farmers in villages of Dhar district in Madhya Pradesh.
5. Computer aided administration of registration department of Andhra Pradesh.

Chapter 15
Grey Hair, Grey Matter, and ICT Policy in the Global South: The Ghana Case

Lloyd G. A. Amoah
Ashesi University, Ghana

ABSTRACT

By exploring the case of Ghana, this chapter examines the often cited linkages between good governance, ICTs, and development in developing societies. Though some significant ICT-related infrastructural development projects have been undertaken in Africa, the empirics indicate that the region, compared to other regions, such as Asia, has yet to experience the magic expected. Using an e-government project at the presidency in Ghana as a case study, this chapter attempts to understand why the vast potential benefits of ICTs have not been realized in countries like Ghana. The argument put forward by the author is that e-government and by extension ICT policy outcomes in developing polities must be understood as partly a reflection of the world view of policy elites, which is at best generally antagonistic, ambivalent, and even apprehensive of the very notion of a cyber society. The chapter concludes with recommendations relevant to Ghana and other developing polities.

INTRODUCTION

By the last two decades of the twentieth century, the ubiquitous presence and impact of Information and Communication Technologies (ICTs) on societies across the globe had become a difficult to ignore reality. Inspired in part by the dot com boom, especially in the West and parts of Asia, African policymakers began to appreciate the potentially revolutionary value of ICTs with regard to fast tracking development. Not be left out, Ghana joined the race to leverage ICTs for development in Africa (Anyimadu & Falch, 2003).

By 2003, the ICT4AD (Information and Communication Technology for Accelerated Development) policy document had been drawn up by the Government of Ghana. Preceding this was the rapid emergence of major players in the mobile telephony and software development sectors and increasing Internet awareness, access, and usage (Anyimadu, et al., 2005). In spite of all this, it is difficult to describe Ghana as an economy and

DOI: 10.4018/978-1-4666-1909-8.ch015

society strategically exploiting, harnessing, and leveraging the developmental potential of ICTs with significant demonstrable results. In other words, public policy in Ghana has not responded adequately to the emergence of the networked society of Castells (2000).

This chapter will argue that to understand Ghana's predicament it is important to deconstruct the complicated interplay between the generation gaps ("grey hairs" versus "jet black hairs"), mindsets (grey matter) and ICTs policy formation. This vexatious, convoluted, and often ignored intersection will be unpacked using participant observation insights gleaned from working on a major e-government project[1] at the Ghanaian presidency complemented by other primary and secondary data sources. The pivotal role that ICTs can play in fast tracking development in the global South is an argument that has been made ad nauseam. Indeed the World Bank has drawn attention to the linkages between good governance, ICTs and development in developing polities in the last two decades. Some significant ICTs related infrastructural development has been undertaken in Africa in the last decade; Chinese and Indian companies have for example won and executed lucrative contracts in this regard. The empirics however indicate that overall and in comparison with other regions such as Asia, the vaunted development potential of ICTs has yet to be reflected in developing countries like Ghana.

Using Ghana as a case study, the burden of this chapter will be to understand why the vast potential benefits of ICTs have not been unleashed in developing countries. The argument will be canvassed that ICTs policy outcomes in developing polities in the last two decades need to be understood as partly a reflection of the world view of policy elites which is at best generally antagonistic, ambivalent and even apprehensive of the very notion of a cyber society. Recommendations relevant to Ghana and other developing polities will be offered to round off the discourse.

AFRICA, E-GOVERNMENT, AND THE DIGITAL AGE

The 1980s in Africa were generally marked by economic and social malaise in most of the countries on the continent. This generalized state of socio-economic crises in the 1980s in Africa has been attributed in part to the primary commodity and oil price shocks of the 1960s and 1970s respectively and the ensuing debt crises (Aryeetey, 1996; Nafziger, 2006). Africa's policy and political elites seemed to have responded to these crises by turning to the Bretton Woods institutions and signing onto the Structural Adjustment Programmes (SAPs) of this International Financial Organization (IFIs). African countries, which adopted the SAPs, had to adhere to strict conditionalities, which were initially economic in nature but later on in the 1990s assumed a political character. Kahler (1992) describes conditionalities as "an exchange of policy changes for external financing" (p. 89). Meeting these conditionalities became a quid pro quo for the loans, grants, and technical assistance that these institutions offered (Amoah, 2005; Killick, 1984; Zormelo, 1996). It must be added that arguably the emergence of conservative governments in Europe and America (Thatcher government in the United Kingdom in 1979 and the Reagan Administration in 1980), the disintegration of the Soviet Union and the fall of the Berlin Wall in 1989 provided the geo-political space for the triumphant march of the essentially neo-liberal ideas of government and economic policy formation in developing countries.

The turn to political conditionalities as alluded to above must be seen as a reflection, at least partly, of the growing dissatisfaction of the IFIs with the policy outcomes of economic restructuring efforts in African countries. The emergence of political conditionalities in the 1990s in Africa marked an evolution in IFIs' policy thinking toward the idea that the re-engineering of the institutions, processes and rules of government in Africa is a necessary condition for economic restructuring

(World Bank, 1989). Fundamentally, the claim was that governments in Africa had to become efficient and effective in the allocation and utilization of resources (financial, material, and personnel) if economic growth and development were to be attained. Thus was born the idea of good governance[2].

By the mid-1990s ICTs[3] had spread with lightening speed across the globe (Castells, 2000). If the efficiency and effectiveness logic of the good governance narrative were to be actualized, it had found an ally in ICTs. The e-government policy[4] of the World Bank, in my view, reflects this recombinant welding of governance and information technology in the pursuit of efficiency and effectiveness.

GHANA AND THE SHIFT TO E-GOVERNMENT

In April 1983, the new military regime of Flight Lieutenant Jerry John Rawlings entered into negotiations with the IMF and the World Bank after initially refusing to do so. Subsequently, the government by 1984 had adopted the Structural Adjustment Programme (SAP) and Economic Recovery Programme (ERP) of the IMF and the World Bank. Ghana's dismal economic situation in the early 1980s had made the turn to the Bretton Woods institutions an inviting option. A concatenation of proximate and distal factors among them political instability characterized by five successful military coups from 1966 to 1981, inappropriate macroeconomic and institutional policies, major external shocks such as the 1974-76 dramatic world oil price increases and the unprecedented severe droughts of 1977 and 1983 among others had set the Ghanaian economy on a downward slide by 1983(Anaman, 2006; Aryeetey, 1996).

One of the key policies that Ghana adopted under the SAP/ERP to deal with its hemorrhaging public institutions was the public sector reform program (Bratton, Mattes, & Gyimah-Boadi, 2005; Oluwu, 1999; Owusu, 2006; Tsikata, 2001) with the World Bank as the key partner. The explicit raison d'être for the attempt at public sector reform was that an efficient and effective public sector was critical for Ghana's socio-economic transformation[5].

US$10.8 million credit was approved in 1987 under the World Bank's Structural Adjustment Institutional Support (SAIS) Project for essentially shoring up the institutional and managerial capacities of Ghana's public service 'in support of the ERP' (World Bank, 2001). At the core of this initial attempt (1987-1994) was a trenchant focus on managing the wage bill of the civil service under the Civil Service Reform Programme (CSRP) (Owusu, 2006, p. 695). Such a focus was operationalized as redeployment schemes, a cessation of automatic graduate entry into the civil service, dismissal of over-aged employees, recruitment freeze among other reform measures such as the drive towards setting up a sound administrative, personnel and financial management system.

By 1994, public sector reforms in Ghana had entered a second phase. This second phase was characterized by a relatively broader, holistic and systemic approach to reform (albeit not abandoning entirely the agenda of the first attempt) which reflected the New Public Management (NPM) in vogue especially in countries of the Organisation for Economic Cooperation and Development (OECD). The National Institutional Reform Programme (NIRP) was set-up in 1994 to lead this second stage of Ghana's public sector reform. This stage was marked by ideas articulated by Osborne and Gaebler (1993) for restructuring the public sector such as decentralization, privatization, and contracting out among others. This framework informed the approach to public sector reforms adopted by the New Patriotic Party (NPP) administration, which took office in January 2001[6]. Dissatisfied with earlier attempts at public sector reforms, the NPP administration set up the Ministry of Public Sector Reform in May 2005 to lead a fresh assault. The Ministry[7] sought to create an efficient and effective public sector, which would be responsive to the demands of the public.

The Government of Ghana signaled its desire to harness the development potential of ICTs via its ICTs policy document, the ICT4AD, which was published in March 2003. The central goal of Ghana's ICTs policy (Government of Ghana, 2003, p. 73) is "to engineer an ICT-led Socio-economic Development Process with the potential to transform Ghana into a middle income information-rich, knowledge-based and technology driven economy and society." Clearly Ghana is attempting to morph into a Castellian network society which can function in the global information technology market which was valued at $728 billion in 2007 by the Gartner Group (World Bank, 2005). Various legal, institutional, and regulatory measures in pursuit of this goal have been undertaken in Ghana (see Table 1). Ghana's E-government policy aims at making the public sector the driver and embodiment of such policy measures. In other words, the belief is that ICTs can and indeed must lie at the core of public sector reforms in order for the sector to fulfill its expected dual roles: purveyor of liberal democratic governance and ally of the free market in an informational age (see Figure 1).

As expected the World Bank's financial and technical support has been a critical component of Ghana's e-government policy formation. This support is channeled through the Bank's E-Ghana project. The Bank is clear about the ultimate agenda of E-Ghana project: "supporting concrete initiatives to implement the Government ICT-led development strategy" (World Bank, 2005). The E-Ghana project seeks to attain the following objectives: " increased employment in the IT-Enabled Services (notably Business Processing Outsourcing or BPO) sector, enhancing efficiency, transparency and accountability in selected government agencies and departments, and improving access to information/transaction capabilities for citizens and businesses in selected

Table 1. Key policy measures in support of ICT-led development strategy

Key Measures	Status
Construction of National Fibre Optic Backbone	First phase completed
Electronic Transactions (E-Transactions); Electronic Communications (E-Communications), National Communications Authority, and the National Information Technology Bills	All passed into law in 2008 by the Parliament of the Republic of Ghana
Ghana Investment Fund for Electronic Communication (GIFEC)[10]	Set up
Nationwide Communication Information Centers (CIC)	Ongoing roll out
E-Ghana Project (an e-government project comprising technology based public financial management systems and Business Process Outsourcing elements).	Ongoing; World Bank financed
Pan-African e-Network Project	Ongoing; Education and Health Component completed
Automation of Gov. Revenue Agencies[11]	PPP agreement finalized for roll out
Development of Enterprise Architecture and e-Government Interoperability Framework (EA/e-GIF)	Proposal stage
Development of Standards and Accreditation in ICT Education	Ongoing
Development of Technology Parks	Ongoing; consultants engaged to develop a strategic plan encompassing investment, marketing, and the business model.
Mobile Number Portability Project	Fully operational
Government Assisted PC acquisition Programme for Ghana's citizens	Ongoing

Figure 1. Key policy planks of Ghana's ICTs policy (adapted from GoG, 2003)

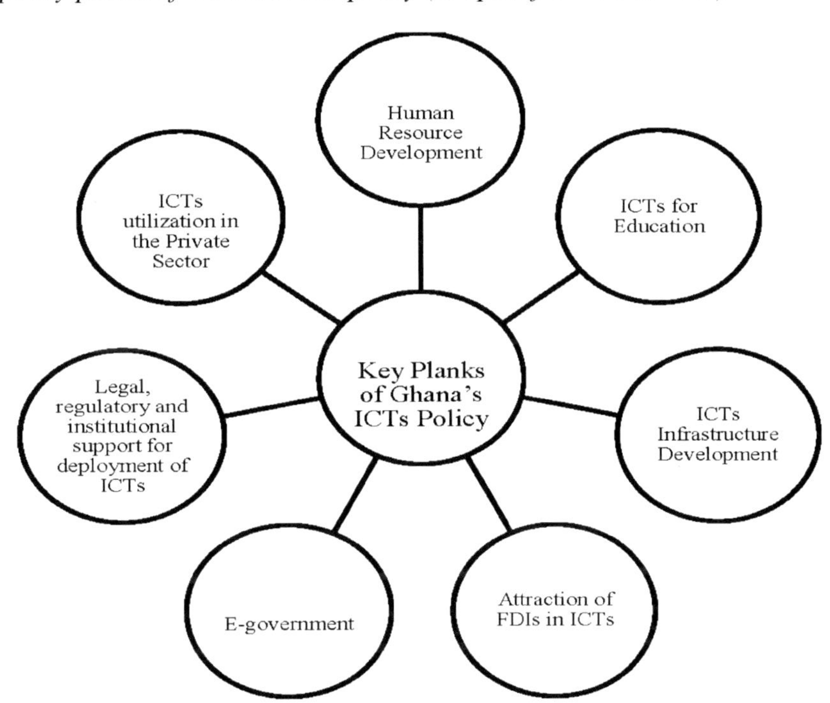

districts" (World Bank, 2005). The deployment of the Government Integrated Financial Management Information System (GIFMIS)[8] to presumably streamline and enhance financial transactions across Ministries, Departments and Agencies (MDAs) in Ghana is a fourth objective that has recently been added. In 2010, the World Bank Board approved $44.7 million from the International Development Agency (IDA) as financing for this fourth objective bringing the total financing approved for the E-Ghana project to $87.4 million[9]. The discourse in this section has sought to tease out the linkages between e-government policies and the broader attempts at ICT-led development strategy in Ghana. Having established this connection the next section will make the claim that the attitudes of policy elites towards e-governance in a developing polity may provide useful clues to their policy beliefs with regard to the overarching issue of ICT-led development strategy (see Figures 2 and 3).

REFLECTIONS ON AN E-GOVERNMENT PROJECT IN THE GHANAIAN PRESIDENCY

From a very provincial conceptualization of ICTs as a tool for enhancing government communications (GoG, 1999), Ghana's policy elites are now evincing a desire to construct a knowledge-based society (GoG, 2010) with ICTs at the hub. Working on a e-government project at the Ghanaian Presidency from 2003-2005 however convinced this researcher that the policy beliefs of Ghana's policy elites at the highest levels of power may undermine the crafting and active operationalization of innovative, system wide, far reaching and consistent policies critical for the construction of a knowledge-based society in the shortest possible time. In other words, the policy beliefs of Ghana's policy elites concerning ICTs may potentially retard the policy changes required to build the much-vaunted knowledge-based society.

Figure 2. Mobile phone mast in Sakumono, Accra

Figure 3. An Internet café signage in Sakumono, Accra

Sabatier (2007) provides very useful insights on the linkages between the policy beliefs of policy elites and policy change. In Sabatier's (2007) influential framework for theorizing policy change, the Advocacy Coalition Framework (ACF), the interaction between the policy beliefs of policy elites and policy change is delineated. The ACF was elaborated by Sabatier (1986, 1987, 1988) working in conjunction with Hans Jenkins-Smith. In fact the theory's organizing concepts were first articulated by Sabatier at a Rotterdam Conference in 1983 (Sabatier, 2007) and has since undergone important revisions. The central concern of this theory is to explain policy change over time within a given policy subsystem[12]. The ACF argues that to explain policy change over time the most important unit of analysis is the policy subsystem (Sabatier, 2007, p. 204) encompassing advocacy coalitions[13] composed of legislators, researchers, bureaucrats, journalists, and interest groups. The hypothesized advocacy coalition reflects the theory's emphasis on the centrality of the policymaking role of policy elites given the complexity of issues of the policy process and with it the salient role of beliefs.

Beliefs and policy change constitute the two dependent variables for the ACF. In the ACF the beliefs of policy elites are ordered in a three tier hierarchy made up of deep core (at a more fundamental level) and policy core beliefs (at a specific and less fundamental level). Deep core beliefs, which cut across most policy subsystems

Involve very general normative and ontological assumptions about human nature, the relative priority of fundamental values such as liberty and equality, the relative priority of the welfare of different groups, the proper role of government vs. markets in general, and about who should participate in government decision making (Sabatier, 2007, p. 194).

These deep core beliefs are the product of the socialization and humanization process that individuals undergo from childhood and thus deemed by the ACF as difficult to change. Policy core beliefs on the other hand are derived from deep core beliefs and extend across a given policy subsystem. According to Sabatier and Jenkins the components of policy core beliefs include

The priority of different policy-related values, whose welfare counts, the relative authority of governments and markets, the proper role of the general public, elected officials, civil servants, experts, and the relative seriousness and causes of policy problems in the sub-system as a whole (Sabatier, 2007, p. 195).

These policy core beliefs are thus related to specific policy subsystems. To the extent that policy core beliefs entail issues of specific policy preferences (e.g. providing or withholding subsidies for Ghanaian farmers) the ACF terms such beliefs policy core policy preferences. In general, the ACF considers policy core beliefs as the most critical nexus linking coalition members. At the bottom of the ACF belief hierarchy are secondary beliefs, which are narrower in scope and deal with instruments for achieving policy ends such as the role of government within a given statute, budgets within a given program, etc. Secondary beliefs are the most amenable to change on account of requiring little evidence (against) and less agreement from coalition members.

The preceding discourse on beliefs is vital essentially because for the ACF change in policy core and secondary beliefs set the basis for major policy change. The ACF argues that the change in both beliefs however must be understood in terms of two critical paths namely policy oriented-learning[14] and external shocks caused by the impact of dynamic external factors[15]; and alternative paths to policy change in the form of internal shocks and negotiated agreements. Policy-oriented learning tends to change secondary beliefs while the other three impact on policy core beliefs (Sabatier, 2007). The ultimate effects of these paths are to alter among other things the notions of the severity of 'wicked problems,' the choice of instruments to deal with them and possible outcomes and thus engender major policy change. In other words, these paths are the causal drivers of major policy change for the ACF. The policy beliefs of Ghana's policy elites along this Sabatierian continuum were on display as I worked on this e-government project.

The Office of the Press Secretary (OPS)[16] was set up at the Ghanaian Presidency[17] during the first term of the Kufuor administration[18]. The OPS was headed by the then press secretary to the president and presidential spokesperson, Kwabena Agyei Agyepong. In creating the OPS, the Kufuour administration was expressing the liberal democracy traditions that the NPP has been heir[19] to since the founding of Ghana. Official documents reviewed by the author indicated clearly that the OPS was seen as a key link (at least on paper) in making the highest levels of government accessible to the Ghanaian citizenry. In other words, the OPS was regarded as a vehicle for democratizing government.

The OPS was made up of three bureaus, namely the Research Bureau, the Local Media Relations Bureau, and the International Media Relations Bureau. The Press Secretary's Secretariat was at the apex of this tripartite structure. I was in charge of the International Media Relations Bureau (IMRB). The key brief of this bureau was to ensure that the daily official engagements of the Ghanaian president were represented accurately in the concourses of the international media. This meant interacting with the major global media conglomerates such as the British Broadcasting Corporation (BBC), Cable News Network (CNN), Reuters, Agénce France Presse (AFP), among others. In this connection, the IMRB had to deal with hundreds of foreign journalists directly or through their diplomatic missions in Accra. To his credit, Kwabena Agyepong had a clear idea of the kind of OPS he wanted to create. Mr. Agyepong wanted a proactive, dynamic, state-of-the-art technology-driven OPS. He had prepared a master plan for this which drew heavily on the White House Communications structure and which he showed to heads[20] of the bureau at our initial rounds of meetings in early 2003. Creating an official online presence[21] for the presidency was part of this master plan, and I was tasked to lead the effort.

Ironically and counter-intuitively, it became clear that the presidency did not consider this particular project top priority, and the absence of funds for it underscored this view. My contacts within the American Embassy in Accra indicated some interest in the project. After a series of meetings involving the political officer at the American Embassy, Susan Parker Burns, the American Government provided funding for the project after close to a year of convincing. The work (design, training, maintenance, etc.) for the project was awarded to a local Web development firm. On April 15, 2004, a press statement announcing the presence of the website for the presidency, www.ghanacastle.gov.gh, was released. It was the first time in Ghana's modern history that the presidency had acquired a presence in cyber space. The site offered daily updates of the president's schedule, his email address, key information on Ghana and the presidency among others. Part of the long-term plan for the website was to allow journalists and ultimately the citizenry to interact in real time with the press secretary and president. The very small team at the IMRB was responsible for updating the website and responding to the flood of e-mails addressed to the president. The e-mails were in the order of a thousand on a daily basis showing that Ghanaians offered the opportunity would not take lightly their civic responsibility to engage consistently with those who governed them.

The administrative and logistical processes involved in running the website were clearly hamstrung by the mindset, which had starved the project of funds from the outset. Internet traffic at the presidency was slow and irregular. This I found out was because the presidency did not have its own server. We were stretched to the limit because too few hands were manning a major e-government project. The presidency was in no mood to hire extra staff and we resorted to using interns straight out of college. Though that helped the manpower, challenges still remained. One found it difficult to understand the state of affairs in the IRMB given that the administration had indicated its belief in an ICT-led development strategy.

Constant interactions with top staff at the ICT department of the presidency revealed that they felt the angst that had pervaded the OPS. Their training trips abroad just like ours were sponsored by donors; their technical advice were politely listened to but not acted upon; they felt uninspired and adrift. I began to listen and watch carefully to try to unravel in some degree what lay at the base of this lassitude when it came to ICTs in the presidency. I noticed that in the OPS our head, the press secretary, was just forty years of age. Indeed the average age at the OPS was about 37 years. The ICT department also showed such youthfulness. On the contrary, those who held the purse and power strings at the presidency while I was there were on average about 60 years old. The president himself was way into his 60s when he assumed office. His chief of staff Kojo Mpiani was his age contemporary. There clearly was a generation gap in which the appreciation of the new cyber world was different and indeed collided. The president was not known, from the numerous discussions (spanning my three year stint at the presidency) I had with top government and party operatives, security apparatchiks, and my own observations, as a technophile. In a sense, he reflected the worldview of his generation: the pre-independence generation or what I term the "grey hairs."

The "grey hairs" had come of age in an era where industrial society was the great ideal. Those who had become policy elites were trained in the main in the then dominant industrialized societies (these societies have now become essentially post-industrial societies). Arguably, the mores, way of life, values, and worldview of the industrial era[22] constitutes part of the deep core beliefs of such "grey hairs" policy elites who still wield enormous political and economic power in Ghana. As Sabatier (2007) has theorized such deep core beliefs are difficult to change and may therefore be clashing with the imperatives of the knowledge society characterized by its novel ideational foundations and sociology (Drucker, 1969, p. 41), which the post-independence generation (jet-black hairs) have had to live with a greater part of their lives and are therefore arguably habituated to. I will argue that the "grey hairs" have had to craft the ICT-led development strategy (and with it e-government policy) because it is one of the inescapable "games" in the global village. Ghana's multilateral and bilateral donors demand it and are ready to fund it (a great incentive). In other words following Sabatier (2007) one can argue that the policy core and secondary beliefs of Ghana's "grey hairs" policy elites underwent some changes on account of especially forced policy learning (donors' policy conditionalities) and external shocks (the rise of a new global ICTs led economy) leading to major policy change (the crafting and implementation of Ghana's ICTs-led development strategy).

This policy change, however, I will aver is in a sense perfunctory because the industrial society mindset (deep core policy beliefs) of Ghana's "grey hair" policy elites is out of sync with their policy core and secondary beliefs which are attempting to accommodate the values of the knowledge society. Ghana's ICTs policy outcomes underscore this view. E-government in Ghana currently is still essentially at the online presence stage[23]. Transactions between government, the citizenry, and business are very much rooted in the industrial society approach: manual and face-to-face. Telecoms operators have proliferated[24], but consumers have complained about quality of service and costs. The Ghana Consumer Protection Association (CPA) went on a street protest (dubbed the "Ghana Phone off Day") against Ghana's mobile phone operators in Accra on the 27th of May, 2010. The emergence of an Internet culture seems to have been thwarted by among other factors bandwidth size (Bon, 2007), speed, high costs and access. Ghana has yet to exploit fully the economic and social advantages of the Internet like Bangalore and Chennai in India have.

SOLUTIONS AND RECOMMENDATIONS

If Ghana's transition to e-government within a knowledge-based society is to be safeguarded it is essential that the policy beliefs of Ghana's policy elites are not ignored in ICTs policy formation process. A decidedly deterministic approach (*Info*Dev/World Bank, 2009) has tended to attend the transition towards e-governance in developing polities initially marked by a focus on deploying technology. This deterministic approach assumed that once technology is deployed e-government policy outcomes sought will automatically be realised. The failures of the focus on technology have led to a shift towards pump priming broader public sector governance processes and structures for e-government. To be expected a key component of this from the World Bank's perspective is the role of leadership at the highest echelons of government in developing countries. To the extent that leadership is important as has been demonstrated in countries like Rwanda, the issue of policy beliefs of such policy elites needs to be critically looked at by interested stakeholders in the e-government agenda in developing countries.

It is not enough for policy elites to express the desire to lead (through for example speeches and policy documents) the e-government agenda; their deeply held policy beliefs (deep core beliefs) matter as well. It must be said that among the "grey hairs," policy elites are those who genuinely believe that e-government is the way to go in an increasingly digitized world. Such individuals will in all probability be tweeting; will be on Facebook; actively using their notebooks and iPads; responding on time to their emails; and have some conversance with commonly used word processing and other software; etc. All these are concrete signs of a perceptive awareness of the reality of the 21st century knowledge society. All stakeholders in the e-government agenda should begin to find ways to indentify and engage such policy elites to lead the process. Crucially this must be wedded to finding ways of bringing younger people into elite policy circles. China did this under Deng Xiaoping in the 1990s (Li, 1998) in relation to major public sector and policy reforms and so did Singapore (Lee, 2000) as the country sought to become a services led economy.

FUTURE RESEARCH DIRECTIONS

If Castells (2000) is right on the informational society as the defining global social reality of this century, the necessity of developing countries finding ways to accommodate this is imperative. In this regard, the transition to e-government becomes crucial not only as end in itself (democratizing government by catalyzing greater participation, access, transparency, and efficiency) but also as a means to an end (the pursuit of ICT-led development). Policy elites have tended to drive the process in developing countries. The Ghana experience however seems to suggest that the deeply held beliefs of policy elites may have a mediating influence on their commitment to and policy choices regarding e-government in particular and ICTs policy in general. Theorizing the linkages between such policy beliefs, an active and visibly expressed commitment to operationalizing ICTs policies and policy choices in developing polities presents a compelling direction for future research. I will suggest a multi- and intra-disciplinary approach that will elaborate a framework that will attempt to trace out other salient causal/explicatory paths and capture the key variables. An exploratory attempt has been made in this chapter.

CONCLUSION

This chapter attempted to draw attention to the challenges and prospects of e-government and ICTs policies in developing polities using an e-government project in Ghana as a case study. The discussions on e-government and ICTs policies was located within the overall framework of public sector reforms as a vital component of the ongoing neo-liberal state construction project promoted in the main by the Bretton Woods institutions in the developing world. The discourse drew heavily on the experiences of the writer who worked on a major and historic e-government project at the Ghanaian presidency from 2003-2005.

The seminal claim of this chapter is that to understand the rather stunted state of affairds regarding e-government in particular and ICTs policy outcomes in general one has to factor in the policy beliefs of the policy elites of this West African nation. In connection with this, the position was canvassed that the policy beliefs of older (pre-independence generation) policy elites (who still hold key leadership positions) runs counter to the organising notions of an informational society which a younger generation of policy elites (post-independence generation who hold less power) tend to subscribe to. The corollary of this tension is a half-hearted, lugubrious, and even contradictory approach to e-government and ICTs policies in Ghana. This may offer lessons for other developing polities. Ultimately, therefore, this chapter argues that public administration's path to the stylized digital *nirvana* in the global South is as much a question of technology and government processes as it is of deeper, vexing ontological matters.

It must be added that the policy narratives on e-governance specifically and ICTs generally in the developing world (which institutions like the World Bank also utilize) reify ICTs as tools "out there" to be used when necessary and ignored at will does not help matters. New narratives are called for which perforce alert the citizenry to ICTs as a virtually inescapable existential reality; the digital age as a kind of condition that one lives with rather than a "tool" one chooses to use or not. This point to the ontological questions that lie at the heart of public administration in the digital age in developing countries this century.

REFERENCES

Amoah, L. G. A. (2005). *Democratization and development aid: Ghana's experience (1989-2003).* (Unpublished Master of Arts Thesis). University of Ghana. Legon, Ghana.

Anaman, K. A. (2006). *Factors that enabled successful economic reform and recovery in Ghana from 1983 to 2006.* Accra, Ghana: Institute of Economic Affairs.

Anyimadu, A., Esseloor, S., Frempong, G., & Stark, C. (2005). Ghana . In Gillward, A. (Ed.), *Towards an African e-Index: Household and Individual ICT Access across 10 African countries.* Johannesburg, South Africa: RIA.

Anyimadu, A., & Falch, M. (2003). Tele-centers as a way of achieving universal access - The case of Ghana. *Telecommunications Policy, 27*, 21–29. doi:10.1016/S0308-5961(02)00092-7

Aryettey, E. (1996). *Structural adjustment and aid in Ghana.* Accra, Ghana: Fredrich Ebert Stiftung.

Bon, A. (2007). *Internet bandwidth problems: Special case of Ghana.* Paper presented at the Web 2.0 for Development Conference. Rome, Italy.

Bratton, M., Mattes, R., & Gyimah-Boadi, E. (2005). *Public opinion, democracy and market reforms in Africa.* Cambridge, UK: Cambridge University Press.

Castells, M. (2000). *The rise of the network society.* Oxford, UK: Blackwell Publishing Limited.

Drucker, P. F. (1969). *The age of discontinuity: Guidelines to our changing society.* New York, NY: Harper and Row.

Government of Ghana. (2003a). *The ICT for accelerated development (ICT4AD) policy.* Accra, Ghana: Government of Ghana.

Government of Ghana. (2003b). *Ghana growth and poverty reduction strategy (GPRSII) (2006-2009).* Accra, Ghana: Government of Ghana.

Government of Ghana. (2005). *Budget statement and economic policy.* Accra, Ghana: Government of Ghana.

Government of Ghana. (2010). *Budget statement and economic policy.* Accra, Ghana: Government of Ghana.

Institute for Public-Private Partnerships. (2009). *Public-private partnerships in e-government: Handbook.* Washington, DC: *info*Dev / World Bank.

Kahler, M. (1992). External influence, conditionality, and the politics of adjustment . In Haggard, S., & Kaufmann, R. R. (Eds.), *The Politics of Economic Adjustment* (pp. 89–136). Princeton, NJ: Princeton University Press.

Killick, T. (Ed.). (1984). *The quest for economic stabilization: The IMF and the third world.* London, UK: ODI.

Kunstelj, M., & Vintar, M. (2004). Evaluating the progress of e-government-A critical analysis. *Information Polity*, *9*(3-4), 131–148.

Lee, K. Y. (2000). *From third world to first: The Singapore story (1965-2000)*. Singapore, Singapore: Times.

Li, D. (1998). Changing incentives of the Chinese bureaucracy. *The American Economic Review*, *88*(2), 393–397.

Mbah, C. (2006). The seductive discourses of development and good governance . In Smith, M. S. (Ed.), *Beyond the African Tragedy: Discourses on Development and the Global Economy* (pp. 121–137). Hampshire, UK: Ashgate.

Nafzinger, E. W. (2006). *Economic development*. Cambridge, UK: Cambridge University Press.

Oluwu, B. (1999). Redesigning African civil service reforms. *The Journal of Modern African Studies*, *37*(1), 1–23. doi:10.1017/S0022278X99002943

Osborne, D., & Gaebler, T. (1993). *Reinventing government: How the entrepreneurial spirit is transforming the public sector*. New York, NY: Plume.

Owusu, F. (2006). Differences in the performance of public organizations in Ghana: Implications for public-sector reform policy. *Development Policy Review*, *24*(6), 693–705. doi:10.1111/j.1467-7679.2006.00354.x

Ronaghan, C. I. (2002). *Benchmarking e-government: A global perspective*. New York, NY: United Nations/DPEPA.

Sabatier, P. A. (1986). Top-down approach and bottom-up approaches to implementation research: A critical analysis and suggested synthesis. *Journal of Public Policy*, *6*, 21–28. doi:10.1017/S0143814X00003846

Sabatier, P. A. (1987). Knowledge, policy-oriented learning, and policy change. *Knowledge*, *8*, 649–692.

Sabatier, P. A. (1988). An advocacy coalition framework for policy change and the role of policy-oriented learning therein. *Policy Sciences*, *21*, 129–168. doi:10.1007/BF00136406

Sabatier, P. A., & Weible, C. M. (2007). The advocacy coalition framework: Innovations and clarifications . In Sabatier, P. (Ed.), *Theories of the Policy Process* (pp. 189–220). Boulder, CO: Westview Press.

Toeffler, A. (1981). *The third wave*. New York, NY: Bantam Books.

Tsikata, Y. M. (2001). *Owning economic reforms: A comparative study of Ghana and Tanzania.* Tokyo, Japan: United Nations University (UNU)/WIDER.

World Bank. (1989). *Sub-Saharan Africa: From crisis to sustainable development*. Washington, DC: World Bank.

World Bank. (2001). *Designing technical assistance projects: Lessons from Ghana and Uganda.* Retrieved in February, 23, 2007 from http://www.worldbank.org

World Bank. (2005). *Project information document (PID): E-Ghana project*. Report No. AB1654. Retrieved in February, 23, 2007 from http://www.worldbank.org

World Bank. (2009). *e-Government primer*. Washington, DC: *info*Dev/World Bank.

Zormello, D. (1996). *Is aid conditionality consistent with national sovereignty?* London, UK: ODI.

ADDITIONAL READING

Ahone, T., & O'Reilly, J. (2007). *Digital Korea: Convergence of broadband internet, 3G cell phones, multiplayer gaming, digital TV, virtual reality, electronic cash, telematics, robotics, e-government and the intelligent home*. London, UK: FutureText.

Al-Hakim, L. (Ed.). (2007). *Global e-government: Theory, applications and benchmarking*. Hershey, PA: IGI Global.

Carrillo, C. J. (Ed.). (2006). *Knowledge societies: Approaches, experiences and perspectives*. Burlington, MA: Elsevier.

Charkravartty, P., & Schiller, D. (2010). Neoliberal newspeak and digital capitalism in crisis. *International Journal of Communication*, *4*, 670–692.

Friedman, T. (2006). *Understanding globalization: The lexus and the olive tree*. New York, NY: Anchor Books.

Friedman, T. (2006). *The world is flat: A brief history of the twenty-first century*. New York, NY: Farrar, Straus, and Giroux.

Fuchs, C. (2010). Class, knowledge and new media. *Media Culture & Society*, *32*(2), 141–150. doi:10.1177/0163443709350375

Ghesquiere, H. (2007). *Singapore's success: Engineering economic growth*. Singapore, Singapore: Thomson.

Gordon, D. M. (1988). The global economy: New edifice or crumbling founadations? *New Left Review*, *168*, 24–64.

Ingram, H., & Schneider, A. L. (2007). Social construction and policy design . In Sabatier, P. A. (Ed.), *Theories of the Policy Process* (pp. 93–126). Boulder, CO: Westview Press.

Jones, A., Kovacich, G. L., & Luzwick, P. G. (Eds.). (2002). *Global information warfare: How business, governments and others achieve objectives and attain competitive advantages*. Boca Raton, FL: CRC Press. doi:10.1201/9781420031546

Kushchu, I. (Ed.). (2007). *Mobile government: An emerging direction in e-government*. Hershey, PA: IGI Global. doi:10.4018/978-1-59140-884-0

Levy, R. (2010). New public management: End of an era. *Public Policy and Administration*, *25*(2), 234–240. doi:10.1177/0952076709357152

Low, L. (2001). The Singapore developmental state in the new economy and polity. *The Pacific Review*, *14*(3), 411–441. doi:10.1080/09512740110064848

McQuire, S. (2008). *The media city: Media, architecture and urban space*. London, UK: Sage.

Miah, A., & Rich, E. (2008). *The medicalization of cyberspace*. London, UK: Routledge.

Morçöl, G. (2002). *A new mind for policy analysis: Toward a postpositivist epistemology and methodology*. Westport, CT: Praeger.

Proux, S., & Morris, J. (2009). Can the use of digital media favour citizen involvement? *Global Media and Communication*, *5*(3), 293–307. doi:10.1177/1742766509348672

Qiang, C. Z. (2007). *China's information revolution*. Washington, DC: The World Bank. doi:10.1596/978-0-8213-6720-9

Rabin, J., Hildreth, W. B., & Miller, G. (Eds.). (2006). *Handbook of public administration*. New York, NY: Taylor and Francis.

Schiller, D. (2009). Actually existing information society. *Television and Media*, *10*(1), 147–148. doi:10.1177/1527476408325108

Seifert, J. W., & Jongpil, C. (2008). Using e-government to reinforce government citizen relation relationships: Comparing government reforms in the United States and China. *Social Science Computer Review*, *27*(1), 3–23. doi:10.1177/0894439308316404

Sreekumar, T. T., & Sanchez, M. R. (2008). ICTs and development: Revisiting the Asian experience. *Science, Technology & Society*, *13*(2), 159–174. doi:10.1177/097172180801300201

Stiglitz, J. (2003). *Globalization and its discontents*. New York, NY: W.W. Norton and Company.

Weijun, W., Yanhui, L., Zhao, D., Li, Y., Hongxiu, L., & Xiaoxi, Y. (Eds.). (2007). *Integration and innovation orient to e-society* (*Vol. 2*). New York, NY: Springer.

ENDNOTES

1. This project was indeed a major component of an attempt to promote open governance via the deployment of ICT at the Ghanaian Presidency. One of the key aims was to allow citizens of Ghana to interface on a regular basis with key officers of the Ghanaian Republic including the president.
2. It must be stressed that some scholars contest the claims and ends of the concept of good governance as it relates to Africa. One such scholar is Cosmas Mbuh who interprets good governance (ideally a useful concept) as a seductive discourse that "works as a form of legitimation that affirms a particular hegemonic order, rather than democratic politics" (Mbuh, 2006, p. 123). I share Mbuh's caution given that the good governance narrative encapsulates Euro-American notions of government and social organization that are not universally held and need not be. It must be added that this caution points up the hegemonizing tendency, which attends the deployment of the good governance idea across the developing world on account of the asymmetries of power between the richer and poorer parts of the world.
3. I follow Castells (2000) in understanding ICT as the anchor of contemporary informational society. The term informational (Castells, 2000, p. 21) "indicates the attribute of a specific form of social organization in which information generation, processing, and transmission become the fundamental sources of productivity and power because of new technological conditions emerging in this historical period." In this spirit, I, therefore, privilege the informational effects of ICT in this analysis.
4. The World Bank's e-government policy is reflected in the activities across the world of its Global Information and Communication Technologies Department and the Information for Development Program (InfoDev). The Bank defines e-government as:
   the use of information and communications technologies by governments to enhance the range and quality of information and services provided to citizens, businesses, civil society organizations, and other government agencies in an efficient, cost-effective, and convenient manner, making government processes more transparent and accountable and strengthening democracy (Institute for Public-Private Partnerships, 2009, p. 1).
5. This point has been stressed in major public policy documents. For example, various Ghana Government documents such as Vision 2020 and the Growth and Poverty Reduction Strategy II (GPRSII).
6. Under the 1992 constitution of the Republic of Ghana, political power has alternated between the two dominant political parties

(the New Patriotic Party and the National Democratic Congress) via periodic competitive multiparty elections.

7 The ministry is now defunct after the NPP lost to the NDC in the 2008 General Elections. A minister of state at the presidency has been tasked with overseeing public sector reforms in Ghana.

8 An earlier attempt had been undertaken under what was known as Public Sector Financial Management Reform Programme (PUFMARP).

9 The World Bank had already approved $40 million for financing the E-Ghana project in 2006. It must be stated that other international organizations such as the United Nations and Ghana's bilaterals support her E-Government agenda and ICT-led development strategy. In this discussion, however, the focus is on the World Bank.

10 GIFEC is funding and expected to finance programs such as the Last Mile Project (extending Internet access to underserved areas); the School Connectivity Project; Rural Payphone Project (EQUATEL); the Post Office Connectivity Project; the Libraries Connectivity Project; and the Prisons Connectivity Project among others.

11 These revenue agencies include the Internal Revenue Service (IRS); Value Added Tax Secretariat (VATS) (both the IRS and VAT are now statutorily under the Ghana Revenue Authority set up in December 2009); and Registrar and Accountant-General's Department (RAGD).

12 Here, Sabatier is concerned about the need for the propositions and concepts of theories of the policy process to be among others logically consistent, empirically falsifiable, and delineate clear causal paths. In evidence is Sabatier's self-admitted penchant for scientific methodology even though he recognizes that other methodologies are equally valid to the extent that one is sufficiently " clear enough to be proven wrong" (Sabatier, 2007, p. 3).

13 A policy subsystem is construed as an arena of interaction between groups and individuals in a specific area of interest (foreign affairs, economy, water, ICTs policy, etc.) in a given territory (Zafonte & Sabatier, 1998).

14 Sabatier and Jenkins's (1999, p. 123) explication of policy-oriented learning which in effect is the impact of feed-back on members of the advocacy coalition(s) is instructive: "relatively enduring alterations of thought or behavioral intentions that result from experience and/or new information and that are concerned with the attainment and revision of policy objectives."

15 For the ACF, these include effects from other policy subsystems, change in socio-economic conditions, changes in public opinion, and changes in systemic governing coalition, etc.

16 The OPS is now essentially defunct, a victim it seems of very public feuding between President Mills's former Press Secretary, Mahama Ayariga, and the Director of Communications at the Presidency, Koku Anyidoho. Clearly an archetypical case of institutional destruction, which, ironically, public sector reforms in Ghana have sought to replace with an institution building ethic.

17 The Ghanaian Presidency is also known informally as the Castle because it is housed in a former fort, the Christiansbourg Castle, built by the Danes in the 17th century on the coast of the Atlantic Ocean's Gulf of Guinea in the capital city of Accra.

18 John Agyekum Kufuor, the Oxford educated lawyer, was a two four-year term president of the Republic of Ghana from January, 7,2001 to January, 7, 2009. Kufuor led the New Patriotic Party (and the tradition the party belongs to) to the presidency—after it

lost power in 1971 after a coup d'etat staged by a Ghanaian military officer, Colonel Ignatius Kutu Acheampong—and ensured the tradition's longest stay in power in Ghana's political history.

[19] The NPP traces its political lineage to the Dr. J. B. Danquah led United Gold Coast Convention(UGCC), which in the early years of the Ghanaian Republic was considered in its politics as a counterpoise to the radical, leftist orientations of Kwame Nkrumah's Convention People's Party.

[20] The Local Media Relations Bureau was headed by Augustine Boadu. The head of the Research Bureau, Kofi Owusu-Bempah, joined the team a few months later.

[21] The literature (Kunstelj, 2004; Hughes, 2004; Ronaghan, 2002) suggests a series of pathways towards the creation of a total e-governance structure from an online official presence for government to a state where there is a total and seamless integration of ICTs-driven government services across administrative boundaries.

[22] Industrial era ideas will include large, hierarchical, bureaucratic organizations where power is centralized, heavy industries characterized by mass production and powered by the internal combustion engine and electricity; the linearization of time (Drucker, 1969; Toefler, 1981).

[23] This online presence seems to be approached sloppily. In my research for this work, I found that some government websites were not updated regularly, if updated at all without efficiency (typos, grammatical, and factual errors abounded). A typical example is the Ministry of Communications website, which I visited on the 12th and 13th of July, 2010. The site is at least 7 years late on information about the telecoms operators in Ghana. The site lists on its home page Ghana's main telecom operators as follows: GT One Touch, Areeba, Mobitel, Kasapa. With the exception of Kasapa, all the companies have changed their names. GT One Touch is now Vodafone Ghana; Areeba in now MTN Ghana; Mobitel in now TIGO. Other telecoms have come on board, such as Zain, which has recently been purchased by Bharti Airtel of India. The industry and information statistics was last updated at 18:23 hrs on the 23rd of November, 2005.

[24] Ghana Government (2010) figures put mobile penetration rates at 63.6% as of July 2009.

Chapter 16
A Positive Hegemony?
Arguing for a Universal Knowledge Regime led by an e-Governance 'Savvy' Global Knowledge Enterprise!

Amlan Bhusan
Bhusan Taylor Ltd, UK

ABSTRACT

Amlan Bhusan raises important questions in "A Positive Hegemony? Arguing for a Universal Knowledge Regime led by an e-Governance 'Savvy' Global Knowledge Enterprise!" To him, there is a growing academic consciousness, regarding the use of e-governance, to deliver social goods in a better way. This voice advocates that more needs to be done by public institutions, governments, and more importantly, the academia, to develop e-governance as an enabler for social efficiency. Such developments would help reach debates and discussions on this area to the grassroots of the policy system. His chapter is neither a commentary of the application of e-governance to deliver social change nor a study of how different governments have handled this area around the world. Rather, it is a practicing consultant's views of the power of e-governance to refine public choice and social decision making and how this process was enriched by a more vigorous role of the academia. Taking specific examples from the education sector, particularly universities, this chapter is a comment on some of the ways in which e-governance 'can' be handled across the education system and how lessons from the developed countries can be used to inspire similar revolutionary changes to the status quo in the developing world. His objective is to promote a greater role for the academia in the public policy making process. The idea is to support a more constructive engagement of the academia with the more vulnerable parts of the social system. Above all, he argues for the benefits of spreading the values of information democracy, right to access to information, among the people. He envisages that the power of a more vocal and active academia would be profound in how it could positively affect the information apartheid affecting many large sections of the developing world. He proposes greater research and development on the means of engaging with e-governance and to establish the mechanisms to enhance, converge, simplify, homogenize, and structuralize the knowledge and information enterprise of the global political and social systems.

DOI: 10.4018/978-1-4666-1909-8.ch016

INTRODUCTION

Basu (2004) refers to e-governance as use of information technologies by government agencies to transform relations with their citizens, businesses, and their other arms of governance. Singla (2002) asserts that, key requirements for establishing e-governance were anticipation, transparency, and accountability. Therefore, it can be said that, to establish a regime of e-governance, any policy system needs to accommodate optimal public participation in decision-making and ensure that all have free and easy access to information. The role of the academia, the research communities and also, the practitioners and consultants, in such a process, deserves due attention here. It can be said that without a positive collaboration between these various factions, a constructive engagement between the people and the governmental machinery would be hard to achieve. The premise of ideal governance being engrained in an active relationship between the government and the society, it will be impossible for any governmental or bureaucratic machinery, to deliver its maximum potentials unless a well-orchestrated collaboration was crafted between the people and the system: the role of the academia and practitioners to help achieve this process assumes extraordinary importance.

Further, the fruits of e-governance cannot and should be confined to the developed world. Examples can be drawn from the successful experiences wherever they take place, to enrich the lives of the more vulnerable sections of the society, particularly in the developing world. The creation of a universally applied e-governance regime, aided by a more vocal and active academia and practice communities, is an ideal outcome of the global knowledge renaissance. Hegemony it is, of knowledge, and this chapter argues in its favour.

BACKGROUND

The role of academia, the universities and research establishments, in shaping the intellectual direction of a society cannot be neglected (Goddard, et al., 2006). The way in which these institutions assist in co-production of new knowledge and invention of new ways of doing research is an extremely important contribution in aide of the knowledge enterprise. Irrespective of their national, socio-cultural, and economic differences, institutions around the world shared their responsibilities to continue to play a major role in developing, fuelling, and engineering a collective pursuit for new knowledge and new ways of doing things. E-governance is a tool, which can help fuel the knowledge building capabilities of a society. By helping integrate and operationalise the knowledge creation enterprise of a functional society, e-governance as a field deserves to be studied, researched, and taught with a greater and a renewed interest.

It is important to note here that mere knowledge provision and teaching alone will not affect desired social outputs. As such, there was no simple correlation between university education and economic growth (Wolf, 2001). There was, however, a strong interlink between 'top' universities and economic growth of the country or region they were based in. This is an indication that academic excellence could not be translated by merely being in the business of knowledge creation. But, efforts to reach knowledge and information to the wider sections of the society, and thereby attempting social change, was what established an institution's social responsibility. This positive correlation is particularly visible in case of several US universities (EU Commission, 2006), and this can be attributed to their growing commitments to being tech savvy and their appetite for using technology to aide their efforts to achieving a well-rounded excellence.

Therefore, it can be noted that institutions could succeed more by effectively enhancing their e-governance potentials to conduct rigorous benchmarking of their performance under various metrics, i.e., research output, quality of research produced, comparative worth, and so on. 'Top' universities are those institutions that were the most efficiently managed ones and had a positive engagement with E-Governance. Interestingly, it can be argued that the examples set forth by the academia can inspire the wider public and private sectors and influence them to imbibe similar values and principles that could engender a positive knowledge culture in their respective policy systems.

This attitude towards efficiency, needs to be implemented more universally by the developed countries to show direction to their developing counterparts in ways to engaging with e-governance in a revolutionary way. This process will demand that more institutions in the developed countries test their capacity to handle e-governance in their administrative and knowledge-sharing processes and create examples to inspire those within the developing systems.

Equally, institutions in the developing countries will have to realise their full potentials to doing knowledge by involving more and engaging with their knowledge communities and draw from the various successful experiences from around the world to establish a culture of assessments and learning within their institutions and processes. This will demand that they increasingly understood the various ways in which assessments and evaluation of policy processes have led to evolution and innovations of ways of doing knowledge in the developed systems. Following with the knowledge sector, the case for e-governance practices within the higher education systems in the developed economies deserve a discussion here. Such a discussion will reveal how these institutions could encourage and aide proliferation of the E-Governance agenda by setting examples within their own sector.

TESTING E-GOVERNANCE WITHIN EDUCATIONAL SYSTEMS IN DEVELOPED SYSTEMS

Raposo et al. (2006) argue that universities should be able to justify their social responsibility by adding sensible workforce and a culture of efficiency to the society. They argue that this was done when universities could effectively integrate education and learning, research commercialisation, achieve effective industrial outreach, and demonstrate extended contacts with other institutions. The determinants used to test the e-governance achievements of a system are effective adoption of public policies, participatorial and democratic decision-making, development of collaborative networks, and provision of access to an open system for governance (Leitner, 2003). Clearly, this requires greater inward looking, an attitude of reflexivity, and a high sense of self improvement. Universities in the developed systems have, therefore, had to adopt an appetite for benchmarking their learning processes in order to assess their comparative strengths and weaknesses and to empower their structure for self regulation by enhancing their texture of critical outlook (Jackson & Lund, 2000). The idea of assessment, evaluation, and benchmarking, assumed tremendous significance for these institutions.

EMPLOYING BENCHMARKING BY ACADEMIA

Several methodologies exist to test benchmarking practices of universities. Notable few include, Mckinnon *et al.* (2000), Charles and Benneworth (2000), and Garlick and Pryor (2004). Mckinnon *et al.* (2000) created a system of benchmarking for Australian universities in that it opened ways to assume an alternative form of governance and a process to test its features when in comparison to other institutions. Similarly, Charles and Benneworth (2000) offered a benchmarking approach to test the effect of higher education institutions in the United Kingdom. They sought a connection between progressive university systems to regional development. They emphasized how universities could develop and regenerate the social capital of a region. However, Garlick and Pryor (2004) advocated that the more important reason for benchmarking was to evaluate quality and performance of organizational processes and practices when compared and contrasted with reference organizations on a global scale (See, Raposo, et al., 2006). This demands that universities progressively employed a more universally applied methodology to establish and test their performances.

Common examples and exercises to benchmark are seen among British patterned education systems including the Performance Based Research Funding Model in New Zealand (See, Parliament Library of NZ, 2004, pp. 7/1-6), Research Assessment Exercises in the UK, similar models in Hong Kong and manual or template form evaluation techniques, as discussed by McKinnon et al. (2000) in the Australian system. Similarly, the European Union also has its own share of evaluation regimes, as outlined under the Lisbon Plan, and particularly greater regional knowledge integration system at the outset of the much-famed Bologna agreement. The application of E-Governance to integrate a common credit system and to establish a system of research and development across the region is noteworthy here (See, GronLund, 2003; Leitner, 2003, 2005; Traunmuller, 2003; OBSEREGIO project, See, Raposo, et al., 2006). But, howsoever orchestrated these various processes were, their outcomes continue to be sporadic and to qualify all results as a perfect example of e-governance led organizational efficiency could be shortsighted. There are several challenges in implementing a universally applicable benchmarking regime and these deserve attention here.

CHALLENGES

There is a considerable debate among academia on the state of underfunding among most universities in the EU and other developed systems including Australia and New Zealand. The common argument is that the remnants of a largely state-controlled and funded tertiary sector continues to impede the intellectual inertia for progress, causing institutional unresponsiveness to the demands of the new generation of students and researchers (EU Commission, 2006). The new generation of students and researchers are hungry for more out of their engagements with the tertiary sector, and these demands cannot be fulfilled by institutions that are increasingly vocal of their resource crunch, affecting their ability to make important decisions fast (EU Commission, 2006). The reliance of universities on public funding exposes them to the economic moods of the countries they were based in. The wider effects of these various conditions include increased dependence of the tertiary sector on consolidated funding, mostly emerging out of the public purse, inability to create decisive and long-term policy changes readily, and inability to easily accommodate revolutionary research and development agenda. This leads to a more risk aversive attitude, and the general effect is the creation of a realm of academic frustration, retarding productivity in the overall knowledge enterprise.

The interesting development to this end has been how academic institutions have addressed these many challenges through a concerted evolution of e-governance aided knowledge production. Such a move was revolutionary for it opened many new avenues to doing knowledge efficiently. For example, it afforded a move towards e-publishing and open access, sorting out many old challenges to access by generating greater administrative efficiency through integration of services and resources. It ensured optimal utilisation of resources, thereby reducing operational costs. E-Governnace, therefore, addressed the old challenges better by presenting immense opportunities for growth, especially by revolutionising ways to advancing, integrating, and operationalising knowledge building and knowledge sharing processes. This was characterised by a series of policy innovations, particularly in the EU, to use the momentum therein, to drive a wider agenda of inclusive social change, thereby generating many important opportunities for the wider region.

OPPORTUNITIES

One of the most important benefits of applying e-governance across university environments is that it promises a concrete evaluation model that can be readily applied across the bureaucratic machinery of the university systems. Benchmarking models utilised by universities often yield an assessment of certain performance indicators that shed light on the existing practices, the critical factors, and the overall evaluation mechanisms in such systems (Ribiero, 2004). For example, to enhance the sustainability of a university its e-governance regime can be studied to ascertain if and how the long term strategies were appropriately integrated and applied within such systems. Further, to demonstrate efficient fiscal prudence, universities are required to establish the consequences of the social investments and public spending on them by creating socially productive research and by producing an educated workforce for ready consumption by a talent and skill-hungry corporate sector (Barosso, 2004). Also, e-governance affords institutions to devolve more authority, thereby extending knowledge platforms to be integrated in an institutional information framework, thereby aiding evolution and greater managerial control and establishing best practices across the breadth of the knowledge systems (Raposo, et al., 2006).

These practices collectively ensure that universities are then capable of securing benchmarking tools to test and positively improve their competencies and capabilities for an informed policymaking, efficient managerial control and implementation, and a whole of administration appetite for growth and inclusiveness. The ultimate improvement occurs in terms of the universities' abilities to create new business avenues, enhance their human capital development protocols, and achieve in establishing an efficient innovation regime (Raposo, et al., 2006). This is an encouraging trend forward!

WAY FORWARD

Orthofer and Traunmüller (2002) comment that E-governance is perhaps the better way to enable organizational innovation and to foster a culture of excellence. For a university system, this would translate in terms of integrating and synchronising the knowledge distribution systems spread across the university systems. This will demand that the end users (students, researchers, faculty, and beyond) will have to be provided with easy access to information from multiple areas. The voluminous data, the multiple software applications, and the complex system architectures have to be carefully managed to prevent asymmetry of information and confusions among the end users (Orthofer & Traunmuller, 2002). The EU system of e-governance in the tertiary education technology has led to an integration process, reflective of this progressive agenda.

As Brusa *et al.* (2007) point out that, starting with the Lisbon agenda, e-governance had emerged progressively across the EU policy on a Pan EU information society marked by the famed Lisbon agreement. Ever since, many action plans have been introduced to foster the climate of e-governance. Two such plans included the Action Plans eEurope 2002 and eEurope 2005, which created a unified voice for an EU information society. Led by the research community, such moves were to encourage the public sector institutions, to draw from the tertiary sector, and accept e-governance and its capacity to deliver public services with readiness and creativity (Orthofer & Traunmoller, 2002). Progressing on this very agenda, the latest such action plan comes in the form of the Information Society Strategy i2010, which explicitly calls for a single EU information space that allows economically viable and secure high bandwith communication, Industry University interlink fostering high quality research through the use of advanced information and communication technology, and the creation of an information society that accelerates achieving benefit for all (Orthofer & Traunmoller, 2002).

In all instances, the role of the universities, the tertiary sector, researchers, and consultants is of immense importance. Through their collective efforts in embracing e-governance, a progressive agenda of inclusive growth could be pushed for. Increasingly, e-governance could be seen as an aide to a whole of administration improvement and an enabler for a regime of evaluation and assessments. This is a tremendous success for any social system.

The EU model of greater integration of e-governance across the tertiary sector to influence and establish a wider social agenda of inclusive public policy making and implementation is indeed an example for other systems to follow. Sure, there were challenges in terms of funding constraints, integration of the vast and different systems of decision making, the differences in socio-cultural and demographic distinctions among the member countries. Considerable challenges came from the vast data and the requirement to make them available under one system of a pan-EU governance. But, through a serious of national, regional, and pan-EU projects, the evolution of task forces and actions plans, the lively academic discussions emerging out of the research corridors, mainly facilitated by the universities and researchers as well as practitioners, have elevated the mood of governance research to that of e-governance research.

The most apt example of this process of a move beyond governance to e-governance came in through the agenda of i20 Government Action plan. This plan declared and espoused growth for all agendas across Europe. The idea is to ensure that no citizen suffers from information challenges, that efficiency in public procurement and services was achieved by enhancing end user satisfaction and by providing extraordinary customer services, enabling greater public participation in the governance process.

Undoubtedly, the major part of these activities will be achieved through a highly vocal and active research environment and the role played by the tertiary sector will be of extreme importance. By engaging with cross sectional groups, by inculcating greater discussions into this subject, and by developing curricula for teaching e-governance, academic institutions will have to play a pivotal role to ensure that the mood for research and development extends to a culture of E-Savvyness among the public institutions and, more so, the end users. The progress and success of the e-governance agenda will depend on the noises made from within the university research atmospheres. This is indeed a very interesting aspect of this process.

The idea of affecting social change, establishing a realm of inclusive, research led public policy making, is a great achievement for the developed countries. By challenging issues on funding, by developing a constructive engagement with the public private interfaces, by enhancing the spirit of intellectual enquiry into managing better through e-governance, and by creating a greater intellectual awareness of how academic institutions and the research community can affect social change, institutions in the developed countries are already on the road to success. Their orientation for success, as demonstrated through various action plans as discussed in the preceding passages, should inspire the developing countries in their approach to governance.

LESSONS FOR THE DEVELOPING WORLD

Having discussed the various challenges and opportunities within the knowledge development and management sectors, particularly the tertiary sector, the above discussions have established the extreme importance of a wider debate as to the role of e-governance in shaping social change. It is clear that the move to achieve better Governance through E-Governance is already in force in the developed world. It is therefore important to discuss the complexities and challenges for efficient e-governance in developing economies and then to comment as to how the education systems in these countries can play a role in revolutionising their appetite for e-savvyness.

Challenges and Opportunities

Bhatnagar (2006) reports that the Asia Pacific region is generally composed of many developing countries with some level of conversation with e-governance. For example, Sri Lanka has plans to develop national e-development programs, while India and Cambodia boast such programmes in place. While China has certain selected avenues for E-Governance in place, Philippines has a remarkable case for use of mobile telephony for advancing public participation in decision-making practices. Hong Kong has examples of the nonprofit sector taking an advanced interest in extending the IT literacy of the senior citizens. India, Sri Lanka, and Mongolia also have appetites for using e-governance to provide essential services to rural areas by developing rural call centers and by espousing greater connectivity across the rural belts (Bhatnagar, 2006; Zussman, 2002; Signore, et al., 2005).

There is awareness among most countries in the developing world, that the merits of e-governance cannot be neglected. Bhatnagar (2006) reports that even if many countries did not have national programs, they did report of some e-government programmes spread out among different sectors. The challenge lies not necessarily in their intentions; it lies in their capacity for change and their competency to adopt their existing systems to such changes. To illustrate further, China has many public procurement procedures patterned on and based in e-governance, but a closer study indicates that very few websites allowed transactions, and information available through such sites were relatively static and not routinely maintained and updated.

Further, developing countries of this region have less than 10% of public agencies that have been computerized. This in itself is a stark question on the reality on ground (Bhatnagar, 2006). If this was a case for most public agencies, the case for penetration of IT across the breadth of the higher education landscapes of the developing countries seems to be very bleak. Heavily bureaucratized administrative landscapes, over dependence on public funds, very little or no autonomy to decide on the flow of the knowledge discourse, often centrally regulated tertiary education systems (e.g. UGC in India, Bangladesh, and other developing countries), along with the rumblings of a highly political environment to work within, academia in the developing world continue to suffer from an acute sense of deprivation and neglect, and this is detrimental to the cause of an open knowledge society that E-Governance should afford them with.

Under such a clout of ambivalence, there needs to be a fresh and loud academic noise challenging the status quo and mooting for a pan-Asian, pan-Latin American, or pan-developing world campaign for e-governance led implementation and service delivery. Universities need to work out strategies to affect social policy making, pointing out that the discourse surrounding e-governance has to touch heavily on the topic of the vulnerable parts of the society whose needs deserve great attention. As Bhatnagar points out, e-governance, even in the developing world, has not really reflected the conditions of the vulnerable parts of the society. He further adds that such countries continue to face troubles to clearly iterate their pro-poor agenda, ineffective in reaching social deliverables across the rural populace, unable to join public private interfaces on the lines of development and shared growth. This is clearly indicative of their inherent lack of capacity to evaluate any process or practice in the interests of greater public discussions.

Way Forward

From a practitioner's point of view, the broader challenges of implementing e-governance reigns on several issues. This is more applicable to developing set ups, where issues of infrastructure, connectivity, and implementation were acute and had considerable effects on the success of any E-Governance regime. Other factors, including poverty, illiteracy of the populace, absence of a well-crafted pro-poor strategy by the government of the day, and the lack of a well-crafted implementation strategy, further impede scopes for the success of the e-governance agenda in developing countries.

In terms of the academic voice pressing for the change to the status quo, the role of academia stretches beyond iteration. A pro e-governance academic voice, will help foster a spirit of evaluation within the social system, generate an educated debate as to the use and viability of an e-governance agenda, propose to enervate various social problems by advocating greater use of technology, and create an atmosphere of public participation and decision-making where different voices can be heard and responded to.

As explained in the Introduction to this chapter, the success of a government lies in its capacity to establish communication between its governance and its citizens. While implementation and service delivery of public policies could occur through the natural outlets of public institutions, departments, and agencies, their ability to perform to social needs and their capacity to deliver social change need to be assessed and evaluated in transparency. That is possibly done only within a well-connected information society that has the capacity and the competency to voice needs, but also to evaluate the ways in which such needs had been addressed by the governmental and administrative machineries.

As Layne (2001) described, fully functional e-governance could be achieved through a four stage process of cataloguing, transaction, vertical, and horizontal integration. Such multi-layered approaches require greater participation of the research community to aide greater discussions and intellectual debates, ultimately delivering information to the common people. As discussed earlier, any e-governance agenda, particularly in developing economies, need to be heavily grounded in awareness of and consideration for the vulnerable parts of the society. Therefore, it can be assumed that the ideal outcome for a successful e-governance campaign was the provision of easy and safe access for all citizens, information on their rights and responsibilities. Such a system would guarantee that citizens were aware of their obligation to the state, and as such, hold their representatives accountable for their performace. Clearly, successful e-governance is a way to give power back to the people.

The role of the universities, the research communities, and in particular, practitioners and consultants engaged in e-governance is of tremendous importance. The EU model has demonstrated how integration and development of cross-country data could create a regime of open access and deliver a wealth of information for the ordinary citizen. The vigorous engagement of the public institutions, the academia, and the research and practice communities in EU states and in most developed systems have many things to teach to the wider developing world.

Not withstanding the challenges inherent in the developing systems, poverty, illiteracy, lack of infrastructure, and corruption, among many stand to retard opportunities for governance in many ways. Lack of transparency, accountability, and responsiveness among the institutions and the citizens compound problems further. To address this situation, there needs to be a refreshing discourse on governance. The move to embracing technology and communication systems, to bridge the gap between the rich and the poor by creating a level playing field of equal opportunity for information, is perhaps a very important step to achieving the millenium development goals set by the UN many years ago. To this effect, wider participation of universities, institutions, and research communities will help support a successful e-governance agenda and bring forth, a new realisation to affect the much needed social changes.

FUTURE RESEARCH DIRECTIONS

The idea of establishing the academia at the centre stage of public policy making is indeed a very powerful suggestion. Not withstanding the inherent biases of the so-called "research wall confinement" of the academics, the sheer depth of their intellectual discourse stands to benefit the often short-sighted approach of governmental organizations operating in seclusion or in a ceremonial engagement with other parts of the society, and this is particularly applicable to developing countries. The idea of creating open access and a common information system among the universities of these emerging nations will help enrich and broaden their knowledge-building exercise and deepen their growing voice for a change to the status quo. This, in addition, will assist in creating greater integration of people of different kinds, producing knowledge of different forms, and giving the academia a platform to share and inform among its constituting groups (e.g. scholars, researchers, commentators, and practitioners) their distinct discoveries. How to afford and establish the establishment of such level playing fields in complex environments within the developing world is an area ripe with opportunities for further research. Further, how the academia can play a pivotal role in shaping more efficient and progressive public policy and improve delivery of social services is also an area for future research.

CONCLUSION

This chapter started by exploring the role of e-governance within the academic landscapes. It commented on how universities and the tertiary sectors in the developed countries managed to move for an information society that was conversant with the demands of the new age citizens. Such progressive agenda had their limitations and challenges, and these were mentioned. Later, drawing from the various measures at e-governance in the developed world, academic landscapes were discussed. Following, an argument was developed to bring about similar structures and systems in the developing world. It was argued that a move from traditional governance systems into a tech savvy, IT enabled approach would help emerging economies establish a greater sense of awareness of rights and responsibilities among their citizens. Such measures will undoubtedly establish systems of transparency and accountability within their social and political spheres, thereby enriching the texture of governance within such systems. This chapter strongly advocated the role that can be played by the academia to help enrich and develop greater public consciousness on public policy. It described how a more vigorous academic voice was needed in collaboration with a progressive practitioners' community to collectively move for an e-governance agenda in emerging economies. This chapter, therefore, envisaged that there was a strong possibility for future research in understanding how academic institutions could work in greater collaboration with the public institutions to deliver social change. The connections between a more e-savvy governance, an engaging academia, and a vocal and responsive populace is indeed an area that can be and should be studied at greater lengths in future works.

REFERENCES

Basu, S. (2004). E-government and developing countries: An overview. *International Review of Law Computers & Technology*, *18*(1), 109–133. doi:10.1080/13600860410001674779

Bhatnagar, S. (2004). *E-government: From vision to implementation: A practical guide with case studies*. New Delhi, India: Sage Publications.

Bhavnagar, S. (2000). Social implications of information and communication technology in developing countries: Lessons from Asian success stories. *The Electronic Journal of Information Systems in Developing Countries*, *1*(4), 1–9.

Charles, D., & Benneworth, P. (2000). *The regional contribution of higher education: A benchmarking approach to the evaluation of the regional impact of a HEI*. Newcastle, UK: Newcastle University.

European Commission. (2002). *An information society for all: Action plan prepared by the council and the European commission for the Feira European council*. Brussels, Belgium: European Commission.

European Commission. (2006). *Developing a knowledge flagship: The European institute of technology*. Brussels, Belgium: European Commission.

Garlick, S., & Pryor, G. (2004). *Benchmarking the university: Learning about improvement*. Canberra, Australia: Australian Government.

Gieber, H., Leitner, C., Orthofer, G., & Traunmüller, R. (2007). Taking best practice forward . In Larson, C. (Ed.), *Digital Government: Advanced e-Government Research and Case Studies*. London, UK: Springer.

Goddard, J., Etzkowitz, H., Puukka, J., & Virtanen, I. (2006). *Supporting the contribution of higher education institutions to regional development. Peer Review Report*. Jyväskylä, Finland: Government Printing Office.

Jackson, N., & Lund, H. (2000). *Benchmarking for higher education*. Buckingham, UK: The Society for Research into Higher Education and Open University Press.

Layne, K. J. L., & Lee, J. (2001). Developing fully functional e-government: A four stage model. *Government Information Quarterly*, *18*(2), 122–136. doi:10.1016/S0740-624X(01)00066-1

Leitner, C. (Ed.). (2003). *eGovernment in Europe: The state of affairs*. Maastricht, The Netherlands: European Institute of Public Administration.

Mahoney, P. (2004). Performance based research funding. *Background Note: Information Briefing Service for Members of Parliament*, *7*, 1–14.

Mckinnon, K., Walker, S., & Davis, D. (2000). *A manual for Australian universities*. Canberra, Australia: Department of Education, Training and Youth Affairs.

Raposo, M., Leitão, J., & Paco, A. (2006). *E-governance of universities: A proposal of benchmarking methodology*. Unpublished.

Signore, O., Chesi, F., & Pallotti, M. (2005). *E-government: Challenges and opportunities*. Paper presented at the CMG Italy - XIX Annual Conference. Rome, Italy.

Singla, M. L. (2002). Transforming the national bone marrow. *Journal of Management Research*, *2*(3), 165–175.

Traunmuller, R., & Lenk, K. (Eds.). (2002). Electronic government. In *Proceedings of the First International Conference*. EGOV.

Wolf, M. (2002). *How to save British universities*. Paper presented at the Singer and Friedlander Lecture. Oxford, UK.

Zussman, D. (2002). *Paper.* Paper presented at the Public Policy Forum Commonwealth Centre for Electronic Governance Integrating Government with New Technologies: How is Technology Changing the Public Sector? Ottawa, Canada.

Chapter 17
Successfully Applying "e" to Governance

Evangelia Mantzari
Athens University of Economics and Business, Greece

Evanthia Hatzipanagiotou
Ministry of Finance, Greece

ABSTRACT

The challenge of better public service offering and the expectations of modern citizens and businesses, as well as the poor past practices of public organizations, bring forward the need to design and implement new systems. These systems are based on current information and communication technologies and are points of reference on the path towards e-government "enlightenment." However, the transition from the traditional processes to the modern ones can be long and strenuous, if relevant projects are not carefully implemented. Therefore, in order to successfully apply electronic practices and methods to public systems of governance, a step-wise approach needs to be formulated starting from traditional standards, leading to transitional procedures, and finally achieving simplification and increased quality of public service by exploiting previous experiences, overcoming past limitations, and applying the lessons learned.

INTRODUCTION

The effort of incorporating electronic methods and introducing modern technological applications to the offering of governmental services can prove to be extremely difficult either due to administrative issues, due to practical limitations, or due to "the fear of the unknown." Usually, the effort to convert the traditional process to an electronic one derives from the necessity to face long-term problems and the aspiration to provide higher quality of service to citizens and to all the involved parties. However, the plan of "electronizing" a governance model can plunge to failure if certain areas are not fully covered and mistakes from the past—which practice and theory both recognize as common—are not taken under careful consideration. Therefore, this chapter focuses on identifying some of the issues that stem from the limitations of the traditional system and may re-

DOI: 10.4018/978-1-4666-1909-8.ch017

strain the transition towards the new e-government systems. In addition, it describes the most critical subjects for consideration and management, while being in the transitional process, and concludes by providing insights on generic rules that lead to successfully applying "e" to Governance.

BACKGROUND

The gradual involvement of Information and Communication Technologies (ICTs) to everyday activities and business practices has transformed the managerial and administrative methods and raised the public's expectations concerning the levels and the quality of service provision. Accordingly, the trend in public administration has evolved in order to respond to modern market requirements and citizens' demands with the development of electronic government systems. The *e-government* revolution offers the potential to reshape the public sector and remake the relationship between citizens and government (Saxena, 2005).

Similarly, *e-governance* can be identified as a second revolution, following the movement of new public management, which may transform not only the way in which most public services are delivered, but also the fundamental relationship between government and citizen (The Economist, 2000) and become a pre-requisite for development (Sen, 1999). Governance is the outcome of politics, policies, and programs that concerns long-term processes and not immediate decision-making (Kettl, 2002). Thus, before attempting to pass over to the "electronic era," public organizations have to re-consider, adapt, and modernize all of their traditional old-fashioned processes.

FROM TRADITION...

The path that leads to the application of e-government methods and systems is marked by long-term routines, mistakes, and unresolved issues. In order to gain the ability to transform them, all involved parties should recognize, realize, and respect the conditions of today before envisioning the reality of tomorrow.

Some of the most critical issues of *traditional governmental practices* are:

- *Bureaucracy (piles of paperwork) and procedural plurality (complex and stiff procedures).*

"The type of an organization designed to accomplish large-scale administrative tasks by systematically coordinating the work of many individuals is called bureaucracy." The basic characteristics of bureaucratic organizations include specialization, hierarchy of authority, a system of rules, and impersonality.

However, "if we want to utilize efficient bureaucracies, we must find democratic methods of controlling them lest they enslave us" (Blau, 1956) and the Internet provides a powerful tool for reinventing governments and establishing democratic methods. It encourages transformation from the traditional bureaucratic paradigm, which emphasizes standardization, departmentalization, and operational cost-efficiency, to the "e-government" paradigm, which emphasizes coordinated network building, external collaboration, and customer services and supports the adoption of "one stop shopping" and customer-oriented principles in design, emphasizing on external collaboration and networking in the development process rather than technocracy (Tat-Kei Ho, 2002).

- *Decentralized agencies all over the country, lacking communication and exchange of information.*

The great number of similar services, dispersed all over the country, deprives of the possibility to communicate and exchange information at the crucial time when they should be ensured. In Greece, for example, before 1998, 300 tax offices and 500 land registry departments used to function per district and without interconnection. At those days, if a tax office employee wanted to be informed about a debtor's real property possession and take measures, he should correspond with 500 land registry departments by sending the equal number of letters! Furthermore, if a citizen owed to a tax office, he could simply state to have lost his ID card and, after getting a new one, submit a new declaration to a different tax office, using a different address, in order to cover his tracks.

- *Personalization or misinterpretation of legislation.*

At the public services of traditional culture, legal interpretations, calculations, display methods and documentation, type of certificates, information provision to citizens and, in general, the employees' behavior are matters of personal choice. Due to the lack of appropriate education, each employee works based on his own understanding of the relevant legislative acts and if someone has additional questions, he will look for answers from the person working next to him or a supervisor. This differentiated level of service often makes citizens to choose a specific employee in order to feel safer or get their job done correctly, even if they have to stand in longer queues and wait much more time.

- *Lack of central monitoring and control for the quality of the offered services to citizens.*

Central administration is absent from the management and control that decentralized units provide to the citizens. This monitoring is randomly exerted by inspectors, who visit the decentralized units once a year, or even not at all. This phenomenon minimizes the potential of central monitoring and decreases the chances of homogeneous service offerings of a standard quality level.

Much of the existing work on the evolution of e-government has explored it from a supply-side perspective, such as evidence presented from surveys of what governments offer online to their citizens. E-government currently has evolved into two identifiable stages. The first stage is the information dissemination phase, while governments catalogue information for public use. The second phase is transaction-based e-government, when e-services, such as paying taxes online, are delivered for public use. The information and transaction phases are closely intertwined with the street-level bureaucracy. One notable finding was that Internet improved the ability for e-citizens to interact with the government, acknowledging some initial movement from street-level to system-level bureaucracies (Reddick, 2004).

- *Absence of an MIS system, which drastically improves the decision-making both centrally and locally.*

The meaning of the terms Management Information System (MIS) and Information System (IS) are identical and interchangeable in an organizational context. They refer to the system providing technology-based information and communication services in an organization. They also refer to the organizational function that manages the system. The concept of a management information system enlarges the scope of information processing to encompass not only applications for transactions and operations, but also applications that support administrative and management functions, support organizational communications and coordination, and add value to products and services (Davis & Olson, 1985).

Above a certain size of an organization, either private or public, management expects "digested" information to support its decision-making process, in the form of exact and relevant data or reports. These data are a reflection of the most important strategic and operational elements of the specific organization, providing an overview of the activities and results for planning, production, marketing, sales, customer relations, cash flow, etc.—depending on the organization's core activity. The quality of those data is of the utmost importance, since the top management of large organizations derives its high-level decisions mostly from such "digests," instead of examining in the depth the core activity components. Since Information Technology as a supporting tool has evolved significantly over the last decades, and business intelligence solutions were developed, more and more attention was drawn to these technologies from companies of the private sector and large or central public administration institutions as well.

Madon and Lewis (2003) also note that "the information system is influenced by both the organizational context in terms of its strategies, structures, politics and culture, and by the wider political socio-economic, cultural and technological climate with which organizations exist."

- *Inefficiency and delays in gathering, arranging, and categorizing facts and data, which are essential for the application of the governmental policies and the short-term and long-term planning (i.e. annual budgeting and reporting).*

An indicative example can be derived from the case of preparation of the Official Greek annual budgeting report before a decade, when tax offices used to work manually and the records of debts were kept on paper files for each debtor. Each year, employees would accumulate the uncollected payments for the balance of each debt (per code) and then they would copy the records to big report sheets and send them to central management. When all the reports were gathered from each regional tax office, a unified report sheet would be created, including all the balances to a national level. Several months after the unified report was prepared, and relatively checked, all the data were used as a basis for the budget report of the following year. It goes without saying that this process contained a big number of mistakes and inaccuracies.

After the implementation of an IT tax system that interconnected all the tax offices and the central unit, the aforementioned data could be collected in a single day and become immediately available to central management.

- *Formulation of long queues in public offices and lack of one-stop shop services.*

Obscure procedures, long queues, re-entering the same information that is already held by the administration, are all practices that are increasingly criticized.

Citizens are gradually getting used to even faster response times and even higher quality of products and services from the private sector, and they expect the same performance from the public sector as well. The provision of public services is rather fragmented, necessitating people to go from one counter to the other. Companies and citizens would much benefit from public services that are seamlessly provided in one-stop shops or online and for which they do not need to know the different departments involved. They also prefer to have services and information tailored to their needs and requirements. However, personalized services require integration and sharing of processes and knowledge across departments and institutions.

Definitions of e-service and e-government are considered in seeking to contextualize the discussion, and a distinction is drawn between e-government and e-public service. While profit motivation has little relevance in the public sector, homogeneity of consumers, definability of tasks and finite and measureable outcomes can serve as likely conditions of success in e-public service. It is proposed that there is a continuum of public sector organizations based on the complexity of task. Initial evidence suggests that e-service delivery has greater potential for success in public sector tasks that have low or limited levels of complexity (Buckley, 2003).

Accordingly, public managers are looking for ways to fully exploit the advantages of Web services and Web services' orchestration technology for improving service delivery, by the creation of citizen-centric, cross-agency processes. Based on pilot projects in the Netherlands, the use of technology requires the introduction of new process orchestration roles and an evolution from hierarchical to agreement-driven relationships (Janssen, et al., 2006).

- *Limitations for the direct tracking—and cracking—of the inconsistent citizens (i.e. tax evasion).*

The inability to cross-reference information in a traditional manual system that does not interconnect with other systems becomes a critical barrier for the tracking of data about inconsistent citizens with regards to their obligations towards the state. Specifically in the area of tax evasion, the IT systems' potential to exchange data and interconnect different services can increase public revenues and expand the taxable base, by tracking tax evaders and taxable material. A relevant example to provide better understanding of the subject concerns the case of a citizen receiving an income from property rentals and avoiding to declare it, in order to evade the income tax payment. With a tax IT system, this citizen's tax evasion could easily be found, by checking the income tax declaration of his tenant - who prefers to declare the amount payable for rent for tax deduction purposes.

- *Lack of business practices, planning, and processes concerning internal auditing*

Internal auditing is an independent, objective assurance and consulting activity designed to add value and improve an organization's operations. It helps an organization accomplish its objectives by bringing a systematic, disciplined approach to evaluate and improve the effectiveness of risk management, control, and governance processes (IIA, 2011). Internal control procedures are policies implemented to provide reasonable assurance of the organization's success in achieving its objectives related to reliable financial reporting, operating efficiency, and compliance with laws and regulations (Corporate Governance, 2011). However, in the public sphere auditing practices are mainly based on financial reporting, while the understanding of improvement margins for processes and practices can make the difference for e-government implementation.

- *Mismanagement of public personnel.*
- *Limited and Experiential Training:* Training in the public sector is relatively limited, especially for innovative practices, and usually takes has a character of emergency response to the issues raised. This lack of complete and formal training is also noticed in many departments, where laws and regulations change rapidly, forcing the civil servants to learn from others' experience or past mistakes.
- *Excess Specialization:* Excess specialization of an organization's personnel is a phenomenon that mostly appears to traditional models of public administration. It presents several drawbacks, such as knowledge withholding for personal gain, difficulty of transferring an employee to a different position or subject, inability of civil servants to realize the overall benefit for the organization due to limited organizational knowledge and narrow vision.
- *Lack of Interest due to Continuous Repetition and Limited Motivation:* Civil servants demand more interesting jobs, with better opportunities for self-development and personal interaction. Since, job security in terms of lifetime employment, benefits and performance assessment are increasingly being eroded, the attractiveness of the government as an employer is at stake.
- *Severe Hierarchical Administration:* The public sector is based on a severe hierar-

chical administration scheme that does not leave much space for individuals that are interested in e-government, innovative and adequately educated to come forward and take charge.

- *No Rewards for Initiative, Creative, or Innovative Ideas:* When an organization does not focus on its intellectual capital, it does not reward innovative ideas and suggestions and keeps its employees' on the sidetracks, then it can never become an organization that learns and transforms. As Warren Bennis quotes, "Innovation—any new idea—by definition will not be accepted at first. It takes repeated attempts, endless demonstrations, and monotonous rehearsals before innovation can be accepted and internalized by an organization. This requires 'courageous patience.'"

THROUGH TRANSITION...

During the design, planning and implementation of an e-government project, certain risks arise, the size of which are usually underestimated on early predictions, and to which one should pay great attention. Some of the most common are:

- *Organizational changes are very difficult to adopt. The employees resist to the introduction of new technology, either because they are afraid of losing some of their tasks, or because they do not want to learn new things and new ways for doing their job.*

Resistance to change emerges when the employee considers the perceived benefit to be low in contrast to an actual high cost (*high cost + low benefit = resistance*). Resistance can be either active or passive. In the first case, active resistance to change may take the form of direct refusal, intimidating / threatening or manipulating tactics, the selective use of facts, criticism, argument, blocking, sabotaging or rumoring. In the case of passive resistance, one can classify reactions that suggest agreement with no response (*Yes – but no effort*), that lead to withholding information, feigning ignorance or standing by and allowing change to fail.

At this point, it should be mentioned that a critical role for the successful introduction of change plays its leader, who must communicate the vision to appropriately selected people that will become change agents. They should be individuals enjoying wide acceptance from the group of people they try to influence, and possess the necessary power and persuasion. Change agents are also the ones that will create the nurturing conditions for the welcoming of change; support it and act as role models for spreading its vision (Cripe, 1993).

- *The adjustments and changes to the legislation are inconsistent and often incompatible with the fast implementation of a flexible IT system for e-government.*

The legislation that is adapted to the processes of a traditional manuscript system cannot be supported by an IT system that aims at simplification and bureaucracy reduction, so it is necessary to change. Any legislative or institutional incompatibilities have to be diagnosed early, ever since the initial design phase of the IT system, in order to make enough time for their transformation and implementation. Synchronization is critical for this process, since the introduction of the IT system should be legally supported and ensured.

The legislative environment, as represented by law practitioners, is difficult to apprehend, realize, and recognize the need for modifications based on a future system, which is not yet tangible, visible, or easily understandable. Thus, the agents suggesting the modifications during the design phase have to be able to familiarize with and justify the whole spectrum of upcoming changes. Legislative changes can make it easier to do the right thing from the start by addressing opportunities and/or providing incentives. This can be a very cost-effective way to increase compliance (Fiscalis, 2010).

- *The parallel operation of both the new IT system and the traditional system cause confusion and overlapping (e.g. legislation of two speeds) that can make the procedures more complex.*

During the establishment of a complete IT system to several common public services (e.g. tax offices), it is unavoidable that both systems, the old and the new, will coexist and co-function over a period of time. This fact suggests the need of a legal framework that will combine and legitimize simultaneously all the available procedures. For this period, only careful strategic design can prevent perplexity and possible overlapping from occurring.

It is often noticed that, during the period of parallel system functioning and until the total transition to the new system, procedures might become more complex than before. As a result, in the short term, the system users might respond with skepticism to the feasibility and value of the adoption of an IT solution that is meant to simplify procedures and improve the effectiveness of their work.

- *The synchronization of activities is critical*

The establishment of a complete IT system to the public services is accompanied by a series of activities that precede, happen simultaneously or follow, and a single delay of an activity can alter the whole timeframe (scheduling) of a project.

To support our point, it is worth mentioning the example of the Greek Integrated Tax System (TAXIS), which was implemented to 300 tax offices distributed over the country. The initial plan expected the gradual establishment of the system to 10 tax offices per week. Before the implementation, a group of experts had to visit the unit and survey the space availability and complete a study concerning their adequacy. Because of the results of these surveys, one third of the tax offices had to relocate and all the rest had to reconstruct or repair their buildings and re-order the spatial location of several departments. Next, the implementation team should oversee the electrical installation, order new office furniture for the computers, move or destroy old equipment, install the hardware and the software, train the personnel (to a training center or on-site at the beginning of the operational phase), enter new data to the system, check the proper functioning of the system, inform the citizens through the local press, and provide technical support and user on-site (for the first week of operations) or distant (call-center) support, as well as continuous training for the employees that would change their scope of work. In conclusion, if one of the aforementioned activities was delayed (e.g. relocation), the whole timetable of implementation would have to be planned again, a process that lead to finalizing the implementation phase with a delay of four months (due to parallel processing of different tasks).

- *Cases of delays with vendors, mismanagement, internal fighting against the new concept in its introduction will most likely appear. Deadlines are sometimes over-run and budgets exceeded.*

For an IT public project, many contractors are responsible for different subsections, as for example hardware equipment, application software, supply and installation of furniture, electrical networks and installations, data entry, user training, etc. Every contractor is signing a different agreement with the government, which includes time schedules that are agreed upon the original master plan and describes the technical and operational guidelines. These contracts can happen in a serial or parallel manner and they can be controlled by the same or different governmental teams operating in the public's interest.

More often than not, delays, missed deadlines and failures happen from a contractor, that lead to new agreements, penalties and timetables for the contractors, whose work follows or depends on previous steps. Then, the allocation of responsibility is not easily detected, the delay/rescheduling of the rest of the project's subsections increase the overall budget and cause frustration to the responsible supervising and managerial teams.

On the other side, the phenomenon of internal (in-service) dispute is observed, concerning the solution a sub-team has chosen, affecting the work of another team. For example if a choice of the hardware equipment team comes in contrast to the solution of the software team, an incompatibility emerges for which each team blames the other. This results to a significant conflict and confusion among the managerial governmental team and allocation of responsibility becomes a difficult-to-resolve issue.

These problematic areas have been diagnosed over the years and experience has led to collaborating schemes for the contractors (usually IT companies), that apply for public procurement projects in full (packages of subsections) and then use sub-contractors. From the governmental aspect, previous experience provides the reasoning to select a single coordinating project management team.

- *Managing external providers and partners can prove to be like "learning to speak a new language."*

Open public procurement tenders, especially in the area of informatics, offer the possibility of participation to foreign companies. In such cases, the English language is mainly selected for oral and written communication, because many technical terms are established in practice. Therefore, the first issue of understanding and communication that usually arises for the involved parties lies to the need of triple translation with the use of a—in-between common—language different than their native. This issue is accentuated by the fact that all of the foreign stakeholders, even the ones who live in the same country and speak the same language, are not always familiar to the specific terminology used by public services.

Furthermore, the language barrier can appear to the difference of understanding between business experts and technical analysts. Particularly during the phase of the *as-is* analysis and the description of user needs for an IT project, a strenuous process of making clarifications has to be undertaken between business experts and technical analysts, that results in more time spending for the familiarization with terms. The outcome of this process is a new "language" of communication and understanding that includes the pre-agreed terms and it is used every time a previous issue comes forward or a deliverable is prepared.

The project partners have to speak an in-between coded language that will allow technical analysts understand a process or an operation, and business experts understand technical and technological terms. It is advisable that all of these people remain on board to the project for its completion and maintenance, so as to be exploited as expert users and consultants for the transference of knowledge.

For example, in TAXIS, whenever a need arose for the adaptation of the IT system to new legislative regulations, the administrative tax official would inform the knowledge tax expert of TAXIS for the new regulation coming to force and, in his turn, the tax expert would inform the technical analyst to implement the change by using his know-how and the agreed-upon coded language. In this way, a theoretical change was transformed to new user requirements for the specific IT system.

- *Training of a potentially high number of differentiated or specialized employees, who may be dispersed all over the coun-*

try—proper scheduling and extreme organizing are very critical for success.

E-government projects usually involve a great number of participating users, geographically dispersed, and highly specialized. This is the reason why, extensive training is almost always a prerequisite for success and must be applied in different speed / timing and with multiple methods. A complex IT project requires the design of several training programs, depending on the extent of specialization, the users' technical expertise, PC literacy and self-efficiency, the levels of the system's accessibility by users, the managerial structuring, as well as the availability of training centers, distant learning, and user support (e.g. call-centers) infrastructures for continuous training, on-site training schemes by experts during the initial implementation, and the affordability to consult relevant case studies, user guides, and training material. The users of new technologies have to be trained quickly and efficiently, and since they are usually distributed to different remote locations, Web-based training is the preferred, and sometimes the only, process for employee training (Chatzoglou, 2009).

Thus, training for an e-government IT project demands becomes a critical success factor, which depends heavily on proper planning, good timing based in the project's implementation and availability of financial resources, infrastructure, logistical equipment and appropriate trainers. The final outcome is based on the effective and on-time implementation of the training program and the level of knowledge consolidation by the system users.

- *Difficulty in recruiting well-qualified talents of IT or business experts, because of public administration codes and regulations. In addition, incentives for attracting those people are low; low wages, loss of prestige, and mundane duties associated with the public service, have led the majority to the private sector.*

Recruitment of qualified talents for the development of e-government projects is a challenging case, because it relates to a series of activities that each public organization should follow in order to locate and attract the appropriate job candidates. In every case, a matching model must be applied to correlate the employee's contribution to the organizational capacities. The employee's contribution is a factor dependent on a total of skills, knowledge, dedication, specialization, creativity, positive attitude and interaction, innovation and commonality of vision. Accordingly, the organizational capacities relate to the rewarding system, the provision of benefits, the availability of continuous training and knowledge enrichment, as well as to the existence of opportunities for personal development, inspiration, and career advancement. As stated by Armstrong (1996), organizational capacities refer to "whatever allows the organization to acquire, manage, develop, and reward the individuals that possess the necessary abilities to maximize their contribution and achieve organizational objectives."

The *sources of recruitment* can either be internal from the organization's workforce, or external from the private sector.

- *Internal sources* of recruitment refer to the tracing of experts that will be involved in the description of the as-is situation and provide user requirements. In addition, it can refer to IT personnel that will be accordingly trained to monitor the technical and functional design of the project, in order to be able to maintain and improve it after implementation. Finally and foremost, internal sources are exploited to locate the persons that will be responsible for project management, programming and smooth operationalization of activities, and suitable for the collaboration and monitoring of the private partners' contractual obligations. The organization's executives should possess the necessary drive and knowledge, but also have specific skills like good judgment, integrity, maturity, organizational flexibility, imagination, rational thinking, inspiration, vision, focus to the project's purpose and ability to people development. They have to be in the position of recognizing and hierarchizing important issues, overcoming organizational limitations and barriers, interacting effectively with business partners, realizing the power and policies of the public organization, and, finally, tracing the appropriate persons or the key factors for decision-making.

- *External sources* of recruitment refer to the careful selection of individuals that will work for the project's development, mainly IT people and administrative executives from the private contractors that will collaborate with the public organization's personnel. In many cases, recruitment can be based to academic researchers or to external consultants that specialize to market conditions and/or research methods. In addition, it can relate to the recruitment of seasonal staff, for data entry needs. The executives from the private sector should possess the ability to combine their knowledge/skills with technology and focus on results, efficiency, and effectiveness.

Internal recruitment presents significant advantages for the organization, such as lower costs, better identification to organizational goals, higher reliability and limited adjustment period, but also some critical disadvantages, like the reproduction of organizational mentality, the danger of meritocracy, and the resistance to prompt commands. On the other side, external recruitment offers the advantages of specialized knowledge, of the introduction of new ideas, of the limited need for training and of the positive contribution to change processes, while it presupposes the disadvantages of higher cost, increased selection risk and more time for adjustment.

The drawback of the public sector lies to its difficulty to locate and attract executives from its internal sources, due to low motivation (often people that leave their previous position to participate to innovative and inspirational projects are left behind promotional schemes and career advancements) and to administrative barriers (e.g. time consuming procedures for departmental transference, relocation and re-positioning). As a result, the public sector tends to address the private sector for staffing, because the private sector enjoys more timely response activation and better flexibility.

TO IMPLEMENTATION!

After recognizing all the critical points and realizing the existing functional and systemic conditions, it is possible to come across a proposal of smoother transition to the electronic era. However, in order to reach a successful conclusion for any public project, the selected e-solution should also respect the following basic principles:

1. Sustainability and Maintainability

Sustainable development 'meets the needs of the present without compromising the ability of future generations to meet their own needs' (Brundtland, 1987, p. 8). So, the implementation of any IT e-government system should follow the principles of sustainable business practices as a guide to the transition of a sustainable economy (Esty & Winston, 2006), while the system's maintainability and adaptation to modern expectations are necessary for guaranteeing its longevity and usefulness.

2. Scalability: Upgrading and Adding New Applications

In software engineering, scalability is the ability of a system, network, or process, to handle growing amounts of work in a graceful manner or its ability to be enlarged to accommodate that growth (Bondi, 2000). Scalability can be measured in various dimensions, such as:

- **Administrative scalability:** The ability for an increasing number of organizations to easily share a single distributed system.
- **Functional scalability:** The ability to enhance the system by adding new functionality at minimal effort.
- **Geographic scalability:** The ability to maintain performance, usefulness, or usability regardless of expansion from concentration in a local area to a more distributed geographic pattern.
- **Load scalability:** The ability for a distributed system to easily expand and contract its resource pool to accommodate heavier or lighter loads. Alternatively, the ease with which a system or component can be modified, added, or removed, to accommodate changing load.

In the case of e-government projects, scalability is very important because the IT systems will have to grow in direct relation to the market's growth, to the increased need for connectivity and support of different public agencies and to the citizens' expectations.

3. Flexibility and Connectivity with Other Systems: Synergies to be Derived

Synergies require that the IT systems of various participating administrative bodies are interconnected and information and administrative processes are 'joined up,' i.e. they must 'interoperate.'

One Member State's administration should be able to access an information resource of another Member State's administration to validate the status of a company or to check the eligibility for social welfare of a citizen from another Member State, with the same ease as it would do for nationally registered companies or citizens. Similarly, the technical and semantic interoperability of geographic information for example, would enhance trans-border intra-agency co-operation, environmental monitoring, and the co-ordination of disaster relief for the private sector.

However, interoperability is not just a technical issue of linking up computer networks. It also concerns organizational issues, such as coordinating processes that span not only intra-organizational boundaries, but also co-operate with partner organizations that may have different internal organization and operations.

Failure to put in place interoperable e-Government systems will induce both economic and social costs. These include: static unresponsive public administrations that are expensive to run and are unable to implement policy timely; inability to develop value added e-Government services; higher costs, greater administrative burden and competitive disadvantage relative to local firms (Liikanen, 2003).

From the *operational aspect*, the following actions should be taken:

- *Set the terms of the systemic research and design, and place contracts with partners.*

Many brainstorming sessions should be organized and best practices from international experience should be studied, in order to compose the master plan of the systems' implementation. This procedure includes the governmental vision and aims, the analysis of the current situation, the suggested models and solutions, as well as the functional and technical planning of the most suitable solution following the feasibility study. As soon as this study is completed and approved by the leadership of the organization, the project is proclaimed to an open public tender, the contractor(s) is selected and the agreement is signed.

- *Secure that the project fulfills the criteria for receiving funding (national or other).*

Funding usually concerns a complex and strenuous procedure, which should be defined at an early stage and surveyed at every stage of implementation, to ensure that the final product abides by the terms and the criteria set for the disbursement of funds. Often, the sources of funding in e-government projects are more than one, e.g. public investment by the national government, or a European funding (aid) to the member states of the Eurozone. In the second case, the project should ensure the criteria of funding that the European Union sets and not only the national government.

- *Propose adjustments and changes for the simplification of tax legislation. The new philosophy and the new norms and procedures brought by the IT, have to be accommodated by the existing rigid legislation system.*

The team working on system development and implementation is responsible for the submission of the proposals to the public organization's leadership with regard to the modification of the legal framework and the simplification of procedures of the new operational environment formed after its operationalization. It is inconceivable for an IT environment to be supported by traditional and often old-fashioned procedures and institutions, which basic purpose was to support mainly manual activities, to solve problems and eliminate the lack of communication between different departments/ points of service of the organization.

Furthermore, the same team should organize its activities, in order to raise the senior executives' awareness, because the adoption of any institutional proposals is depending on their decisions. Often the sensibility of those who are responsible for the projects is underestimated, leading to the confrontation of obstacles, delays, and negativity by the legislatures who experience a knowledge gap concerning the decisions referred due to lack of relative information.

- *Create a number of internal business processes streamlining and re-engineering provisions, aiming at better quality of service.*

The technology relevant to IT projects developed on behalf of organizations for the public sector should aim on the provision of qualitative services to citizens and the limitation of people's and businesses complaints about governmental requirements. This aim is achievable with the invention of automatic internal processes and networks using either the same or different IT systems, in order to make all the information available to the user, irrespectively of where it has been initially registered. This means that the customer/citizen would not need to adduce piles of documents. At this point, one could refer to the example of the Citizens' Service Centers of the Greek Ministry of Internal Affairs that are located in every municipal department of Greece. At these Service Centers, citizens have -among others- the opportunity to submit tax statements, request and receive copies and certificates from the majority of public authorities, either through on-line connection of the Center with the network of each authority or through an internal process systemized for this purpose, functioning via telephone, e-mail, or fax.

- *Design teams; allocate tasks and duties to capable and willing employees. A critical factor for success is selecting the right person for the right job.*

The recruitment process for the project's main team and sub-teams depends initially on individuals' willingness to participate in the project and, secondly, on their personal characteristics, innate or acquired from knowledge and experience, which the project team examines. Furthermore, emphasis should be given to the working mentality and the new understanding of job requirements. This means that the members of each group should embrace the vision and aims of the project and try to contribute according to the role they are expected to play. Thus, the change of mentality consists in the transition from "my job" to "my role."

The organizational structure of IT projects usually includes three stages: the project's managerial team, the team responsible for the subsystems or units of the project and finally the technical analysts, the business experts and other specialists, who work on the development of every subsystem or unit.

The managerial team often includes roles such as: coordinator manager, functional manager, technical manager, and human resources manager. Additionally, the sub-teams that refer to subsystems or units of the project may share a horizontal or vertical integration. As an example of vertical integration, one could mention design, software development, control, implementation, and proper functioning of a project's subsystem, while in the case of horizontal completion one could refer to the design team and implementation of training.

- *Complete the necessary technical assignments to prepare the infrastructure (hardware, software, buildings, offices, etc.) accommodating the new e-processes.*

The development and establishment of a new service or system demands a set of conditions and procedures for the preparation and the creation of the necessary infrastructures, which completion follows the implementation schedule. Specific examples include: the control of buildings' adequacy or the need for services' relocation, the re-organization of departments and offices, the installation of electrical equipment and networks, the delivery and arrangement of furniture and mechanical equipment, the sufficiency control of air-conditioning supplies, the removal of outdated equipment and office supplies, the delivery of new office supplies and expendables, the setting and functioning of the appropriate software. Thus, the most critical success factors are timing and team coordination.

- *Prepare databases and set data-entry rules (when applicable).*

A necessary condition, preceding the software application to the operational points of service, is the preparation of databases and tables of reference. A critical decision for the system's proper function is if and what historical information will be imported to the system and how important they can be for its completeness. Due to significant complexity, data entry costs for the replacement of handwritten records and increased possibility of mistakes, most modern IT systems are designed so as to demand the minimum—only necessary—historical data. In every case, a set of rules and a set of programs must be prepared for the process of data entry, which would define the mandatory fields of the system to insert specific information. These fields can then be used as control keys / points for an operation, setting the acceptable value range for each field according to codifications or logical limitations of imported characters and the default prices when there is missing information for the specific control key. For example, based on the required personal information fields of a citizen, when the year of birth is not known during data entry, an indicative year (1900) can be used to signify the missing value and alert the end user to find the actual date and correct the entry later.

Data entry programs should be designed to be open (with no systemic controls) and easy to use (with not many screen transitions), in order to facilitate the speed of recording. In addition, the need for designing a special handwritten form should not be neglected, because the authorized department has to capture, check, and validate the data that are copied from the handwritten records, or create specific transition programs for the collection of data from older computerized records.

- *Run system tests and pilot applications.*

Testing the system's good operation is consisted of running a series of documented usability scenarios related to real life situations, which purpose is to examine whether the system responds in the predicted manner. So, technical analyst and expert users that are responsible for the documentation of scenarios and the design of the system's tests, have to exhaust their imagination in predicting any probable or improbable handling of the IT system and put their selves in the position of the most ignorant or the most clever and cunning end user.

For instance, in a tax office during full operation and in service of dozens of citizens, a security officer, who was annoyed by the fan noise coming out of the main server, pressed the power button to switch off the fan. As a result, the system's application ended abruptly, so the entire task team and the tax office personnel had to spend days to retract the data that might have been lost. This anecdotal case brought forward the need for extra security measures for the safety of the main server.

A pilot application is an especially safe procedure, mainly for long range IT projects that are dispersed to many geographical points. A service or a unit is selected out of the group and all the procedures are followed for the implementation and operationalization of the system. Pilot testing under actual—real life—conditions is a method that allows for future problems—that could not be identified during lab testing—to emerge and be corrected before full scale implementation of the system. Prepare databases and set data-entry rules (when applicable).

- *Train end users, public officers, and IT operational administrators.*

Training of end users and of the operational administrators, who are responsible of the system's on-site maintenance, should finalize shortly before the beginning of the system's operation. Specifically, the month before the launch is a safe time span, given that local support will be provided for the first days of operation, covering also the need for job training. For more complex information systems, which have many points implementation, high degree of staff specialization, different subsystems and multiple rights of access, it's difficult to respect the aforementioned one month rule—unless correct prediction, proper planning, and appropriate training scheduling precede the implementation phase.

- *Continue training after the implementation, and set up help desks for providing support to the end users both locally and centrally.*

Continuous training processes are an essential element of the sustainability and improvement of the system's effectiveness. Novice users, after a few months of familiarization with the system, desire to learn more tools or operational subsystems. They tend to have new demands and adjust more tasks or functions according to this new work method, therefore new questions occur, which the project team has difficulty to address via telephone communication.

Additionally, end users might be promoted or transferred to positions with different operational requirements; new employees might be hired and trained from the beginning, while there is always a need for updated training for additional new operations and systems. An effective and economic method to resolve all these issues, besides the solution of seminars, is distant learning or e-learning that diminishes any relocation difficulties that might arise.

A help desk can be a simple or a more complex service of particular significance, for the response to "hot" problems that demand an immediate attention. It usually consists of a call center that initially receives each question, and, if the call center is unable to provide the answer, it transfers the question to the responsible project sub-team.

Furthermore, responses to frequently asked questions can provide a relevant Web site and a call service, which gradually obtains autonomy and extensive expertise.

A more compound solution can offer the existence of a supporting network, locally and centrally, with persons and task forces able to provide direct intervention on-site at a minimum response time. This kind of solution is usually applied at the beginning of a system's operation, when there is a great dispersion rate for a limited amount of time and a large number of on-site transactions with citizens.

- *Inform citizens and "market" the new philosophy internally and externally (AIDA [Awareness, Interest, Desire, Action])*

Raising citizens' awareness for their potential when transacting with governmental services should be based on a targeted communication plan, focused on promotion, acceptance and adoption of the new philosophy. More specifically, citizens' interest for the use of innovative services can be reinforced, if the public emphasize separately to each sector, company, or individual on the benefits provided by change. Then the citizens' motivation and desire to adopt the new system and comply with modern processes can increase, resulting in affirmative actions and positive responses.

- *Continuously review and improve the system.*

Technology evolves rapidly and so do the users' needs of an information system. As end users pass through the technology adoption stages, they ask for more and more information and functions, relevant to citizens' demands for new services. From the moment the system's implementation is completed, the supporting project team needs to observe the system and constantly suggest improvements. Continuously updated services, of shorter or longer range, can keep public motivation active and offered services modernized.

LESSONS LEARNED

- **Think big, start small, scale fast:** Start with rudimentary easily understandable processes. Then coordinate those activities into an overarching strategic vision. Sustain people's interest by earning little wins.
- **Face governmental facts:** Public sector organizations operate in settings very different from the private sector. Change is inherent in implementing public policies. Laws are changed, priorities shift, and implementation accordingly has to adjust. In addition, small policy changes can require major changes in IT structures. Similarly, the time allowed for legislation to come into effect, is often too short for proper IT systems to be built and launched.
- **Avoid emerging technologies:** New technologies are tempting because they often promise better solutions and fascinating possibilities for business change. The risk of failure can be reduced by using well-proven approaches or—even better—standard software, although this will often imply that business processes have to be adapted to the possibilities offered by the IT system. The application of common commercial practices, rather than "customized software," has proven time and time again to be the most successful solution.
- **Identify and manage risks:** Risk identification and management are paramount features of successful IT project management.

Independent consultants from outside the administration can help identify risks and offer clairvoyant solutions. Knowledge management and management control systems adapted to the culture must be implemented.

- **Strengthen leadership and accountability, focus on business change:** Leadership is a key issue of project management. Clear lines of responsibility and accountability are needed for the successful management of any IT project, because there is no such thing as an IT project in isolation. Rather than, every system should be seen as a tool and means to other ends -notably a change in processes. IT projects are thus business projects that need to be *led* by top management and not simply *run* by IT or business experts.
- **Manage knowledge and human resources:** A recurrent problem is the lack of IT skills in the public sector. Against the background of a very tight IT labor market and an ever-increasing demand for high-qualified staff, the competitiveness of the public employer has to be visibly strengthened. Ways to do that include differentiated payment systems, provision of initiatives for creativity and innovation, and overall better knowledge and human resources management.
- **Involve end-users:** the potential impact of IT initiatives on people must be anticipated. Thus, effective and appropriately timed training, user support and change management procedures should be implemented, in order to positively predispose the end-users. On the other side, to increase positive attitude towards e-Government services, governmental agencies should develop implementation strategies that emphasize the usefulness of e-Government services, work-style compatibility, and user trust (Hung, et al., 2006). In conclusion, the main factors for success rely upon the sensitivity, ability, and flexibility of the human resources who participate in the transformation of big projects from vision to reality.

REFERENCES

Alter, S. (2002). *Information systems: The foundation of e-business* (4th ed.). Upper Saddle River, NJ: Prentice Hall.

Armstrong, M. (1996). *A handbook of personnel management practice*. London, UK: Kogan Page.

Bennis, W. (2005). *Quote*, Retrieved 23/7/2011, from http://asideconsulting.blogspot.com/2005/08/warren-bennis-quote.html

Blau, P. (1956). *Bureaucracy in modern society*. New York, NY: Crown Publishing Group/Random House.

Bondi, A. B. (2000). Characteristics of scalability and their impact on performance. In *Proceedings of the 2nd International Workshop on Software and Performance*, (pp. 195-203). Ottawa, Canada: IEEE.

Brundtland, G. H. (1987). *Our common future: Report of the World commission on environment and development*. Oxford, UK: Oxford University Press.

Buckley, J. (2003). E-service quality and the public sector. *Managing Service Quality*, *13*(6), 453–462. doi:10.1108/09604520310506513

Ceriello, V. R. (1991). *Human resource management systems –Strategies, tactics and techniques*. Lexington, VA: Lexington Books.

Chatzoglou, P. D., Sarigiannidis, L., Vraimaki, E., & Diamantidis, A. (2009). Investigating Greek employees' intention to use web-based training. *Computers & Education*, *53*, 877–889. doi:10.1016/j.compedu.2009.05.007

Choudrie, J., & Dwivedi, Y. K. (2004). Investigating the socio-economic characteristics of residential consumers of broadband in the UK. In *Proceedings of the Tenth Americas Conference on Information Systems*, (pp. 1558-1567). New York, NY: IEEE.

Cripe, E. J. (1993). How to get top-notch change agents. *Training & Development*, *47*(12), 52–54, 57–58.

Davis, G. B., & Olson, M. H. (1985). *Management information systems: Conceptual foundations, structure, and development* (2nd ed.). New York, NY: McGraw-Hill, Inc.

Duboc, L., Rosenblum, D. S., & Wicks, T. (2006). A framework for modelling and analysis of software systems scalability. In *Proceeding of the 28th International Conference on Software Engineering ICSE 2006*, (pp. 949-952). ICSE.

Esty, D. C., & Winston, A. S. (2006). *Green to gold: How smart companies use environmental strategy to innovate, create value, and build competitive advantage*. New Haven, CT: Yale University Press.

Governance, C. (2011). *Wikipedia*. Retrieved from http://en.wikipedia.org/wiki/Corporate_governance

Hill, M. D. (1990). What is scalability? *ACM SIGARCH Computer Architecture News*, *18*(4), 18–21. doi:10.1145/121973.121975

Holliday, I., & Kwok, R. C. W. (2004). Governance in the information age: Building e-government in Hong Kong. *New Media & Society*, *6*, 549. doi:10.1177/146144804044334

Hughes, O. E. (2003). *Public management and administration* (3rd ed.). Hampshire, UK: Palgrave Macmillan.

Hung, S.-Y., Chang, C.-M., & Yu, T.-J. (2006). Determinants of user acceptance of the e-government services: The case of online tax filing and payment system. *Government Information Quarterly*, *23*, 97–122. doi:10.1016/j.giq.2005.11.005

IIA. (2011). *Wikipedia*. Retrieved from http://en.wikipedia.org/wiki/Institute_of_Internal_Auditors

Janssen, M., Gortmaker, J., & Wagenaar, R. W. (2006). Information systems management. *Web Service Orchestration in Public Administration: Challenges, Roles, and Growth Stages*, *23*(2), 44–55.

Kegan, R., & Laskow-Lahey, L. (2001). *The real reason people won't change. Harvard Business Review OnPoint*. Boston, MA: Harvard Business School Publishing.

Kettl, D. F. (2002). *The transformation of governance*. Baltimore, MD: John Hopkins University Press.

Kotter, J. P. (1995, March/April). Leading change: Why transformation efforts fail. *Harvard Business Review*.

Lawrence, P. R. (1954, May/June). How to deal with resistance to change. *Harvard Business Review*.

Liikanen, E. (2003). *e-Government for Europe's public services of the future*. Retrieved from http://www.uoc.edu/dt/20336/index.html

Madon, S., & Lewis, D. (2003). *Information systems and non- governmental development organizations (NGOs): Advocacy, organizational learning and accountability in a southern NGO*. Working Paper. London, UK: London School of Economics.

Reddick, C. G. (2005). Citizen interaction with e-government: From the streets to servers? *Government Information Quarterly*, *22*(1), 38–57. doi:10.1016/j.giq.2004.10.003

Report, F. E. (2010). *Helping taxpayers get it "right from the start"*. Stockholm, Sweden: Swedish Tax Agency.

Saxena, K. B. C. (1996). Reengineering public administration in developing countries. *Long Range Planning*, *29*(5), 704–712. doi:10.1016/0024-6301(96)00064-7

Scalability. (2011). *Wikipedia.* Retrieved from http://en.wikipedia.org/wiki/Scalability

Sen, A. (1999). *Development and freedom*. Oxford, UK: Oxford University Press.

Tat-Kei Ho, A. (2002). Reinventing local governments and the e-government initiative. *Public Administration Review*, *62*(4), 434–444. doi:10.1111/0033-3352.00197

Teicher, J., Hughes, O., & Dow, N. (2002). E-government: A new route to public sector quality. *Managing Service Quality*, *12*(6), 384–393. doi:10.1108/09604520210451867

Turban, E., McLean, E., & Wetherbe, J. (2002). *Information technology for management: Transforming business in the digital economy* (3rd ed.). New York, NY: Wiley.

UN/ASPA. (2002). Benchmarking e-government: a global perspective. Retrieved from http://unpan1.un.org/intradoc/groups/public/documents/un/unpan021547.pdf

Compilation of References

Abbate, J. (1999). *Inventing the internet*. Cambridge, MA: MIT Press.

Abramson, M. A., & Harris, R. S. (2003). *The procurement revolution*. Sterling, CT: Rowman and Littlefield.

Accenture. (2002). *E-government leadership – Realizing the vision*. Retrieved from http://www.accenture.com/xd/xd.asp?it=enWeb&xd=industries%5Cgovernment%5Cgove_welcome.xml

Acevedo, B., & Common, R. (2006). Governance and the management of networks in the public sector. *Public Management Review*, *8*(3), 395–414. doi:10.1080/14719030600853188

Acevedo, B., & Common, R. (2010). Governance, networks and policy change: The case of cannabis in the United Kingdom. In Osborne, S. P. (Ed.), *The New Public Governance?* (pp. 394–412). New York, NY: Routledge.

Acquisti, A., & Gross, R. (2006). Imagined communities: Awareness, information sharing, and privacy on the Facebook. In P. Golle & G. Danezis (Eds.), *Proceedings of 6th Workshop on Privacy Enhancing Technologies,* (pp. 36–58). Cambridge, UK: Robinson College.

Adams, B. (2004). Public meetings and the democratic process. *Public Administration Review*, *64*(1), 43–54. doi:10.1111/j.1540-6210.2004.00345.x

Agarwal, V., Mittal, M., & Rastogi, L. (2003). *Enabling e-governance – Integrated citizen relationship management framework – The Indian perspective*. Retrieved from http://www.e11online.com/pdf/e11_whitepaper2.pdf

Akther, M. S., Onishi, T., & Kidokoro, T. (2007). E-government in a developing country: Citizen-centric approach for success. *International Journal of Electronic Governance*, *1*(1), 38–51. doi:10.1504/IJEG.2007.014342

Al-Kibsi, G., Boer, K., Mourshed, M., & Rea, N. (2001). Putting citizens on-line, not in line. *The McKinsey Quarterly*, *2*, 65–73.

Allen, R. (2001). Workflow: An introduction. In Fischer, L. (Ed.), *The Workflow Handbook 2001* (pp. 15–38). Lighthouse Point, FL: Future Strategies.

Alter, S. (2002). *Information systems: The foundation of e-business* (4th ed.). Upper Saddle River, NJ: Prentice Hall.

American Society for Quality. (2008). Basic concepts, glossary: Quality. *American Society for Quality.* Retrieved November15, 2008, from http://www.asq.org/glossary/q.html

Amoah, L. G. A. (2005). *Democratization and development aid: Ghana's experience (1989-2003).* (Unpublished Master of Arts Thesis). University of Ghana. Legon, Ghana.

Anaman, K. A. (2006). *Factors that enabled successful economic reform and recovery in Ghana from 1983 to 2006.* Accra, Ghana: Institute of Economic Affairs.

Anderson, K. (1999). Reengineering public sector organizations using information technology. In Heeks, R. (Ed.), *Reinventing Government in the Information Age*. London, UK: Routledge.

Andresani, G., & Ferlie, E. (2006). Studying Governance within the British public sector and without. *Public Management Review*, *8*(3), 415–431. doi:10.1080/14719030600853220

Anttiroiko, A. (2005). Towards ubiquitous government: The case of Finland. *E-Service Journal*, *4*(1), 65–70. doi:10.2979/ESJ.2005.4.1.65

Anyimadu, A., Esseloor, S., Frempong, G., & Stark, C. (2005). Ghana. In Gillward, A. (Ed.), *Towards an African e-Index: Household and Individual ICT Access across 10 African countries*. Johannesburg, South Africa: RIA.

Anyimadu, A., & Falch, M. (2003). Tele-centers as a way of achieving universal access - The case of Ghana. *Telecommunications Policy*, *27*, 21–29. doi:10.1016/S0308-5961(02)00092-7

Apfel, A. L. (2002). *The total value of opportunity approach*. Washington, DC: Gartner Research.

Apfel, A. L., & Murphy, T. (2007). *Five perspectives beyond ROI: A process for scoring and prioritizing projects and program*. Washington, DC: Gartner Research.

Archer, N. P., & Ghasemzadeh, F. (1999). An integrated framework for project portfolio selection. *International Journal of Project Management*, *17*(4), 207–216. doi:10.1016/S0263-7863(98)00032-5

Armstrong, M. (1996). *A handbook of personnel management practice*. London, UK: Kogan Page.

Artto, K. A., Martinsuo, M., & Aalto, T. (2001). *Project portfolio management: Strategic business management through projects*. Helsinki, Finland: Project Management Association Finland.

Aryettey, E. (1996). *Structural adjustment and aid in Ghana*. Accra, Ghana: Fredrich Ebert Stiftung.

Atkinson, R. (2003). *Network government for the digital age*. Washington, DC: Progressive Policy Institute.

Australian Public Service Commission. (2004). *Connecting government: Whole of government responses to Australia's priority challenges*. Report No. 4. Canberra, Australia: Management Advisory Committee.

Azuma, M. (2001). SQuaRE: The next generation of the ISO/IEC 9126 and 14598 international standards series on software product quality. In *Proceedings of the European Software Control and Metrics - Escom 2001*. London, UK: ESCOM.

Backus, M. (2001). *E-governance and developing countries: Introduction and examples*. Research Report, No. 3. The Hague, The Netherlands: International Institute for Communication and Development. Retrieved from editor.iicd.org/files/report3.doc

Backus, M. (2001). *E-governance in developing countries*. Research Brief No. 1. New Delhi, India: International Institute for Communication and Development (IICD).

Baev, V. (2005). Social and philosophical aspects of e-governance paradigm formation for public administration. In *Razon Y Palabra, 42*.

Bakvis, H., & Juillet, L. (2004). *The horizontal challenge: Line departments, central agencies and leadership*. Ottawa, Canada: Canada School of Public Service.

Barkatullah, T. M. (2008). ICT human capacity building. In *Country Report: Bangladesh*. Dhaka, Bangladesh: Bangladesh Computer Council.

Basu, S. (2004). E-government and developing countries: An overview. *International Review of Law Computers & Technology*, *18*(1), 109–133. doi:10.1080/13600860410001674779

Baum, C., & Maio, D. (2000). *Gartner's four phases of e-government model*. Washington, DC: Gartner's Group.

Baumol, W. J., Litan, R. E., & Schramm, C. J. (2007). *Good capitalism, bad capitalism and the economics of growth and prosperity*. New Haven, CT: Yale.

BCC. (2009). *About us*. Retrieved from http://www.bcc.net.bd/

Bennis, W. (2005). *Quote,* Retrieved 23/7/2011, from http://asideconsulting.blogspot.com/2005/08/warren-bennis-quote.html

Berner, M. (2001, Spring). Citizen participation in local government budgeting. *Popular Government*, 23–30.

Berry, F. S., Brower, R. S., Choi, S. O., Goa, W. X., Jang, H., Kwon, M., & Word, J. (2004). Three traditions of network research: What the public management research agenda can learn from other research communities. *Public Administration Review*, *64*(5), 539–552. doi:10.1111/j.1540-6210.2004.00402.x

Bhatnagar, S., & Chawla, R. (2004). *Online delivery of land titles to rural farmers in Karnataka, India*. Paper presented in 'Scaling up Poverty reduction: A Global Learning Process and Conference'. Shanghai, China.

Bhatnagar, S. (2003). E-governance and access to information. In Hodess, R., Inowlocki, T., & Wolfe, T. (Eds.), *Transparency International Global Corruption Report 2003: Special Focus Access to Information* (p. 28). London, UK: Pluto Press.

Bhatnagar, S. (2004). *E-government: From vision to implementation: A practical guide with case studies*. New Delhi, India: Sage Publications.

Bhatnagar, S. C. (1999). *E-government: Opportunities and challenges*. Ahmadabad, India: Indian Institute of Management.

Bhatnagar, S. C. (2001). *Philippine customs reform*. Washington, DC: World Bank.

Bhattacharjya, J., & Chang, V. (2006). *An Exploration of the Implementation and Effectiveness of IT Governance Processes in Institutions of Higher Education in Australia*. Australia: Curtin University of Technology.

Bhavnagar, S. (2000). Social implications of information and communication technology in developing countries: Lessons from Asian success stories. *The Electronic Journal of Information Systems in Developing Countries*, *1*(4), 1–9.

Bhogle, S. (2008). E-governance. In Anttiroiko, A.-V. (Ed.), *Electronic Government: Concepts, Methodologies, Tools and Applications*. Hershey, PA: IGI Global.

Blau, P. (1956). *Bureaucracy in modern society*. New York, NY: Crown Publishing Group/Random House.

Bon, A. (2007). *Internet bandwidth problems: Special case of Ghana*. Paper presented at the Web 2.0 for Development Conference. Rome, Italy.

Bondi, A. B. (2000). Characteristics of scalability and their impact on performance. In *Proceedings of the 2nd International Workshop on Software and Performance*, (pp. 195-203). Ottawa, Canada: IEEE.

Borge, R. (2005). La Participación Electrónica: Estado de la Cuestión y Aproximación a su Clasificación. *IDP: Revista de Internet* [UOC]. *Derecho y Política*, *1*, 1–15.

Bouckaert, G., & Pollitt, C. (2005). *Public management reform: A comparative analysis*. Oxford, UK: Oxford University Press.

Boyd, D., & Ellison, N. (2008). Social network sites: Definition, history and scholarship. *Journal of Computer-Mediated Communication*, *13*(1), 210–230. doi:10.1111/j.1083-6101.2007.00393.x

Bratton, M., Mattes, R., & Gyimah-Boadi, E. (2005). *Public opinion, democracy and market reforms in Africa*. Cambridge, UK: Cambridge University Press.

Bretschneider, S. (2003). Information technology, e-government, and institutional change. *Public Administration Review*, *63*(6). doi:10.1111/1540-6210.00337

Breznitz, D. (2007). *Innovation and the state: Political choice and strategies for growth in Israel, Taiwan, and Ireland*. New Haven, CT: Yale.

Brinkerhoff, P. C. (2000). *Mission-based management: Leading your not-for-profit in the 21st century*. New York, NY: John Wiley & Sons.

Broadbent, M., & Weill, P. (1993). Improving business and information strategy alignment: Learning from the banking industry. *IBM Systems Journal*, *32*(1), 162–179. doi:10.1147/sj.321.0162

Brown, D. (1999). Information systems for improved performance management: Development approaches in U.S. public agencies. In Heeks, R. (Ed.), *Reinventing Government in the Information Age*. London, UK: Routledge.

Brown, D. (2005). Electronic government and public administration. *International Review of Administrative Sciences*, *71*(2), 241–254. doi:10.1177/0020852305053883

Brundtland, G. H. (1987). *Our common future: Report of the World commission on environment and development*. Oxford, UK: Oxford University Press.

Brynjolfson, E. (1992). *The productivity paradox of information technology: Review and assessment*. Retrieved 2004-02-09, from http://ccs.mit.edu/papers/CCSWP130/ccswp130.html

Brynjolfson, E., & Hitt, L. M. (1998). Beyond the productivity paradox - Computers are the catalyst for bigger changes. *Communications of the ACM*. Retrieved from http://ebusiness.mit.edu/erik/bpp.pdf

Brynjolfson, E., & Hitt, L. M. (2003). *Computing productivity: Firm-level evidence*. Cambridge, MA: MIT.

Brynjolfson, E., & Kahin, B. (Eds.). (2000). *Understanding the digital economy*. Cambridge, MA: MIT Press.

BTRC. (2009). *About us*. Retrieved from http://www.btrc.gov.bd/

Buckley, J. (2003). E-service quality and the public sector. *Managing Service Quality*, *13*(6), 453–462. doi:10.1108/09604520310506513

Bulletin, C. B. I. (2006). *Prime minister Dr Manmohan Singh delivering speech on the occasion of foundation stone laying ceremony*. New Delhi, India: Government Publication.

Burrell, G., & Morgan, G. (1979). *Social paradigms and organization analysis*. New York, NY: Heinemann.

Calder, A., & Moir, W. (2006). *The Calder-Moir IT governance framework*. Retrieved March 28. 2008, from http://www.itgovernance.co.uk

Callahan, K. (2007). *Elements of effective governance: Measurement, accountability and participation*. New York, NY: Taylor and Francis.

Campbell, J. L., Hollingworth, J. R., & Lindberg, L. N. (Eds.). (1991). *Governance of the American economy*. Cambridge, UK: Cambridge University Press. doi:10.1017/CBO9780511664083

Canadian Council for Refugees. (1998). *Best settlement practices*. Retrieved October 2008, from http://ccrweb.ca/bpfinal.pdf

Canadian Council for Refugees. (2000). *Canadian national settlement service standards framework*. Retrieved October 2008, from http://ccrweb.ca/standards.pdf

Cap Gemini & Ernest and Young. (2001). *Web-based survey on electronic public services*. Retrieved from http://europa.eu.int/information_society/eeurope/news_library/documents/bench_online_services.doc

Cariño, L. V. (2006). From traditional public administration to governance: Research in NCPAG, 1952-2002. *Philippine Journal of Public Administration*, *50*(1-4), 1–24.

Caseley, J. (2004, March 13). Public sector reform and corruption: CARD façade in Andhra Pradesh. *Economic and Political Weekly*, 1152–1155.

Caseley, J. (2006). Multiple accountability relationships and improved service delivery performance in Hyderabad City, South India. *International Review of Administrative Sciences*, *72*(4), 531. doi:10.1177/0020852306070082

Cassidy, J. (2002). *Dot.con - The real story of why the internet bubble burst*. London, UK: Penguin.

Castells, M. (2000). *The rise of the network society* (2nd ed.). Oxford, UK: Blackwell Publishers.

Castells, M. (2001). *The internet galaxy - Reflections on the internet, business, and society*. Oxford, UK: Oxford University Press.

Castelnovo, W. (2009). Supporting innovation in small local government organizations. In *Proceedings of the 3rd European Conference on Information Management and Evaluation,* (pp. 99-106). IEEE.

Ceriello, V. R. (1991). *Human resource management systems –Strategies, tactics and techniques*. Lexington, VA: Lexington Books.

Chadwick, A. (2003). Bringing e-democracy back in. *Social Science Computer Review*, *21*(4), 443–455. doi:10.1177/0894439303256372

Chadwick, A., & May, C. (2003). Interaction between states and citizens in the age of internet: E-government' in the United States, Britain and European Union. *Governance: An International Journal of Policy, Administration and Institutions*, *16*(2). doi:10.1111/1468-0491.00216

Chan, Y. E. (2002). Why haven't we mastered alignment? The importance of the informal organization structure. *MIS Quarterly Executive*, *1*(21), 76–112.

Chan, Y., Huff, S., Barclay, D., & Copeland, D. (1997). Business strategic orientation, information systems strategic orientation, & strategic alignment. *Information Systems Research*, *8*(1), 125–150. doi:10.1287/isre.8.2.125

Chan, Y., & Reich, B. (2007). IT alignment: What have we learned? *Journal of Information Technology*, *22*, 297–315. doi:10.1057/palgrave.jit.2000109

Charles, D., & Benneworth, P. (2000). *The regional contribution of higher education: A benchmarking approach to the evaluation of the regional impact of a HEI*. Newcastle, UK: Newcastle University.

Charmaz, K. (2007). *Constructing grounded theory*. Thousand Oaks, CA: Sage Publications.

Chatzoglou, P. D., Sarigiannidis, L., Vraimaki, E., & Diamantidis, A. (2009). Investigating Greek employees' intention to use web-based training. *Computers & Education, 53*, 877–889. doi:10.1016/j.compedu.2009.05.007

Chircu, A. M., & Lee, D., & Hae-Dong. (2005). E-government: Key success factors for value discovery and realization. *Electronic Government, 2*(1), 11–24. doi:10.1504/EG.2005.006645

Choudrie, J., & Dwivedi, Y. K. (2004). Investigating the socio-economic characteristics of residential consumers of broadband in the UK. In *Proceedings of the Tenth Americas Conference on Information Systems*, (pp. 1558-1567). New York, NY: IEEE.

Ciborra, C., & Navarra, D. D. (2005). Good governance, development theory, and aid policy: Risks and challenges of e-government in Jordan. *Information Technology for Development, 11*(2), 14–15. doi:10.1002/itdj.20008

Citizenship and Immigration Canada. (1996). *Consultations on settlement renewal: Finding a new direction for newcomer integration*. Ottawa, Canada: Citizenship and Immigration Canada.

Citizenship and Immigration Canada. (2002). *Immigration-contribution accountability measurement system: Security requirements for service provider organizations*. Retrieved October 2008, from http://integration-net.ca/english/ini/caf-cipc/doc/r300-index.htm

Coallier, F., & Gérin-Lajoie, R. (2006). *Open government architecture: The evolution of De Jure standards, consortium standards and open source software*. CIRANO Project Report Num. 2006rp-02. Montreal, Canada: CIRANO.

Cohn-Berman, B. (2005). *Listening to the public: Adding the voices of the people to government performance measurement and reporting*. New York, NY: The Fund for the City of New York.

Cole, M., & Partson, G. (2006). *Unlocking public value: A new model for achieving high performance in public service organizations*. Hoboken, NJ: Wiley Publishing.

Collins, N., & Butler, P. (2003). When marketing models clash with democracy. *Journal of Public Affairs, 3*(1), 52–62. doi:10.1002/pa.133

Commons Group. (2004). *OCASI: Settlement sector database needs study: Final report*. Toronto, Canada: Commons Group.

Considine, M., & Lewis, J. (1999). Governance at ground level: The frontline bureaucrat in the age of markets and networks. *Public Administration Review, 59*(6), 467–480. doi:10.2307/3110295

Cooper, T. L., Bryer, T. A., & Meek, J. W. (2006). Citizen-centered collaborative public management. *Public Administration Review, 66*(s1), 76–88. doi:10.1111/j.1540-6210.2006.00668.x

Cortada, J. W., Gupta, A. M., & Le Noir, M. (2007). *How the most advanced nations can remain competitive in the information age*. Palo Alto, CA: IBM.

Council of the European Union. (2002). *Evolution of e-government in the European Union*. Retrieved from http://www.map.es/csi/pdf/egovEngl_definitivo.pdf

Criado, J. I. (2009). *Entre sueños utópicos y visiones pesimistas*. Madrid, Spain: Instituto Nacional de Administración Pública.

Cripe, E. J. (1993). How to get top-notch change agents. *Training & Development, 47*(12), 52–54, 57–58.

Curtin, G. G., Sommer, M. H., & Vis-Sommer, V. (2003). *The world of e-government*. New York, NY: The Haworth Press.

Cuthill, M., & Fein, J. (2005). Capacity building facilitating citizen participation in local governance. *Australian Journal of Public Administration, 64*(4), 63–80. doi:10.1111/j.1467-8500.2005.00465a.x

Dada, D. (2006). The failure of e-government in developing countries: A literature review. *The Electronic Journal of Information Systems in Developing Countries, 26*(7), 1–10.

DANIDA. (2006). *Business opportunity study within the IT and telecommunication industry in Bangladesh*. Copenhagen, Denmark: Ministry of Foreign Affairs.

DARPG. (2006). *Background note: Improving public service delivery – Management aspects and technology applications*, (pp. 3-4). Retrieved June 8, 2007 from http://darpg.nic.in/arpg-website/conference/chiefsecy-conf/csc.doc

David, P. A. (1990). The dynamo and the computer: A historical perspective on the modern productivity paradox. *The American Economic Review*, *1*(2), 355–361.

Davis, G. B., & Olson, M. H. (1985). *Management information systems: Conceptual foundations, structure, and development* (2nd ed.). New York, NY: McGraw-Hill, Inc.

Davis, J. (2004). Corruption in public service delivery: Experience from South Asia water and sanitation sector. *World Development*, *32*(1), 56–61. doi:10.1016/j.worlddev.2003.07.003

De Coito, P., & Williams, L. (2000). *Setting the course: A framework for coordinating services for immigrants and refugees in the peel region*. Mississauga, Canada: Canadian Government Printing Office.

De Miguel, M. (2010). E-government in Spain: An analysis of the right to quality e-government. *International Journal of Public Administration*, *33*(1), 1–10. doi:10.1080/01900690903178454

De Mul, J., Müller, E., & Nusselder, A. (2001). *ICT de baas? Informatietechnologie en de menselijke autonomie*. Utrecht, The Netherlands: Center for Public Innovation.

Deane, A. (2003). *Increasing voice and transparency using ICT tools: E-government, e-governance*. Washington, DC: World Bank.

Debraceny, R. S. (2006). Re-engineering IT internal controls - Applying capability maturity models to the evaluation of IT controls. In *Proceedings of the 39th Hawaii International Conference on System Sciences*. Hawaii, HI: IEEE.

Dekker, A., & Kleinknecht, A. H. (2003). *Flexibiliteit, technologische vernieuwing en de groei van de arbeidsproductiviteit* (No. A203). Delft, The Netherlands: TUDelft, Faculteit Techniek, Bestuur en Management.

Deleon, L. (2005). Public management, democracy, and politics. In Ferlie, Lynn Jr., & Pollit (Eds.), *The Oxford Handbook of Public Management*. Oxford, UK: Oxford University Press.

Demmers, J., Jilberto, A. E. F., & Hogenboom, B. (2004). Good governance and democracy in a world of neoliberal regimes. In Demmers, J., Jilberto, A. E. F., & Hogenboom, B. (Eds.), *Good Governance in the Era of Global Neoliberalism: Conflict and De-Politicization in Latin America, Eastern Europe and Africa* (pp. 1–19). London, UK: Routledge. doi:10.4324/9780203478691_chapter_1

Derzsi, Z., & Gordijn, J. (2006). A framework for business/IT alignment in networked value constellations. In T. Latour & M. Petit (Eds.), *18th International Conference on Advanced Information Systems Engineering, CAiSE 2006*, (pp. 219-226). Namur, Belgium: Namur University Press.

Dey, B. K. (2000). Challenges and opportunities- A future vision. *The Indian Journal of Public Administration*, *46*(3), 305.

Di Maio, A. (2003a). *Value for money is not enough in public sector IT projects*. Washington, DC: Gartner Research.

Di Maio, A. (2003b). *Traditional ROI measures will fail in government*. Washington, DC: Gartner Research.

Di Maio, A. (2003c). *New performance framework measures public value of IT*. Washington, DC: Gartner Research.

Di Maio, A. (2003d). *How to measure the public value of IT*. Washington, DC: Gartner Research.

Donna, E., & Yen, D. C. (2006). E-government: Evolving relationship of citizens and government domestic, and international development. *Government Information Quarterly*, *23*(2), 207–235. doi:10.1016/j.giq.2005.11.004

Drucker, P. F. (1969). *The age of discontinuity: Guidelines to our changing society*. New York, NY: Harper and Row.

Duboc, L., Rosenblum, D. S., & Wicks, T. (2006). A framework for modelling and analysis of software systems scalability. In *Proceeding of the 28th International Conference on Software Engineering ICSE 2006*, (pp. 949-952). ICSE.

Duffy, J. (2001). Maturity models: Blueprints for e-volution. *Strategy and Leadership*, *29*(6), 19–26. doi:10.1108/EUM0000000006530

Duffy, J. (2002). *IT/business alignment: Is it an option or is it mandatory*. Framingham, MA: IDC.

Dugdale, A. (2004). *E-governance: Democracy in transition*. Paper presented at the Annual Meeting of the International Institute of Administrative Sciences. Washington, DC.

Dunleavy, P., Margetts, H., Bastow, S., & Tinkler, J. (2006). *Digital era governance: IT corporations, the state, and e-government*. Oxford, UK: Oxford University Press.

Dutton, W. H. (Ed.). (1996). *Information and communication technologies: Visions and realities*. Oxford, UK: Oxford University Press.

Dwivedi, O. P., & Mishra, D. S. (2007). Good governance: A model for India. In Farazmand, A., & Pinkowski, J. (Eds.), *Handbook of Globalization, Governance and Public Administration* (pp. 701–741). New York, NY: Taylor & Francis.

Economist. (2000, June 22). A survey of government and the internet: The next revolution. *The Economist*. Retrieved from http://www.economist.com/node/80746

Economist. (2000, June 24). The next revolution – A survey of government and the internet. *The Economist*, 3.

e-Government Media. (2009, November 7). The progress of e-government initiatives in Brunei Darussalam. *Borneo Bulletin*, p. 67.

Ehsan, M. (2004). Origin, ideas and practice of new public management: Lessons for developing countries. *Administrative Change*, *31*(2), 69–82.

EIPA. (2003). *eGovernment in Europe: The state of affairs*. Brussels, Belgium: Publishing Atlanta.

Eisenhardt, K. M. (1989). Building theories from case study research. *Academy of Management Review*, *14*(4), 532–550.

Ellison, N., Steinfield, C., & Lampe, C. (2007). The benefits of Facebook friends: Social capital and college students' use of online social network sites. *Journal of Computer-Mediated Communication*, *12*(4), 1143–1168. doi:10.1111/j.1083-6101.2007.00367.x

E-Readiness. (2012). *Wikipedia*. Retrieved from http://en.wikipedia.org/wiki/E-readiness

Esty, D. C., & Winston, A. S. (2006). *Green to gold: How smart companies use environmental strategy to innovate, create value, and build competitive advantage*. New Haven, CT: Yale University Press.

EU Commission. (2000, March 23). E-Europe: An information society for all. *Lisbon*.

EU. (2004). *eGovernment research in Europe*. Retrieved from http://europa.eu.int/information_society/programmes/egov_rd/text_en.htm

European Commission. (2000). *eEurope action plan*. Retrieved from http://www.europa.eu.int

European Commission. (2002). *An information society for all: Action plan prepared by the council and the European commission for the Feira European council*. Brussels, Belgium: European Commission.

European Commission. (2006). *Developing a knowledge flagship: The European institute of technology*. Brussels, Belgium: European Commission.

European Commission. (2007a). *i2010 - Annual information society report 2007*. Brussels, Belgium: European Commission.

European Commission. (2007b). *The user challenge benchmarking: The supply of online public services*. Brussels, Belgium: Capgemini.

Fang, Z. (2002). *E-government in digital era: Concept, practice, and development*. Bangkok, Thailand: School of Public Administration, National Institute of Development Administration (NIDA).

Farber, D. (2008, November 16). *Obama appoints YouTube (Google) as secretary of video*. [Web log comment and video file]. Retrieved from http://news.cnet.com/obama-appoints-youtube-google-as-secretary-of-video/

Farr, D. (2005). *Service mapping, service Ontario*. Paper presented at the Provincial Government Conference. Lake Carling, Canada.

Federal Architecture Working Group. (2000). *Architecture alignment and assessment guide*. Ottawa, Canada: Federal Architecture Working Group.

Fenn, J. (2008). *Gartner's hype cycle special report for 2008*. Retrieved November 20, 2008, from http://www.gartner.com

Ferlie, E., & Steane, P. (2002). Changing developments in NPM. *International Journal of Public Administration*, *25*(12), 1459–1470. doi:10.1081/PAD-120014256

Finger, M., & Pécoud, G. (2003). *From e-government to e-governance? Towards a model of e-governance*. Paper presented at the 3rd European Conference on e-Government. Dublin, Ireland.

Finger, M., & Langenberg, T. (2007). Electronic governance. In Anttiroiko, A. V., & Malkia, M. (Eds.), *Encyclopedia of Digital Government* (*Vol. 2*). Hershey, PA: IGI Global.

Finger, M., & Pecoud, G. (2003). From e-government to e-governance? Towards a model of e-governance. *Electronic. Journal of E-Government*, *1*(1), 5.

Finnegan, R. (1996). Using documents. In Sapsford, R., & Jupp, V. (Eds.), *Data Collection and Analysis* (pp. 138–151). Thousand Oaks, CA: Sage Publications.

Flanagan, J., & Nicholls, P. (2007). *Public sector business cases using the five case model: A toolkit*. Retrieved April 4, 2008, from http://www.hm-treasury.gov.uk/d/greenbook_toolkittemplates170707.pdf

Flinders, K. (2010, June 11). *Will LinkedIn reshape the recruitment sector?* [Web log comment]. Retrieved from http://www.computerweekly.com/Articles/2010/06/14/241559/Will-LinkedIn-reshape-the-recruitment-sector.htm

Florida, R. (2002). *The rise of the creative class*. New York, NY: Basic Books.

Floyd, S. W., & Wooldridge, B. (1990). Path analysis of the relationship between competitive strategy, information technology, and financial performance. *Journal of Management Information Systems*, *7*(1), 47–64.

Fountain, J. (2001). *Building the virtual state: Information technology and institutional change*. Washington, DC: Brookings Institution.

Fransman, M. (2002). *Telecoms in the internet age - From boom to bust to.?*Oxford, UK: Oxford University Press.

Frederickson, H. G. (2005). Whatever happened to public administration? Governance, governance everywhere. In Ferlie, Lynn Jr., and Pollit (Eds.), *The Oxford Handbook of Public Management*. Oxford, UK: Oxford University Press.

Freeman, C. (1998). Lange wellen und arbeitslosigkeit. [Long waves and unemployment] In Thomas, H., & Nefiodow, L. A. (Eds.), *Kondratieffs Zyklen der Wirtschaft - An der Schwelle neue Vollbeschäftigung?*Herford, Germany: BusseSeewald.

Freeman, C., & Louçã, F. (2001). *As time goes by - From the industrial revolutions to the information revolution*. Oxford, UK: Oxford University Press.

Freeman, C., & Soete, L. (1997). *The economics of industrial innovation* (3rd ed.). Cambridge, MA: The MIT Press.

Freeman, L. C. (2004). *The development of social network analysis: A study in the sociology of science*. Vancouver, Canada: Booksurge.

Freeman, R. E. (1994). The politics of stakeholder theory: Some future directions. *Business Ethics Quarterly*, *4*(4), 409–421. doi:10.2307/3857340

Gant, J., & Gant, D. (2002). *Web portal functionality and state government e-service*. Paper presented at the 35th Hawaii International Conference on System Sciences. Hawaii, HI.

Garlick, S., & Pryor, G. (2004). *Benchmarking the university: Learning about improvement*. Canberra, Australia: Australian Government.

Ghapanci, A., Albadavi, A., & Zarei, B. (2008). A framework for e-government planning and implementation. *Electronic Government: An International Journal*, *5*(1), 71–90. doi:10.1504/EG.2008.016129

Giacalone, R. A., & Rosenfeld, P. (1989). *Impression management in the organization*. Hillsdale, NJ: Lawrence Erlbaum Associates.

Gieber, H., Leitner, C., Orthofer, G., & Traunmüller, R. (2007). Taking best practice forward. In Larson, C. (Ed.), *Digital Government: Advanced e-Government Research and Case Studies*. London, UK: Springer.

Go, A., Inkster, K., & Lee, P. (1996). *Making the road by walking it: A workbook for re-thinking settlement*. Toronto, Canada: CultureLink.

Goddard, J., Etzkowitz, H., Puukka, J., & Virtanen, I. (2006). *Supporting the contribution of higher education institutions to regional development. Peer Review Report*. Jyväskylä, Finland: Government Printing Office.

Golden, W., Hughes, M., & Scott, M. (2003). The role of process evolution in achieving citizen centered e-government. In D. Galletta & J. Ross (Eds.), *9th Americas Conference on Information Systems*, (pp. 801-810). Atlanta, GA: Association for Information Systems.

Golubeva, A., & Merkuryeva, I. (2006). Demand for online government services: Case studies from St. Petersburg. *Information Polity: The International Journal of Government & Democracy in the Information Age*, *11*(3/4), 241–254.

Gorla, N. (2008). Hurdles in rural e-government projects in India: Lessons for developing countries. *Electronic Government: An International Journal*, *5*(1), 91–102. doi:10.1504/EG.2008.016130

Gortmaker, J., Janssen, M., & Wagenaar, R. W. (2005). Accountability of electronic cross-agency service-delivery processes. In M. Wimmer, R. Traunmüller, Å. Grönlund, & K. V. Andersen (Eds.), *4th International Conference on Electronic Government, EGOV 2005*, (pp. 49-56). Copenhagen, Denmark: Springer.

Gotembiewsky, R. T. (1977). *Public administration as a developing discipline, part 1: Perspectives on past and present*. New York, NY: Marcel Dekker.

Governance, C. (2011). *Wikipedia*. Retrieved from http://en.wikipedia.org/wiki/Corporate_governance

Government of Bangladesh. (2002). *National information and communication technology (ICT), policy 2002*. Dhaka, Bangladesh: Ministry of Science and Information and Communication Technology.

Government of Bangladesh. (2009). *National information and communication technology (ICT), policy 2009*. Dhaka, Bangladesh: Ministry of Science and Information and Communication Technology.

Government of Ghana. (2003a). *The ICT for accelerated development (ICT4AD) policy*. Accra, Ghana: Government of Ghana.

Government of Ghana. (2003b). *Ghana growth and poverty reduction strategy (GPRSII) (2006-2009)*. Accra, Ghana: Government of Ghana.

Government of Ghana. (2005). *Budget statement and economic policy*. Accra, Ghana: Government of Ghana.

Government of Ghana. (2010). *Budget statement and economic policy*. Accra, Ghana: Government of Ghana.

Government of India. (2007). Fourth report: Second administrative reforms commission. In *Ethics in Governance*, (p. 141). Retrieved August 9, 2007 from http://arc.gov.in

Graham, S., & Marvin, S. (2001). *Splintering urbanism - Networked infrastructures, technological mobilities and the urban condition*. London, UK: Routledge. doi:10.4324/9780203452202

Granovetter, M. (2004). *The social construction of corruption*. Paper presented at the Conference 'The Norms, Beliefs, and Institutions of the 21st Century Capitalism: Celebrating the 100th Anniversary of Max Weber's the Protestant Ethic and Spirit of Capitalism'. Ithaca, NY. Retrieved May 25, 2007 from http://www.stanford.edu/dept/soc/people/faculty/granovetter/documents/The%20Social%20Construction%20of%20Corruption%20June%2005.doc

Greiner, L. (2005). *State of the marketplace: E-government gateways*. Retrieved from http://www.faulkner.com/products/faulknerlibrary/00018297.htm

Grindle, M. S., & Thomas, J. W. (1991). Implementing reform: Arenas, stakes and resources. In *Public Choice and Policy Change: The Political Economy of Reform in Developing Countries* (pp. 149–150). Baltimore, MD: Johns Hopkins University Press.

Groenewegen, J. P. M. (2008). *Personal communication*. Delft, The Netherlands: Academic Press.

Groenewegen, J. P. M. (1989). *Planning in een markteconomie*. Delft, The Netherlands: Eburon.

Gronlund, A. (2007). Electronic government. In Anttiroiko, A. V., & Malkia, M. (Eds.), *Encyclopedia of Digital Government* (*Vol. 2*). Hershey, PA: IGI Global.

Grönlund, A., & Horan, T. A. (2004). Introducing e-gov: History, definitions and issues. *Communications of the Association for Information Systems*, *15*, 713–729.

Guldentops, E. (2004). Governing information technology through COBIT. In Van Grembergen, W. (Ed.), *Strategies for Information Technology Governance*. Hershey, PA: IGI Global.

Gulledge, T. R., & Sommers, R. A. (2002). Business process management: Public sector implications. *Business Process Management Journal*, *8*(4), 364–376. doi:10.1108/14637150210435017

Gunter, B. G. (2008). Aid, debt and development in Bangladesh. *UNDP Policy Dialogue Series, 38*. Dhaka, Bangladesh: UNDP Bangladesh. Retrieved from http://www.bangladeshstudies.org

Hagen, M. (1997). *A typology of electronic democracy*. Retrieved 2010 from http://www.uni-giessen.de/fb03/vinci/labore/netz/hagen.htm

Haigh, N., & Griffiths, A. (2008). E-government and environmental sustainability: Results from three Australian cases. *Electronic Government: An International Journal*, *5*(1), 45–62. doi:10.1504/EG.2008.016127

Halinen, A., Salmi, A., & Havila, V. (1999). From dyadic change to changing business networks: An analytical framework. *Journal of Management Studies*, *26*(6), 779–794. doi:10.1111/1467-6486.00158

Han, S. (2007, July 26). Overcome red tape to be more productive. *The Brunei Times*, p. 1.

Haque, M. S. (2002). E-governance in India: Impact on relations among citizens, politicians and public servants. *International Review of Administrative Sciences*, *68*(2), 246. doi:10.1177/0020852302682005

Harris, R., & Rajora, R. (2006). *Empowering the poor: Information and communication technology for governance and poverty reduction- Study of rural development projects in India, a UNDP-APDIP publication*. New Delhi, India: Elsevier Press. Retrieved May 22, 2007 from http://www.apdip.net/publications/ict4d/EmpoweringThePoor.pdf

Hasan, S. (2003). Introducing e-government in Bangladesh: Problems and prospects. *International Social Science Review*, *78*. Retrieved from www.questia.com/PM.qst?a=o&se=gglsc&d=5002081994

Hazair, H. (2007, September 20). Speed, security stifling online banking growth. *The Brunei Times*, p. 1.

Heeks, R. (2001). Understanding e-governance for development. *IDPM i-Government Working Paper, 11*.

Heeks, R. (2001). *Understanding e-governance for development*. Manchester, UK: University of Manchester. Retrieved June 4, 2007 from http://unpan1.un.org/intradoc/groups/public/documents/NISPAcee/UNPAN015484.pdf

Heeks, R. (2003). *Most egovernment-for-development projects fail: How can risks be reduced?* iGovernment Working Paper Series, No 14. Manchester, UK: IDPM, University of Manchester.

Heeks, R. (2000). *Reinventing government in the information age*. London, UK: Roultedge.

Heeks, R. (Ed.). (2001). *Reinventing government in the information age*. London, UK: Routledge.

Hefetz, A., & Warner, M. (2004). Privatization and its reverse: Explaining the dynamics of the government contracting process. *Journal of Public Administration: Research and Theory*, *14*(2), 171–190. doi:10.1093/jopart/muh012

Henderson, J., & Venkatraman, H. (1993). Strategic alignment: Leveraging information technology for transforming organizations. *IBM Systems Journal*, *32*(1), 472–484. doi:10.1147/sj.382.0472

Henry, N. (1995). *Public administration and public affairs*. New York, NY: Prentice-Hall, Inc.

Hill, C. W. L. (2009). *International business: Competing with the global marketplace* (7th ed.). New York, NY: McGraw-Hill Irwin.

Hill, M. D. (1990). What is scalability? *ACM SIGARCH Computer Architecture News*, *18*(4), 18–21. doi:10.1145/121973.121975

Hipp, L., & Warner, M. E. (2008). Market forces for the unemployed? Training vouchers in Germany and the USA. *Social Policy and Administration*, *42*(1), 77–101.

Ho, A. (2002). Reinventing local government and the e-government initiative. *Public Administration Review*, *62*(4), 434–444. doi:10.1111/0033-3352.00197

Holder, S. (2001). *Settlement service standards: An inventory of work-in-progress and future steps*. Toronto, Ontario: Ontario Council of Agencies Serving Immigrants.

Holliday, I., & Kwok, R. C. W. (2004). Governance in the information age: Building e-government in Hong Kong. *New Media & Society*, *6*, 549. doi:10.1177/146144804044334

Holm, L. M., Kühn P., M. & Viborg A., K. (2006). IT Governance – Reviewing 17 IT Governance Tools and Analyzing the Case of Novozymes A/S. *Proceedings of the 39th Hawaii International Conference on System Science*. Hawaii/USA.

Hood, C. (1983). *The tools of government in the digital age*. New York, NY: Palgrave Macmillan.

Hood, C. (2006). The tools of government in the information age. In Moran, M., Rein, M., & Goodin, R. E. (Eds.), *The Oxford Handbook of Public Policy* (pp. 469–481). Oxford, UK: Oxford University Press. doi:10.1093/oxfordhb/9780199548453.003.0022

Hoque, S. M. S. (2006). E-governance in Bangladesh: A scrutiny from the citizens' perspective. In R. Ahmad (Ed.), *The Role of Public Administration in Building a Harmonious Society*, (pp. 346-365). Napsipag, Manila: Asian Development Bank.

Hosni, A. (2009, November 27). Good response to HSBC online services. *Borneo Bulletin*, p. 4.

Hughes, O. E. (2003). *Public management and administration* (3rd ed.). Hampshire, UK: Palgrave Macmillan.

Hung, S.-Y., Chang, C.-M., & Yu, T.-J. (2006). Determinants of user acceptance of the e-government services: The case of online tax filing and payment system. *Government Information Quarterly*, *23*, 97–122. doi:10.1016/j.giq.2005.11.005

Ibrahim, K. (2007, July 23). Size of BSB to increase 10-fold. *The Brunei Times*, p. 1.

ICF International. (2004, Fall). The business value of CMMI. *Perspective Reports*.

ICMA. (2007). *Customer service and 311/CRM technology in local government: Lessons on connecting with citizens*. Retrieved from http://icma.org/en/results/research_and_development/smart_communities/311

IDA. (2006). *The great campaign 1986–1991*. Singapore, Singapore: Infocomm Development Authority.

IIA. (2011). *Wikipedia*. Retrieved from http://en.wikipedia.org/wiki/Institute_of_Internal_Auditors

Information Systems Audit and Control Association. (2006). *CobIT* (4th ed.). Ottawa, Canada: Information Systems Audit and Control Association.

Institute for Public-Private Partnerships. (2009). *Public-private partnerships in e-government: Handbook*. Washington, DC: *info*Dev / World Bank.

Integrated Settlement Planning Research Consortium. (2000). *Revisioning the newcomer settlement support system*. Ottawa, Canada: Integrated Settlement Planning Research Consortium.

International Monetary Fund. (2009). *World economic outlook database-April 2010*. Geneva, Switzerland: IMF.

Internet.com. (2010). *Web 2.0*. Retrieved from http://www.webopedia.com/TERM/W/Web_2_point_0.html

ISACA. (2005). *Aligning COBIT, ITIL and ISO17799 for business benefit*. Rolling Meadows, IL: ISACA.

ISACA. (2007a). *Control objectives for information and related technology*. Rolling Meadows, IL: ISACA.

ISACA. (2007b). *COBIT 4.1 excerpt*. Rolling Meadows, IL: ISACA.

Islam, M. M., & Ahmed, A. M. S. (2007). Understanding e-governance: A theoretical approach. *Asian Affairs*, *29*(4), 29–46.

ISO 9000. (2005). *Quality management systems – Fundamentals and vocabulary*. Geneva, Switzerland: International Organization for Standardization.

ISO 9001. (2000). *Quality management systems – Requirements*. Geneva, Switzerland: International Organization for Standardization.

ISO/IEC 12119. (1994). *Information technology – Software packages - Quality requirements and testing.* Geneva, Switzerland: International Organization for Standardization.

ISO/IEC 14598-2. (2000). *Software engineering – Product evaluation – Part 2: Planning and management.* Geneva, Switzerland: International Organization for Standardization.

ISO/IEC 14598-3. (2000). *Software engineering – Product evaluation – Part 3: Process for developers.* Geneva, Switzerland: International Organization for Standardization.

ISO/IEC 14598-4. (1998). *Software engineering – Product evaluation – Part 4: Process for acquirers.* Geneva, Switzerland: International Organization for Standardization.

ISO/IEC 14598-5. (1998). *Information technology – Software product evaluation – Part 5: Process for evaluators.* Geneva, Switzerland: International Organization for Standardization.

ISO/IEC 14598-6. (2001). *Software engineering – Product evaluation – Part 6: Documentation of evaluation modules.* Geneva, Switzerland: International Organization for Standardization.

ISO/IEC 15026. (1998). *Information technology – System and software integrity levels.* Geneva, Switzerland: International Organization for Standardization.

ISO/IEC 17598-1. (1999). *Information technology – Software product evaluation – Part 1: General overview.* Geneva, Switzerland: International Organization for Standardization.

ISO/IEC 25000. (2005). *Software engineering – Software product quality requirements and evaluation (SQuaRE) – Guide to SQuaRE.* Geneva, Switzerland: International Organization for Standardization.

ISO/IEC 25001. (2007). *Software engineering – Software product quality requirements and evaluation (SQuaRE) – Planning and management.* Geneva, Switzerland: International Organization for Standardization.

ISO/IEC 25020. (2007). *Software engineering – Software product quality requirements and evaluation (SQuaRE) – Measurement reference model and guide.* Geneva, Switzerland: International Organization for Standardization.

ISO/IEC 25030. (2007). *Software engineering – Software product quality requirements and evaluation (SQuaRE) – Quality requirements.* Geneva, Switzerland: International Organization for Standardization.

ISO/IEC 25051. (2006). *Software engineering – Software product quality requirements and evaluation (SQuaRE) – Requirements for quality of commercial off-the-shelf (COTS) software product and instructions for testing.* Geneva, Switzerland: International Organization for Standardization.

ISO/IEC 25062. (2006). *Software engineering – Software product quality requirements and evaluation (SQuaRE) – Common industry format (CIF) for usability test reports.* Geneva, Switzerland: International Organization for Standardization.

ISO/IEC 9126-1. (2001). *Software engineering – Product quality – Part 1: Quality model.* Geneva, Switzerland: International Organization for Standardization.

ISO/IEC 9126-2. (2003). *Software engineering – Product quality – Part 2: External metrics.* Geneva, Switzerland: International Organization for Standardization.

ISO/IEC 9126-3. (2003). *Software engineering – Product quality – Part 3: Internal metrics.* Geneva, Switzerland: International Organization for Standardization.

ISO/IEC 9126-4. (2004). *Software engineering – Product quality – Part 4: Quality in use metrics.* Geneva, Switzerland: International Organization for Standardization.

ISO/IEC TR 25021. (2007). *Software engineering – Software product quality requirements and evaluation (SQuaRE) – Quality measure elements.* Geneva, Switzerland: International Organization for Standardization.

ITGI. (2003). *Board briefing on IT governance* (2nd ed.). Rolling Meadows, IL: ITGI.

ITGI. (2006a). *IT control objectives for Sarbanes-Oxley* (2nd ed.). Rolling Meadows, IL: ITGI.

ITGI. (2006b). *Enterprise value: Governance of IT investments: The Val IT framework*. Rolling Meadows, IL: ITGI.

ITGI. (2006c). *Enterprise value: Governance of IT investments, the business case*. Rolling Meadows, IL: ITGI.

ITGI. (2006d). *Enterprise value: Governance of IT investments, the ING case study*. Rolling Meadows, IL: ITGI.

ITGI. (2007). *VAL IT case study: Value governance - The police case study*. Rolling Meadows, IL: ITGI.

ITGI. (2008). *Enterprise value: Governance of IT investments, the Val IT framework 2.0 extract*. Rolling Meadows, IL: ITGI.

ITSMF. (2007). *An introductory overview of ITIL® V3*. Berkshire, UK: ITSMF.

Jackson, P., & Curthoys, N. (2001). *E-government: Developments in the US and UK*. Paper presented at the 12th International Workshop on Database and Expert Systems Applications. Munich,Germany.

Jackson, N., & Lund, H. (2000). *Benchmarking for higher education*. Buckingham, UK: The Society for Research into Higher Education and Open University Press.

Janssen, M., Gortmaker, J., & Wagenaar, R. W. (2006). Information systems management. *Web Service Orchestration in Public Administration: Challenges, Roles, and Growth Stages*, *23*(2), 44–55.

Jessop, B. (2005). Capitalism, steering and the state. In *Formen und Felder politischer Intervention: Zur Relevanz von Staat und Steuerung* (pp. 30–49). Munster, Germany: WestfÃlisches Dampfboot.

Johnson, I. (1997). *Redefining the concept of governance*. Gatineau, Canada: Political and Social Policies Division, Policy Branch, Canadian International Development Agency (CIDA).

Jonk, A., & Van Velzen, G. (2002). *De politieke partij in de netwerksamenleving*. Den Haag, The Netherlands: Academic Press.

Jupp, V. (2003). Realizing the vision of egovernment. In Curtin, C. G., Sommer, M. H., & Vis-Sommer, V. (Eds.), *The World of E-Government* (pp. 129–147). London, UK: Routledge.

Kahler, M. (1992). External influence, conditionality, and the politics of adjustment. In Haggard, S., & Kaufmann, R. R. (Eds.), *The Politics of Economic Adjustment* (pp. 89–136). Princeton, NJ: Princeton University Press.

Kakihara, M. (2003). *Emerging work practices of ICT-enabled mobile professionals*. (PhD Thesis). University of London. London, UK.

Kakihara, M., & Sørensen, C. (2002). *Mobility: An extended perspective*. Paper presented at the 35th Hawaii International Conference on System Sciences. Hawaii, HI.

Karim, M. R. A. (2003). Technology and improved service delivery: Learning points from the Malyasian experience. *International Review of Administrative Sciences, 69*.

Kaushik, A. (2010). *Web analytics 2.0*. Indianapolis, IN: Wiley Publishing.

Kearns, G. S., & Lederer, A. L. (2000). The effect of strategic alignment on the use of IS-based resources for competitive advantage. *The Journal of Strategic Information Systems*, *9*(4), 265–293. doi:10.1016/S0963-8687(00)00049-4

Keele, L. (2007). Social capital and the dynamics of trust in government. *American Journal of Political Science*, *51*(2), 241–254. doi:10.1111/j.1540-5907.2007.00248.x

Kegan, R., & Laskow-Lahey, L. (2001). *The real reason people won't change. Harvard Business Review OnPoint*. Boston, MA: Harvard Business School Publishing.

Kenniscentrum. (2006). *Nederlandse overheid referentie architectuur: NORA 2.0 samenhang en samenwerking binnen de elektronische overheid*. The Hague, The Netherlands: Ministry of the Interior and Kingdom Relations.

Kerr, G., & Simard, L. (2002). *Powers, evaluation of the OASIS computerization project: Final report*. Retrieved October 2008, from www.realworldsystems.net

Kettl, D. F. (2002). *The transformation of governance*. Baltimore, MD: John Hopkins University Press.

Kickert, W. J. M. (1993). Complexity, governance and dynamics: Conceptual explorations of public network management. In Kooiman, J. (Ed.), *Modern Governance: New Government-Society Interactions*. London, UK: Sage.

Kifle, H., & Low, K. C. P. (2009). e-Government implementation and leadership – The Brunei case study. *Electronic. Journal of E-Government*, *7*(3), 271–282.

Killick, T. (Ed.). (1984). *The quest for economic stabilization: The IMF and the third world*. London, UK: ODI.

Kim, (2005). Special report on the sixth global forum on reinventing government: Toward participatory and transparent governance. *Public Administration Review*, *65*(6), 646–654. doi:10.1111/j.1540-6210.2005.00494.x

Kim, H. K. (2005). *Ubiquitous government - Dreams and issues*. Paper presented at the 39th International Council for Information Technology in Government Administration (ICA) Conference. Salzburg, Austria.

King, C. S., Feltey, K. M., & O'Neil, B. O. (1998). The question of participation: Toward authentic public participation in public administration. *Public Administration Review*, *58*(4), 317–327. doi:10.2307/977561

Kingsley, C. (2010). *Making the most of social media: 7 lessons from successful cities*. Retrieved from https://www.fels.upenn.edu/sites/www.fels.upenn.edu/files/PP3_SocialMedia.pdf

Kitchenham, B., Pfleeger, S., Pickard, L., Jones, P., Hoaglin, D., Emam, K., & Rosenberg, J. (2002). Preliminary guidelines for empirical research in software engineering. *IEEE Transactions on Software Engineering*, *28*(8), 721–734. doi:10.1109/TSE.2002.1027796

Kleinberg, J. (2008). The convergence of social and technological networks. *Communications of the ACM*, *51*(11), 66–72. doi:10.1145/1400214.1400232

Klein, H., & Myers, M. (1999). A set of principles for conducting and evaluating interpretive field studies in information systems. *Management Information Systems Quarterly*, *23*(1), 67–93. doi:10.2307/249410

Kleinknecht, A. H. (1987). *Innovation patterns in crisis and prosperity - Schumpeter's long cycle reconsidered*. New York, NY: St. Martin's Press.

Klijn, E. H. (2005). Network and inter-organizational management. In Ferlie, Lynn Jr., & Pollit (Eds.), *The Oxford Handbook of Public Management*. Oxford, UK: Oxford University Press.

Klijn, E. H. (2010). Trust in governance networks: Looking for conditions for innovative solutions and outcomes. In Osborne, S. P. (Ed.), *The New Public Governance?* (pp. 303–321). New York, NY: Routledge.

Konana, P. (2007, January 29). Can IT-enabled services lower corruption?. *The Hindu, Thiruvanathapuram Edition*.

Konsynski, B., & McFarlan, W. F. (1990). Information partnerships - Shared data, shared scale. *Harvard Business Review*, *68*(5), 114–120.

Kooiman, J. (2003). *Governing as governance*. London, UK: Sage.

Koppenjan, J. F. M., & Groenewegen, J. P. M. (2005). Institutional design for complex technological systems. *International Journal of Technology. Policy and Management*, *5*(3), 240–257.

Kotler, P. (2004). *El marketing de servicios profesionales*. Barcelona, Spain: Paidós Empresa.

Kotler, P., & Lee, N. (2006). *Marketing in the public sector*. Upper Saddle River, NJ: Wharton School Publishing.

Kottemann, J. E., & Boyer-Wright, K. M. (2010). Socioeconomic foundations enabling e-business and e-government. *Information Technology for Development*, *16*(1), 4–15. doi:10.1002/itdj.20131

Kotter, J. P. (1995, March/April). Leading change: Why transformation efforts fail. *Harvard Business Review.*

Krantz, D. H., Luce, R. D., Suppes, P., & Tversky, A. (1971). Foundations of measurement: *Vol. 1. Additive and polynomial representations*. San Diego, CA: Academic Press Inc.

Kuhn, T. S. (1996). *The structure of scientific revolutions* (3rd ed.). Chicago, IL: The University of Chicago Press.

Kumar, V., Mukerji, B., Butt, I., & Persaud, A. (2007). Factors for successful e-government adoption: A conceptual framework. *The Electronic. Journal of E-Government*, *5*(1), 63–76.

Kunstelj, M., & Vintar, M. (2004). Evaluating the progress of e-government-A critical analysis. *Information Polity*, *9*(3-4), 131–148.

Lam, J. C. Y., & Lee, M. K. O. (2006). Digital inclusiveness – Longitudinal study of internet adoption by older adults. *Journal of Management Information Systems*, *22*(4), 177–206. doi:10.2753/MIS0742-1222220407

Lawrence, P. R. (1954, May/June). How to deal with resistance to change. *Harvard Business Review*.

Lawson, T. (1997). *Economics and reality*. London, UK: Routledge.

Layne, K. J. L., & Lee, J. (2001). Developing fully functional e-government: A four stage model. *Government Information Quarterly, 18*(2), 122–136. doi:10.1016/S0740-624X(01)00066-1

Lee, B. Y. (2007). *Singapore e-government*. Paper presented at the 4th Ministerial e-Government Conference. Lisbon, Portugal.

Lee, K. Y. (2000). *From third world to first: The Singapore story (1965-2000)*. Singapore, Singapore: Times.

Lee, T.-R., Wu, H.-C., Lin, C.-J., & Wang, H.-T. (2008). Agricultural e-government in China, Korea, Taiwan, and the USA. *Electronic Government: An International Journal, 5*(1), 63–70. doi:10.1504/EG.2008.016128

Leitner, C. (Ed.). (2003). *eGovernment in Europe: The state of affairs*. Maastricht, The Netherlands: European Institute of Public Administration.

Lemstra, W. (2006). *The internet bubble and the impact on the development path of the telecommunication sector.* Delft, The Netherlands: TUDelft.

Lenhart, A., & Fox, S. (2009). *Twitter and status updating*. Washington, DC: Pew Internet & American Life Project.

Leskovec, J., Adamic, L., & Huberman, B. (2006). The dynamics of viral marketing. In *Proceedings of the 7th ACM Conference on Electronic Commerce,* (pp. 228-237). ACM Press.

Li, D. (1998). Changing incentives of the Chinese bureaucracy. *The American Economic Review, 88*(2), 393–397.

Liikanen, E. (2003). *e-Government for Europe's public services of the future*. Retrieved from http://www.uoc.edu/dt/20336/index.html

Ling, T. (2002). Delivering joined-up government in the UK: Dimensions, issues and problems. *Public Administration, 80*(4), 615–642. doi:10.1111/1467-9299.00321

Lipsey, R. G., Carlaw, K. I., & Bekar, C. T. (2005). *Economic transformations: General purpose technologies and long term economic growth.* Oxford, UK: Oxford University Press.

Lovelock, C. H. (2008). *Marketing de servicios: Personal, tecnología y estrategia*. Mexico City, México: Pearson Prentice Hall.

Low, K. C. P. (2011). *Typologies of the Singapore corporate culture*. Retrieved from http://www.GlobalTrade.net

Low, K. C. P., & Mohd Zain, A. Y. (2008). *Creating the competitive edge, the father leadership way*. Paper presented at the International Conference on Business & Management - Creating Competitive Advantage in The Global Economy. Bandar Sewi Begawan, Brunei.

Low, K. C. P., Almunawar, M. N., & Mohiddin, F. (2008). *E-government in Brunei Darussalam – Facilitating businesses, creating the competitive edge*. Paper presented at the International Conference on Business & Management - Creating Competitive Advantage in the Global Economy. Bandar Sewi Begawan, Brunei.

Low, K. C. P. (2005). Towards a framework & typologies of Singapore corporate cultures. *Management Development Journal of Singapore, 13*(1), 46–75.

Low, K. C. P. (2007). The cultural value of resilience – The Singapore case study. *Cross-Cultural Management: An International Journal, 14*(2), 136–149. doi:10.1108/13527600710745741

Low, K. C. P. (2008). Core values that can propel e-government to be successfully implemented in Negara Brunei Darussalam. *Insights to a Changing World Journal, 4*, 106–120.

Low, K. C. P. (2009). *Corporate culture and values: The perceptions of corporate leaders of cooperatives in Singapore*. Saarbrucken, Germany: VDM-Verlag.

Low, K. C. P. (2009a). Strategic Maintenance & leadership excellence in place marketing – The Singapore perspective. *Business Journal for Entrepreneurs, 3*, 125–143.

Low, K. C. P., & Ang, S. L. (2011). e-Government should be for the people and owned by the people – The Brunei perspective: Lessons to be learnt from others. *Business Journal for Entrepreneurs, 1*, 117–135.

Luftman, J. (2000). Assessing business-IT alignment maturity. *Communications of AIS, 4.*

Luftman, J. (2003). *Competing in the information age: Align in the sand* (2nd ed.). Oxford, UK: Oxford University Press. doi:10.1093/0195159535.001.0001

Luftman, J. N. (2003). Assessing IT-business alignment. *Information Systems Management, 20*(4), 9–15. doi:10.1201/1078/43647.20.4.20030901/77287.2

Lukensmeyer, C. J., & Torres, L. H. (2006). *Public deliberation: A manger's guide to citizen engagement*. Washington, DC: Center for the Business of Government.

Lynn, L. E. Jr. (2010). What endures? Public governance and the cycle of reform. In Osborne, S. P. (Ed.), *The New Public Governance?* (pp. 105–123). New York, NY: Routledge.

Madon, S. (2004). Evaluating the developmental impact of e-governance initiatives: An exploratory framework. *The Electronic Journal on Information Systems in Developing Countries, 20*(5).

Madon, S., & Lewis, D. (2003). *Information systems and non- governmental development organizations (NGOs): Advocacy, organizational learning and accountability in a southern NGO*. Working Paper. London, UK: London School of Economics.

Maes, R., Rijsenbrij, D., Truijens, O., & Goedvolk, H. (2000). *Redefining business-IT alignment through a unified framework*. White Paper. Amsterdam, The Netherlands: University of Amsterdam.

Mahizhnan, A., & Andiappan, A. (2002). *e-Government – The Singapore case study*. Retrieved from http://www.infitt.org/ti2002/papers/60ARUN.PDF

Mahoney, P. (2004). Performance based research funding. *Background Note: Information Briefing Service for Members of Parliament, 7*, 1–14.

Mansell, R. (Ed.). (2002). *Inside the communications revolution - Evolving patterns of social and technical interactions*. Oxford, UK: Oxford University Press.

Mansell, R., & Steinmueller, W. E. (2002). *Mobilizing the information society - Strategies for growth and opportunity*. Oxford, UK: Oxford University Press.

Marche, S., & McNiven, J. D. (2003). E-government and e-governance: The future isn't what it used to be. *Canadian Journal of Administrative Sciences, 20*(1), 74. doi:10.1111/j.1936-4490.2003.tb00306.x

Marche, S., & Niven, J. D. (2009). E-government and e-governance: The future isn't what it used to be. *Canadian Journal of Administrative Sciences, 20*(1), 74–86. doi:10.1111/j.1936-4490.2003.tb00306.x

Markov, R. (2006). Economical impact of IT investments in the public sector: The case of local electronic government. In *Proceedings of the International Business Informatics Challenge 2006*. Dublin, Ireland: Business Informatics Challenge.

Mathiason, J. (2009). *Internet governance - The new frontier of global institutions*. London, UK: Routledge.

Mbah, C. (2006). The seductive discourses of development and good governance. In Smith, M. S. (Ed.), *Beyond the African Tragedy: Discourses on Development and the Global Economy* (pp. 121–137). Hampshire, UK: Ashgate.

McFarlane, G. (2006). *IT and the balanced scorecard*. Retrieved April 4. 2008, from http://www.deloitte.com/view/en_KZ/kz/services/enterprise-risk-services/b907e60e612fb110VgnVCM100000ba42f00aRCRD.htm

McGinnis, P. (2003). Creating a blueprint for e-government. In Curtin, C. G., Sommer, M. H., & Vis-Sommer, V. (Eds.), *The World of E-Government* (pp. 51–63). London, UK: Routledge.

Mckinnon, K., Walker, S., & Davis, D. (2000). *A manual for Australian universities*. Canberra, Australia: Department of Education, Training and Youth Affairs.

McShea, M. (2006). IT metrics - IT value management: Creating a balanced program. *IEEE Computer Society. IT Professional, 8*(6), 31–37. doi:10.1109/MITP.2006.138

Meadows, D. L. (1972). *Rapport van de club van Rome*. Utrecht, The Netherlands: Spectrum.

Mead, R. (1994). *International management: Cross cultural dimension*. Oxford, UK: Blackwell Business.

Mehdi, A. (2005). Digital government and its effectiveness in public management reform. *Public Management Review, 7*(3), 484.

Mellor, W., & Parr, V. (2002). *Government online: An international perspective annual global report*. Retrieved from http://tnsofres.com/gostudy2002

Mellor, N. (2006). E-citizen - Developing research-based marketing communications to increase awareness and take-up of local authority e-channels. *Aslib Proceedings, 58*(5), 436–446. doi:10.1108/00012530610692384

Ministry of Planning. (2009). *e-Government procurement*. Retrieved from http://www.cptu.gov.bd/EProcurement.aspx

Ministry of Religious Affairs. (2009). *Hajj management information system*. Retrieved from http://www.bangladeshdir.com/webs/catalog/hajj_management_information_system__ministry_of_religious_affairs__bangladesh.html

Mislove, A., Marcon, M., Gummadi, K., Druschel, P., & Bhattacharjee, P. (2007). Measurement and analysis of online social networks. In *Proceedings of the 7th ACM SIGCOMM Conference on Internet Measurement*, (pp. 29-42). San Diego, CA: ACM Press.

Misra, S. (2005). E-governance: Responsive and transparent service delivery mechanism. In Singh, A. (Ed.), *Administrative Reforms: Towards Sustainable Practices* (pp. 290–292). New Delhi, India: Sage Publications.

Misuraca, G. C. (2007). *e-Governance in Africa - From theory to action*. Trenton, NJ: Africa World Press.

Mitra, A. (2005). Direction of electronic governance initiative within two worlds: Case for a shift in emphasis. *Electronic Government, 2*(1).

Mofleh, S. I., & Wanous, M. (2009). Reviewing existing methods for evaluating e-government websites. *Electronic Government: An International Journal*, *6*(2), 129–142. doi:10.1504/EG.2009.024438

Mohiddin, F. (2007). *Information systems success in Brunei: The impact of organisational structure and culture*. (Unpublished PhD Thesis). Curtin University of Technology. Bentley, Australia.

Mohiddin, F., & Low, K. C. P. (2008). *e-Government & national culture – The way forward*. Paper presented at the Human Resources Compendium: Current Global Trends in HR Best Practices. Bandar Sewi Begawan, Brunei.

Möller, K. K., & Halinen, A. (1999). Business relationships and networks: Managerial challenge of network era. *Industrial Marketing Management*, *28*(1), 413–427.

Moon, M. (2002). The evolution of e-government among municipalities: Rhetoric or reality? *Public Administration Review*, *62*(4), 424–433. doi:10.1111/0033-3352.00196

Moon, M. J. (2002). The evolution of e-government among municipalities: Rhetoric or reality? *Public Administration Review*, *62*(4), 424–433. doi:10.1111/0033-3352.00196

Moon, M. J., & Bretschneider, S. (1997). Can state government actions affect innovation and its diffusion? An extended communication model and empirical test. *Technological Forecasting and Social Change*, *54*(1), 57–77. doi:10.1016/S0040-1625(96)00121-7

Moore, G. E. (1965). Cramming more components onto integrated circuits. *Electronics, 38*(8).

Moore, M. H. (1995). *Creating public value - Strategic management in government*. Cambridge, MA: Harvard University Press.

Moore, W. H., & Mukherjee, B. (2006). Coalition government formation and foreign exchange markets: Theory and evidence from Europe. *International Studies Quarterly*, *50*(1), 93–118. doi:10.1111/j.1468-2478.2006.00394.x

Moulaert, F., & Cabaret, K. (2006). Planning, networks and power relations: Is democratic planning under capitalism possible? *Planning Theory*, *5*(1), 51–70. doi:10.1177/1473095206061021

Mueller, M. L. (1993, July). Universal service in telephone history - A reconstruction. *Telecommunications Policy*, 352–369. doi:10.1016/0308-5961(93)90050-D

Mueller, M. L. (2002). *Ruling the root: Internet governance and the taming of cyberspace*. Cambridge, MA: The MIT Press.

Mueller, M. L. (2010). *Networks and states: The global politics of internet governance*. Cambridge, MA: MIT Press.

Murakami, T. (2003). Establishing the ubiquitous network environment in Japan. *Nomura Research Institute Papers, 66*.

Murra, M. E. (2003). *E-government: From real to virtual democracy*. Retrieved from http://unpan1.un.org/intradoc/groups/public/documents/other/unpan011094.pdf#search='egovernment%3Afrom%20Real%20to%20virtual%20democracy

Nafzinger, E. W. (2006). *Economic development*. Cambridge, UK: Cambridge University Press.

National Association of State Chief Information Officers. (2007). *Connecting state and local government: Collaboration through trust and leadership*. Lexington, KY: NASCIO.

Neuman, W. L. (2003). *Social research methods: Qualitative and quantitative approaches* (5th ed.). Boston, MA: Allyn & Bacon.

Norrie, J. L. (2006). *Improving results of project portfolio management in the public sector using a balanced strategic scoring model.* (DPM Thesis). Royal Melbourne Institute of Technology University. Melbourne, Australia.

NWPB. (2009). *Portal home*. Retrieved from http://www.bangladesh.gov.bd/

Nye, J. Jr. (1999). Information technology and democratic governance. In Karmarck, E. C., & Nye, J. Jr., (Eds.), *Democracy.com? Governance in Networked World*. Hollis, NH: Hollis Publishing Company.

O'Reilly, T. (2005). *What is web 2.0: Design patterns and business models for the next generation of software*. [Web log comment]. Retrieved from http://oreilly.com/web2/archive/what-is-web-20.html

OCASI & COSTI. (1999). *The development of service and sectoral standards for the immigrant services sector: Discussion document*. Toronto, Canada: OCASI and COSTI.

OCASI. (2000). *Models of settlement service workshop, co-located with the OCASI annual conference Oct 20, 2000, Geneva Park, Ontario, Canada*. Retrieved October 2008, from http://atwork.settlement.org/downloads/atwork/Models_of_Settlement_Service.pdf

OCASI. (2008). *Online directory of member agencies*. Retrieved October 2008, from http://www.ocasi.org/membership/OCASI_Online_Directory.pdf

OECD. (2001). *Understanding the digital divide*. Paris, France: OECD Publications. Retrieved from http://www.oecd.org/dataoecd/38/57/1888451.pdf

OECD. (2003). *The e-government imperative*. Paris, France: OECD e-Government Studies.

OECD. (2003). *The e-government imperative: Main findings*. Retrieved from http://www.oecd.org/dataoecd/60/60/2502539.pdf

Oja, M.-K., & Schrader, A. (2008). *From internet to internet of things – The evolution of u-government*. Retrieved from http://www.iadis.net/dl/final_uploads/200817C045.pdf

Oluwu, B. (1999). Redesigning African civil service reforms. *The Journal of Modern African Studies*, *37*(1), 1–23. doi:10.1017/S0022278X99002943

O'Mahony, M., & Van Ark, B. (2003). *EU productivity and competitiveness: An industry perspective*. Luxembourg, Luxembourg: European Communities.

Orelli, R. L., Padovani, E., & Scorsone, E. (2010). E-government, accountability, and performance: Best-in-class governments in European union countries. In Reddick, C. G. (Ed.), *Comparative E-Government* (pp. 561–586). London, UK: Springer. doi:10.1007/978-1-4419-6536-3_29

Orszag, P. (2009, December 8). *Memorandum for the heads of executive departments and agencies: Open government directive*. Retrieved from http://www.whitehouse.gov/omb/assets/memoranda_2010/m10-06.pdf

Osborne, D., & Gaebler, T. (1992). *Reinventing government: How the entrepreneurial spirit is transforming the public sector*. Reading, MA: Addison-Wesley.

Osborne, D., & Gaebler, T. (1993). *Reinventing government: How the entrepreneurial spirit is transforming the public sector*. New York, NY: Plume.

Osborne, D., & Plastrik, P. (2003). *Herramientas para transformar el gobierno* [The reinventor's fieldbook]. Barcelona, Spain: Paidós.

Osborne, S. (2009). *The new public governance*. London, UK: Routledge.

Osborne, S. P. (2006). The new public governance? *Public Management Review*, *8*(3), 377–387. doi:10.1080/14719030600853022

Osborne, S. P. (2010a). The (new) public governance: A suitable case of treatment? In Osborne, S. P. (Ed.), *The New Public Governance?* (pp. 1–16). New York, NY: Routledge.

Osborne, S. P. (2010b). Public governance and public services delivey: A research agenda for the future. In Osborne, S. P. (Ed.), *The New Public Governance?* (pp. 413–428). New York, NY: Routledge.

Osborne, S. P., McLaughlin, K., & Chew, C. (2010). Relationship marketing, relational capital and the governance of public services delivery. In Osborne, S. P. (Ed.), *The New Public Governance?* (pp. 185–199). New York, NY: Routledge.

Osimo, D. (2008). *Web 2.0 in government: Why and how? JRC Scientific and Technical Reports*. Paris, France: European Commission.

Owen, T. (1999). *The view from Toronto: Settlement services in the late 1990s*. Paper presented at the Vancouver Metropolis Conference. Toronto, Canada.

Owusu, F. (2006). Differences in the performance of public organizations in Ghana: Implications for public-sector reform policy. *Development Policy Review*, *24*(6), 693–705. doi:10.1111/j.1467-7679.2006.00354.x

Palmer, I. (2003). *State of the world: E-government implementation*. Retrieved from http://www.faulkner.com/products/faulknerlibrary/00018297.htm

Palvia, P., Palvia, S., & Whitworth, J. E. (2002). Global information technology: A meta analysis of key issues. *Information & Management*, *39*(5), 403–414. doi:10.1016/S0378-7206(01)00106-9

Pani, N., Misra, S. S., & Sahu, B. S. (2004). *Modern system of governance: Good governance v/s e-governance*. New Delhi, India: Anmol Publications.

Paré, G., & Elam, J. (1997). Using case study research to build theories of it implementation. In Lee, A., Liebunau, J., & DeGross, J. (Eds.), *Information Systems and Qualitative Research* (pp. 70–100). London, UK: Chapman and Hall.

Parsons, W. (1995). *Public policy - An introduction to the theory and practice of policy analysis*. Cheltenham, UK: Edward Elgar.

Pawson, R., & Tilley, N. (1997). *Realistic evaluation*. London, UK: Sage Publications Ltd.

PCIP. (2002). *Roadmap for e-government in the developing world*. Retrieved from http://www.pacificcouncil.org/pdfs/e-gov.paper.f.pdf

Pengelly, J. (2004). *ITIL foundations*. London, UK: GTS Learning.

Perez, C. (1983). Structural change and the assimilation of new technologies in the economic and social system. *Futures*, *15*, 357–375. doi:10.1016/0016-3287(83)90050-2

Perez, C. (2002). *Technological revolutions and financial capital: The dynamics of bubbles and golden ages*. Cheltenham, UK: Edward Elgar.

Perez, C. (2004a). *The new techno-economic paradigm*. Amsterdam, The Netherlands: Ministry of Economic Affairs DG Telecom & Post.

Perez, C. (2004b). *Various TEP*. Cambridge, UK: Cambridge University Press.

Peterson, R. (2003). Information strategies and tactics for information technology governance. In *W. Van Grembergen (2004), Strategies for Information Technology Governance*. Hershey, PA: IGI Global. doi:10.4018/978-1-59140-140-7.ch002

Pica, D., & Kakihara, M. (2003). *The duality of mobility: Understanding fluid organizations and stable interaction*. Paper presented at the 11th European Conference on Information Systems. Naples, Italy.

Pieterson, W., Teerling, M., & Ebbers, W. (2008). Channel perceptions and usage: Beyond media richness factors. *Electronic Government Proceedings*, *5184*, 219–230. doi:10.1007/978-3-540-85204-9_19

Pina, V., Torres, L., & Royo, S. (2010). Is e-government leading to more accountable and transparent local governments? An overall view. *Financial Accountability & Management*, *26*(1), 3–20. doi:10.1111/j.1468-0408.2009.00488.x

Planning Commission. (2009). *About planning commission*. Retrieved from http://www.plancomm.gov.bd/about.asp

PMO. (2009). *Government digitized forms online*. Retrieved from http://www.forms.gov.bd/

PMO. (2010). *Access to information (A2I) programme, quarterly progress report*. Retrieved from http://www.gov.bd

Porter, M. (1985). *Competitive advantage: Creating and sustaining performance*. New York, NY: Free Press.

Porter, M. (1998). *On competition*. Boston, MA: Harvard Business School Press.

Powell, T. (1992). Organizational alignment as competitive advantage. *Strategic Management Journal*, *13*(2), 119–134. doi:10.1002/smj.4250130204

Prabhu, C. S. R. (2006). *E-governance: Concepts & case studies*. New Delhi, India: Prentice Hall of India.

PriceWaterhouseCoopers. (2004). *Rethinking the European ICT agenda - Ten ICT-breakthroughs for reaching Lisbon goals*. The Hague, The Netherlands: Ministry of Economic Affairs.

Prins, C. (2001). Electronic government: Variations on a concept. In Prins, J. E. J. (Ed.), *Designing E-Government* (pp. 1–5). Dordrecht, The Netherlands: Kluwer Law International.

Qualman, E. (2010, May 5). *Social media 2 (refresh)*. [Web log comment]. Retrieved from http://socialnomics.net/2010/05/05/social-media-revolution-2-refresh/

Rahman, A. (2010). *Digital Bangladesh bank*. Paper presented at the National Seminar on Vision 2021: Challenges for Engineering Profession. Chittagong, Bangladesh.

Rahman, M. H., Low, K. C. P., Almunawar, M. N., Mohiddin, F., & Ang, S. L. (2012). E-government policy implementation in Brunei: Lessons learnt from Singapore. In Manoharan, A., & Holzer, M. (Eds.), *Active Citizen Participation in e-Government: A Global Perspective*. Hershey, PA: IGI Global. doi:10.4018/978-1-4666-0116-1.ch018

Rannu, R., Saksing, S., & Mahlakõiv, T. (2010). *The mobile government: 2010 and beyond*. White Paper. Retrieved from http://m-gov.mobi

Raposo, M., Leitão, J., & Paco, A. (2006). *E-governance of universities: A proposal of benchmarking methodology*. Unpublished.

Raymond, L., Uwizeyemungu, S., & Bergeron, F. (2006). Motivations to implement ERP in e-government: An analysis from success stories. *Electronic Government: An International Journal*, *3*(3), 225–240. doi:10.1504/EG.2006.009597

Reddick, C. (2005). Citizen interaction with e-government: From the streets to servers? *Government Information Quarterly*, *22*, 38–57. doi:10.1016/j.giq.2004.10.003

Reddick, C. G. (2005). Citizen interaction with e-government: From the streets to servers? *Government Information Quarterly*, *22*(1), 38–57. doi:10.1016/j.giq.2004.10.003

Reich, B., & Benbasat, I. (1996). Development of measures to investigate the linkage between business and information technology objectives. *Management Information Systems Quarterly*, *20*(1), 55–81. doi:10.2307/249542

Reinhard, N., Sun, V., & Agune, R. (2006). ICT Spending in Brazilian Public Administration. *Proceedings of the 19th Bled eConference eValues*. Bled, Slovenia.

Report, C. M. S. (2006). *Tracking corruption in India-2005: Towards sustaining good governance.* New Delhi, India: Centre for Media Studies. Retrieved from http://www.cmsindia.org

Report, F. E. (2010). *Helping taxpayers get it "right from the start"*. Stockholm, Sweden: Swedish Tax Agency.

Rhodes, R. A. W. (1996). New governance: Governing without government. *Political Studies*, *44*(4), 652–667. doi:10.1111/j.1467-9248.1996.tb01747.x

Rhodes, R. A. W. (2007). Understanding governance: Ten years on. *Organization Studies*, *28*(8), 1243–1264. doi:10.1177/0170840607076586

Ridgway, B. (2006). *Ubiquitous government: Enabling innovation in a connected world*. Tokyo, Japan: Microsoft Asia-Pacific.

Ridley, G., Young, J., & Carroll, P. (2004). COBIT and its utilization - A framework from the literature. In *Proceedings of the 37th Hawaii International Conference on System Sciences.* Hawaii, HI: IEEE.

Riley, T. B., & Riley, C. G. (2003). E-governance to e-democracy - Examining the evolution. In *International Tracking Survey Report 2003*. Ottawa, Canada: Riley Information Services.

RIPE. (2005). *About RIPE*. Retrieved 2005-09-26, from www.ripe.net/ripe/about.html

Roberts, F. S. (1979). *Measurement theory with applications to decision-making, utility and social sciences*. Reading, MA: Addison Wesley Publishing Company.

Robson, W. (1997). *Strategic management and information systems*. London, UK: Prentice Hall.

Roggenkamp, K. (2004). *Development modules to unleash the potential of mobile government: Developing mobile government applications from a user perspective*. Paper presented at the 4th European Conference on e-Government. Dublin, Ireland.

Romero, R., & Téllez, J. A. (2010). *Voto electrónico, derecho y otras implicaciones*. Mexico City, México: Universidad Nacional Autónoma de México (UNAM).

Ronaghan, C. I. (2002). *Benchmarking e-government: A global perspective*. New York, NY: United Nations/ DPEPA.

Roy, J. (2011). The promise (and pitfalls) of digital transformation. In Leone, R. P., & Ohemeng, F. L. K. (Eds.), *Approaching Public Administration: Core Debates and Emerging Issues*. Toronto, Canada: Edmond Montgomery Publications.

Russell, P. (1983). *The global brain - Speculation on the evelutionary leap to planetary consciousness*. Los Angeles, CA: J.P. Tarcher.

Sabatier, P. A. (1986). Top-down approach and bottom-up approaches to implementation research: A critical analysis and suggested synthesis. *Journal of Public Policy*, *6*, 21–28. doi:10.1017/S0143814X00003846

Sabatier, P. A. (1987). Knowledge, policy-oriented learning, and policy change. *Knowledge*, *8*, 649–692.

Sabatier, P. A. (1988). An advocacy coalition framework for policy change and the role of policy-oriented learning therein. *Policy Sciences*, *21*, 129–168. doi:10.1007/ BF00136406

Sabatier, P. A., & Weible, C. M. (2007). The advocacy coalition framework: Innovations and clarifications. In Sabatier, P. (Ed.), *Theories of the Policy Process* (pp. 189–220). Boulder, CO: Westview Press.

Sabherwal, R., & Chan, Y. (2001). Alignment between business and IS strategies: A study of prospectors, analyzers and defenders. *Information Systems Research*, *12*(1), 11–33. doi:10.1287/isre.12.1.11.9714

Sagheb-Tehrani, M. (2007). Some steps towards implementing e-government. *Journal of ACM. Computers & Society*, *37*(1), 22–29. doi:10.1145/1273353.1273356

Said Ya'akub, I. (2007, December 4). Be single-minded to succeed. *The Brunei Times*, p. 1.

Sallé, M., & Rosenthal, S. (2005). Formulating and implementing an HP IT program strategy using COBIT and HP ITSM. In *Proceedings of the 38th Hawaii International Conference on System Sciences*. Hawaii, HI: IEEE.

Sallé, M. (2004). *IT service management and IT governance: Review, comparative analysis and their impact on utility computing*. Palo Alto, CA: HP Laboratories.

Salomon, L. M. (2002). *The tools of government*. Oxford, UK: Oxford University Press.

Santana Tapia, R. (2006a). *What is a networked business?* Tech. Rep. TR-CTIT-06-23a. Enschede, The Netherlands: University of Twente.

Santana Tapia, R. (2006b). *IT process architectures for enterprises development: A survey from a maturity model perspective*. Tech. Rep. TR-CTIT-06-04. Enschede, The Netherlands: University of Twente.

Santana Tapia, R., Daneva, M., & van Eck, P. (2007). Validating adequacy and suitability of business-IT alignment criteria in an inter-enterprise maturity model. In S. Ceballos (Ed.), *EDOC 2007: Proceedings of the 11th IEEE International Enterprise Distributed Object Computing Conference*, (pp. 202-213). Washington, DC: IEEE Computer Society Press.

Saul, R. (2000). The IT balanced scorecard – A roadmap to effective governance of a shared services IT organization. *Information Systems Control Journal*, *2*, 31–38. doi:10.1023/A:1010041819361

Savedoff, W. D., & Hussmann, K. (2006). Why are health systems prone to corruption? In *Transparency International Global Corruption Report 2006: Special Focus Corruption and Health* (p. 13). London, UK: Pluto Press.

Saxena, K. B. C. (1996). Reengineering public administration in developing countries. *Long Range Planning*, *29*(5), 704–712. doi:10.1016/0024-6301(96)00064-7

Saxena, K. B. C. (2005). Towards excellence in e-governance. *International Journal of Public Sector Management*, *18*(6), 498–513. doi:10.1108/09513550510616733

Scalability. (2011). *Wikipedia.* Retrieved from http://en.wikipedia.org/wiki/Scalability

Schaap, F. (2002). *The words that took us there*. Amsterdam, The Netherlands: Aksant Academic Publishers.

Schurr, S. H., Burwell, C. C., Devine, W. D. Jr, & Sonenblum, S. (1990). *Electricity in the American economy - Agent of technical progress*. New York, NY: Greenwood Press.

Scott, J. K. (2006). E" the people: Do U.S. municipal government web sites support public involvement? *Public Administration Review*, *66*(3), 341–353. doi:10.1111/j.1540-6210.2006.00593.x

SEI. (2006). *CMMI for development version 1.2*. Pittsburgh, PA: Carnegie Mellon University.

SEI. (2008). *CMMI history*. Retrieved March 29, 2008 from http://www.sei.cmu.edu/cmmi/faq/his-faq.html

Sen, A. (1999). *Development and freedom*. Oxford, UK: Oxford University Press.

Senn, A. (2004). *Realing: Tackling business and IT alignment*. CIO Advertising Supplement. New York, NY: Deloitte Development LLC.

Serrano, R. (2003). *What makes inter-agency coordination work? Insights from the literature and two case studies*. Washington, DC: Inter-American Development Bank.

Serrat, O. (2010, May 24). *Social media and the public sector*. [Web log comment]. Retrieved from http://www.globalknowledgeexchange.net/social-media-and-the-public-sector

Sethibe, T., Campbell, J., & McDonald, C. (2007). IT governance in public and private sector organisations: Examining the differences and defining future research directions. In *Proceedings of the 18th Australasian Conference on Information Systems.* Toowoomba, Australia: IEEE.

Shadrach, B., & Ekeanyawu, L. (2003). *Improving the transparency, quality and effectiveness of pro-poor public services using ICTs: An attempt by transparency international*. Paper presented in the 11th International Anti-corruption Conference. Seoul, South Korea.

Shakir, U. (2000). Re-visioning the newcomer settlement support system by ISPR consortium. In *Proceedings of the OCASI Annual Conference Oct 20, 2000*. Geneva Park, Canada: OCASI. Retrieved October 2008, from http://atwork.settlement.org/downloads/newcomer_settlement_support_system.pdf

Shapiro, C., & Varian, H. R. (1999). *Information rules - A strategic guide to the network economy*. Boston, MA: Harvard Business School Press.

Shleifer, A. (1998). State versus private ownership. *Journal of Economic Perspectives, 12*(4), 141.

Signore, O., Chesi, F., & Pallotti, M. (2005). *E-government: Challenges and opportunities*. Paper presented at the CMG Italy - XIX Annual Conference. Rome, Italy.

Silcock, R. (2001). What is e-government. *Parliamentary Affairs*, *54*, 88–101. doi:10.1093/pa/54.1.88

Simonsson, M., & Hultgren, E. (2005). Administrative systems and operation support systems – A comparison of IT governance maturity. In *Proceedings of the CIGRÉ International Colloquium on Telecommunications and Informatics for the Power Industry.* Cuernavaca, Mexico: CIGRE.

Simonsson, M., & Johnson, P. (2006). *Defining IT governance – A consolidation of literature*. Stockholm, Sweden: Royal Institute of Technology (KTH).

Simonsson, M., Johnson, P., & Wijkström, H. (2006). *Model-based IT governance maturity assessment with COBIT*. Stockholm, Sweden: Royal Institute of Technology (KTH).

Singla, M. L. (2002). Transforming the national bone marrow. *Journal of Management Research*, *2*(3), 165–175.

Slater, W. F. (2002). *Internet history and growth*. Retrieved 2003-01-08, 2003, from www.isoc-chicago.org

Smith, A. D. (2008). Business and e-government intelligence for strategically leveraging information retrieval. *Electronic Government: An International Journal*, *5*(1), 31–44. doi:10.1504/EG.2008.016126

Snellen, I. (2005). E-government: A challenge for public management. In Ferlie, Lynn Jr., & Pollit (Eds.), *The Oxford Handbook of Public Management*. Oxford, UK: Oxford University Press.

Sobhan, F., Shafiullah, M., Hossain, Z., & Chowdhury, M. (2004). *Study of e-government in Bangladesh*. Dhaka, Bangladesh: Bangladesh Enterprise Institute.

Solomon, L. M. (2002). *The tools of government: A guide to the new governance*. Oxford, UK: Oxford University Press.

Song, G. (2005). Mobile technology application in city management: An illumination of project nomad in UK. *Municipal Administration and Technology*, *7*(3), 103–106.

Sørensen, C. (2003). *Research issues in mobile informatics: Classical concerns, pragmatic issues and emerging discourses*. Retrieved from http://mobility.is.lse.ac.uk/html/downloads.htm

Srinivas, A., & Nayar, L. (2007). Elephant must remember: Growth story, yes, but there are over 300 million yet to see it. *Outlook, the Weekly News Magazine, 47*(15), 20-22.

Stutzman, F. (2006). An evaluation of identity-sharing behavior in social network communities. In *Proceedings of the 2006 iDMA and IMS Code Conference*. International Digital and Media Arts Journal.

Sward, D., & Lansford, R. (2007). *Measuring IT success at the bottom line*. Palo Alto, CA: Intel Corporation.

Symons, C. (2006). *Measuring the business value of IT*. Cambridge, MA: Forrester Research Inc.

Tapscott, D., & Williams, D. (2006). *Wikinomics: How mass collaboration changes everything*. New York, NY: Portfolio.

Tat-Kei Ho, A. (2002). Reinventing local governments and the e-government initiative. *Public Administration Review*, *62*(4), 434–444. doi:10.1111/0033-3352.00197

Teicher, J., Hughes, O., & Dow, N. (2002). E-government: A new route to public sector quality. *Managing Service Quality*, *12*(6), 384–393. doi:10.1108/09604520210451867

Tidd, J., Bessant, J., & Pavitt, K. (2001). *Managing innovation - Integrating technological, market and organizational change* (2nd ed.). Chichester, UK: John Wiley & Sons.

Timney, M. (1998). Overcoming administrative barriers to citizen participation: Citizens as partners, not adversaries. In King, C. S., & Stivers, C. (Eds.), *Government is Us: Public Administration in an Anti-Government Era* (pp. 88–99). Thousand Oaks, CA: Sage.

Toeffler, A. (1981). *The third wave*. New York, NY: Bantam Books.

Traunmuller, R., & Lenk, K. (Eds.). (2002). Electronic government. In *Proceedings of the First International Conference*. EGOV.

Trommel, W. (1999). *ICT en nieuwe arbeidspatronen*. Den Haag, The Netherlands: Government Press.

Tsikata, Y. M. (2001). *Owning economic reforms: A comparative study of Ghana and Tanzania*. Tokyo, Japan: United Nations University (UNU)/WIDER.

Turban, E., McLean, E., & Wetherbe, J. (2002). *Information technology for management: Transforming business in the digital economy* (3rd ed.). New York, NY: Wiley.

UN. (2008). *E-government survey- From e-government to connected governance*. New York, NY: Division for Public Administration and Development Management.

UN. (2010). *United Nations e-government survey 2010*. New York, NY: Department of Economic & Social Affairs.

UN/ASPA. (2002). Benchmarking e-government: a global perspective. Retrieved from http://unpan1.un.org/intradoc/groups/public/documents/un/unpan021547.pdf

UNCTAD. (2002). *Reports on e-commerce and development of the united nations conference on trade and development*. New York, NY: United Nations.

UNDESA. (2002). *Plan of action - E-government for development*. Rome, Italy: Government of Italy, Ministry for Innovation and Technologies. Retrieved from http://www.palermoconference2002.org

UNDESA. (2003). *World public sector report: E-government at the crossroad*. New York, NY: UN Department of Economic and Social Affairs.

UNDP. (2004). E-governance. *Essentials*, *15*, 2.

United Nations. (2002). *Benchmarking e-government: A global perspective*. New York, NY: UN Publishing.

United Nations. (2010). *E-government survey*. New York, NY: UN Publishing.

UNSCO. (2007). *Education in Brunei Darussalam statistics*. Retrieved from http://stats.uis.unesco.org/unesco/TableViewer/document.aspx?ReportId=121&IF_Language=eng&BR_Country=960

US Congress. (2002). *US 2002 e-government act*. Washington, DC: US Congress Printing Office.

Van Ark, B., & Piatkowski. (2004). *Productivity, innovation and ICT in old and new Europe*. Groningen, The Netherlands: Rijksuniversiteit Groningen.

Van Ark, B., Inklaar, R., & McGuckin, R. H. (2003). *The contribution of ICT-producing and ICT-using industries to productivity growth: A comparison of Canada, Europe and the United States. The Haag*. The Netherlands: Centre for the Study of Living Standards.

Van Bon, J. (2004). *IT service management, an introduction based on ITIL*. Zasltbommel, The Netherlands: Van Haren Publishing.

van Eck, P., Blanken, H., & Wieringa, R. (2004). Project GRAAL: Towards operational architecture alignment. *International Journal of Cooperative Information Systems*, *13*(3), 235–255. doi:10.1142/S0218843004000961

Van Grembergen, W. (2002). Introduction to minitrack: IT governance and its mechanisms. In *Proceedings of the 35th Hawaii International Conference on System Sciences*. Hawaii, HI: IEEE.

Van Grembergen, W., De Haes, S., & Guldentops, E. (2004). Structures, processes and relational mechanisms for IT governance. In Van Grembergen, W. (Ed.), *Strategies for Information Technology Governance*. Hershey, PA: IGI Global. doi:10.4018/978-1-59140-140-7.ch001

Vaníček, J., Vrana, I., & Aly, S. (2009). Fuzzy aggregation and averaging for group decision making: A generalization and survey. *Knowledge-Based Systems*, *22*(1), 79–84. doi:10.1016/j.knosys.2008.07.002

Vietor, R. H. K. (2007). *How countries compete: Strategy, structure, and government in the global economy*. Boston, MA: Harvard Business School.

Vincent, I. (1999). Collaboration and integrated services in the NSW public sector. *Australian Journal of Public Administration*, *58*(3), 50–54. doi:10.1111/1467-8500.00105

Vintar, M., & Nogra˘sek, J. (2010). How much can we trust different e-government surveys? The case of Slovenia. *Information Polity*, *15*, 199–213.

Vittal, N. (1999). *Applying zero tolerance to corruption*. Retrieved June 28, 2007 from http://www.cvc.nic.in/vscvc/note.html

Warland, C., & Ridley, G. (2005). Awareness of IT control frameworks in an Australian state government: A qualitative case study. In *Proceedings of the 38th Hawaii International Conference on System Sciences*. Hawaii, HI: IEEE.

Weerakkody, V., Dwlvedi, Y., & Kurunananda, A. (2009). Implementing e-government in Sri Lanka: Lessons from the UK. *Information Technology for Development*, *15*(3), 171–192. doi:10.1002/itdj.20122

Weick, K. E. (1995). *Sensemaking in organizations*. Thousand Oaks, CA: Sage Publications.

Weill, P., & Ross, J. W. (2005). *A matrixed approach to designing IT governance*. Retrieved March 16. 2008, from http://sloanreview.mit.edu/the-magazine/2005-winter/46208/a-matrixed-approach-to-designing-it-governance/

Weill, P., & Broadbent, M. (1998). *Leveraging the new infrastructure: How market leaders capitalize on information technology*. Boston, MA: Harvard Business School Press.

Weill, P., & Ross, J. W. (2004). *IT governance – How top performers manage IT decision rights for superior results*. Boston, MA: Harvard Business School Press.

Weiser, M. (1991). The computer for the 21st century. *Scientific American*, •••, 94–100. doi:10.1038/scientificamerican0991-94

Welch, E. W., Hinnant, C. C., & Moon, M. J. (2005). Linking citizen satisfaction with e-government and trust in government. *Journal of Public Administration: Research and Theory*, *15*(3), 371–391. doi:10.1093/jopart/mui021

Wentink, V. (2002). *Digital alterego and the new economy*. Zeist, The Netherlands: Comparc.

Wentink, V. (2004). *Een nieuwe horizontalisering*. Zeist, The Netherlands: Comparc.

West, D. (2000). *Assessing e-government: The internet, democracy and service delivery by state and federal governments*. Retrieved from http://www.insidepolitics.org/egovtreport00.html

West, D. (2008a). Improving technology utilization in electronic government around the world. Retrieved from http://www.brookings.edu/reports/2008/0817_egovernment_west.aspx

West, D. (2008b). *State and federal e-government in the United States*. Retrieved from http://www.brookings.edu/reports/2008/0826_egovernment_west.aspx

West, D. M. (2008). *Improving technology utilisation in electronic government around the world, 2008*. Retrieved from http://www.brookings.edu/.../2008/0817_egovernment_west.aspx

West, D. M. (2004). E-government and the transformation of service delivery and citizen attitudes. *Public Administration Review*, *64*(1). doi:10.1111/j.1540-6210.2004.00343.x

Williamson, A. (2007). *Citizen participation in the unified New Orleans plan*. Unpublished Paper. Boston, MA: Harvard University.

Williamson, O. E. (1998). Transaction cost economics: How it works, where it is headed. *The Economist*, *146*(1), 23-58.

Wolf, M. (2002). *How to save British universities*. Paper presented at the Singer and Friedlander Lecture. Oxford, UK.

Wolf, C. Jr. (1990). *Markets or governments - Choosing between imperfect alternatives*. Cambridge, MA: The MIT Press.

Wong, W., & Welch, E. W. (2004). Does e-government promote accountability? A comparative analysis of website openness and government accountability. *Governance: An International Journal of Policy, Administration and Institutions*, *17*(2), 275–297. doi:10.1111/j.1468-0491.2004.00246.x

Working Group on IT for Eleventh Plan Report. (2007). *Recommendations of working group on eleventh five year plan 2007 - 12: Information technology sector*. Retrieved June 7, 2007 from http://planningcommission.nic.in/aboutus/committee/wrkgrp11/wg11_IT.pdf

Working Group. (2002). *Roadmap for e-government in the developing world*. Los Angeles, CA: Pacific Council on International Policy.

World Bank. (1989). *Sub-Saharan Africa: From crisis to sustainable development*. Washington, DC: World Bank.

World Bank. (1997). *World development report: The state in a changing world*. Oxford, UK: Oxford University Press.

World Bank. (1998). *Assessing aid: What works, what doesn't, and why*. Oxford, UK: Oxford University Press.

World Bank. (2001). *Designing technical assistance projects: Lessons from Ghana and Uganda*. Retrieved in February, 23, 2007 from http://www.worldbank.org

World Bank. (2003). *World development report 2004: Making services work for poor people*. Washington, DC: The World Bank.

World Bank. (2005). *Project information document (PID): E-Ghana project*. Report No.AB1654. Retrieved in February, 23, 2007 from http://www.worldbank.org

World Bank. (2006). *Reforming public services in India: Drawing lessons from success*. New Delhi, India: Sage Publications India.

World Bank. (2007). Country brief: Russian federation, economy. Retrieved from http://web.worldbank.org

World Bank. (2007). *Introduction to e-government: What is e-government?* Retrieved from http://worldbank.org/

World Bank. (2008). *World development indicators database*. Washington, DC: World Bank.

World Bank. (2009). *e-Government primer*. Washington, DC: *info*Dev/World Bank.

Yin, R. K. (2003). *Case study research: Design and methods*. Thousand Oaks, CA: Sage Publications.

Zaho, D., & Rosson, M. (2009). How and why people Twitter: The role that microblogging plays in informal communication at work. In *Proceedings of the ACM International Conference on Supporting Groupworth,* (pp. 243-252). New York, NY: ACM Press.

Zormello, D. (1996). *Is aid conditionality consistent with national sovereignty?* London, UK: ODI.

Zussman, D. (2002). *Paper.* Paper presented at the Public Policy Forum Commonwealth Centre for Electronic Governance Integrating Government with New Technologies: How is Technology Changing the Public Sector? Ottawa, Canada.

Zuurmond, A. (1994). *De infocratie*. Den Haag, The Netherlands: Phaedrus.

About the Contributors

Muhammad Muinul Islam teaches Public Administration in Jahangirnagar University. Prior to this, he worked for Asian University of Bangladesh and a non-profit organization. His research interest includes different issues of public policy and governance, public budgeting and finance, and public service ethics. He graduated from University of Dhaka and attended the MPA program in Georgia State University with a Fulbright scholarship. He contributed encyclopedia entries, book chapters, and journal articles in local and international journals. He was a co-editor of the book *Bangladesh: Politics, Public Administration, and Governance* from VDM, Germany.

Mohammad Ehsan is an Assistant Professor of Public Administration at the University of Dhaka, Bangladesh. He has attained academic degrees from University of Dhaka, Bangladesh; University of Bergen, Norway; and Carleton University, Canada. He is scheduled to defend his doctoral dissertation at the Department of Political Science, Dalhousie University, soon. He also taught at Shahjalal University of Science and Technology, Bangladesh, as well as at Dalhousie University, Canada. He has worked for the NGO sector and donor community in various capacities as well. So far, he has published three books, a few book chapters, encyclopedia entries, and several articles in refereed international journals. His current research interests include public sector management, public sector ethics, and anti-corruption policy and program evaluation, e-governance, and higher education. His primary research focus is on South and Southeast Asia. He has memberships in several academic and professional associations, and is the recipient of, among many others, NORAD Fellowship and IDRC Doctoral Research Award.

* * *

Mohammad Nabil Almunawar *Ir. IPB Indonesia, MSc UWO Canada, Ph.D. UNSW, Australia,* is a Senior Lecturer at Faculty of Business, Economics, and Policy Studies, Universiti of Brunei Darussalam (UBD), Brunei Darussalam. Dr. Almunawar has published many papers in refereed journals as well as international conferences. He has many years of teaching experiences in the area of computer and information systems. He was a respected consultant in developing information systems for United Nations (WHO) projects, Central Bank of Indonesia, and some private companies. His overall research interest is application of IT in Management and Electronic Commerce. He is also interested in object-oriented technology, databases, and multimedia retrieval.

Lloyd G. Adu Amoah teaches at Ashesi University College, Berekuso, Ghana. His research interests focus on theories of the policy process and the linkages between philosophy and public administration in developing polities, Africa-China relations, and Africa-BRICS interactions, among others. Dr. Amoah's work has appeared in *Administrative Theory and Praxis*, *Fudan Journal of the Humanities and Social Sciences*, *African Affairs Journal*, and the *Ghana Policy Journal*, among others. His latest work is: "Public Policy Formation in Africa in the Wake of the Global Financial Meltdown: Building Blocks for a New Mind in a Multi-Polar World" in Dietz et al.'s *African Engagements: Africa Engaging an Emerging Multipolar World* (Brill, 2011). He is the executive director of Strategy3, a strategy and public policy think-pad based in Accra, Ghana.

Sik Liong Ang, B. Sc. (Hon. London), M. Phil. (London), MBA, has more than 30 years of experience in Petroleum Engineering and Chemistry, Education, and Chinese Culture. In business, he has some good experience in transforming and reorganising a company unit into an effective and efficient outfit toward an ISO standard. He has also a sound knowledge of contract management and leadership training. Ang, a Ph.D. executive candidate, is currently pursuing his Ph.D. programme on Confucian Leadership and Management, and he is also a research assistant for an e-Government research grant project at the Universiti Brunei Darussalam.

Alberto Asquer is Lecturer of Business and Corporate Strategy at the Faculty of Economics of the University of Cagliari, Italy. His research is mainly focused on the political economy of liberalization, privatization, and re-regulation processes, and occasionally related to other fields such as management, strategic management, public policy, and regulatory reform. Recent publications of his include: "Implementing Fiscal Decentralization: A Case Study of a Regional Tax Agency in Italy" in *Governance*, *23*(4), 609-621; "The Regulatory Reform of Water Infrastructure in Italy: Overall Design and Local Variations" in *Water Policy*, *12*(s1), 66-83; and "Regulatory Reform and Industrial Restructuring: The Cases of Water, Gas, and Electricity in Italy" in *Competition and Regulation in Network Industries*, *11*(1), 85-117.

Amlan Bhusan is a transnational management consultant with a decade of experiences undertaking several highly successful consulting projects in diverse geographic and economic settings. From strategy to diversification, from consolidation of inefficient business units to effective manpower management, recruitment and other core knowledge management functions, Amlan is active across several portfolios. In the recent past, Amlan has been a strategy and diversification specialist running his consulting firm in New Zealand, successfully delivering on several high profile projects within the education, tourism, and applied skills development sectors, before establishing his current consulting practice, Bhusan Taylor Ltd., in London. Amlan has successfully led and managed several multimillion-dollar consulting assignments within strategic diversification, assisting many organizations in UK, Scandinavian EU, Switzerland, Italy, USA, and India, among other South Asian countries. Amlan is a graduate in Public Policy from the School of Government, Wellington, New Zealand, and has a research Masters degree in Business and Economics from Lund University, Sweden. In 2006, he was also nominated to a Doctorate of Letters Honoris Causa. Amlan spends his time between India and the UK, consulting to several global organizations within the education and training sector and is also active in the traditional terrains of consulting work, like technology, people, and resources.

Maya Daneva is an Assistant Professor at the Information Systems Group at the University of Twente, The Netherlands. Her research is focused on business-IT alignment, maturity models, requirements engineering, cost estimation of large systems, and the application of qualitative research methods. Maya also serves as a liaison to the Dutch IT industry. Prior to this, Maya spent 9 years as a Business Process Analyst in the Architecture Group of TELUS, Canada's second largest telecommunication company in Toronto, where she consulted on ERP requirement engineering processes and their evaluation as well as on ERP-supported coordination models and business process design. Maya has authored more than 70 research and experience papers.

Carlotta del Sordo, BA Business Economics, University of Bologna, Italy, Ph.D. Management of the Public Sector, University of Salerno, Italy, visiting scholar at Boston University, US, Accounting Department, Lecturer in Business Economics, Department of Management, University of Bologna, Italy. His current research interests are management accounting theory and management control in the public sector.

Md. Gofran Faroqi graduated from Dhaka University, Bangladesh, in Sociology and started a career in Public Service of Bangladesh and served in both field and secretariat administrations in different capacities. He has been promoted as a Senior Assistant Secretary, and lastly, he was serving as a Upazila Nirbahi Officer (UNO) in Lakhai Upazila of Habiganj District. He completed his Master of Public Administration degree from Flinders University, Australia, in 2009. Currently, he is enrolled in Flinders University, Australia, as a PhD Candidate in the Department of Politics and Public Policy. His major areas of research interests are public management, contemporary issues in public policy, governance and politics, accountability issues in GO/NGO sectors, e-governance, etc.

Keith Hales. Following a successful career in the military and in the private sector, he joined Bond University as an Assistant Professor in the School of Information technology at Bond University. He has published papers in IT Governance, Emergency Management, and Strategic Alignment in the Public Sector. His ongoing research interests are in the deployment of IT in the SME business sector, IT in developing countries, and in supply chain management.

Evanthia Hatzipanagiotou holds a Master Degree in Human Resources Management from Athens University of Economics and Business and a B.Sc. in Economic Science from the Department of Law, Economics, and Political Science of the University of Athens. Evanthia is employed at the Greek Ministry of Finance, in the General Secretariat of Tax and Customs Affairs—General Division of Taxation, as a Director of the Public Revenue Collection Division. Previously, she has worked as Operational Manager of the TAXIS project for the implementation, training, and installation of the first IT taxation system in Greece. Evanthia participates to the Fiscalis EU Program, representing the Greek Ministry of Finance, which aims at enhancing the fight against tax fraud, improving administrative procedures, and ensuring the exchange of information between national tax administrations and traders, through trans-European tax IT projects. Furthermore, Evanthia is the author of many articles about human resource management, tax legislation, and public reform.

Kathryn Kloby is an Assistant Professor in the Department of Political Science at Monmouth University and is the Director of the Public Policy Graduate Program. She teaches undergraduate and graduate courses in public policy, public administration, public management, and research methods. Dr. Kloby's research interests include public sector performance measurement, citizen engagement, and e-governance.

Wolter Lemstra is Senior Research Fellow at the Department Technology, Policy, and Management of the Delft University of Technology, and Senior Lecturer at the Strategy Academy, Rotterdam, in The Netherlands. He thereby links his academic interests to 25 years of experience in the telecom sector. He occupied senior management positions in system engineering and product management at Philips, marketing and business development for Eastern Europe at AT&T. Most recently, he was Vice President at Lucent Technologies, responsible for marketing and business development in the Europe, Middle East, and Africa region. His current research is focused on sector governance, industry structure developments, firm strategic behaviour, and innovation trajectories. He has been a (co)author in projects for the Ministry of Economic Affairs and for the European Commission. His most recent co-authored and co-edited book is *The Innovation Journey of Wi-Fi – The Road to Global Success*. He is a co-founder of the Cognitive Radio Platform NL.

Patrick Kim Cheng Low, Ph.D. (South Australia), Chartered Marketer, Certified MBTI Administrator, Chartered Consultant and Certified Behavioral Consultant (IML, USA), brings with him more than 25 years of combined experience from sectors as diverse as the electronics, civil service, academia, banking, human resource development, and consulting. The once Visiting Professor, Graduate School of Business, the University of Malaya (2007), Prof. Dr. Low, the Deputy Dean, Postgraduate Studies and Research (2009), is teaching in Universiti Brunei Darussalam. An academician-practitioner, a prolific author (author of twelve books, including bestsellers: *Strategic Customer Management*, 2006, 2002, 2000 [one of Borders' top ten in 2001/2]; *Sales Success*, 2006, 2003; *Team Success*, 2003; and *The Power of Relationships*, 2001), and a business coach, Prof. Dr. Low is also the founder of BusinesscrAFT™ Consultancy and an associate of UniSA. His most recent books are *Successfully Negotiating in Asia*, Springer (2010), and *Corporate Culture and Values*, VDM-Verlag (2009).

Evangelia Mantzari is currently a PhD Candidate at Athens University of Economics and Business, in the Department of Management Science and Technology, and a Research Officer in the ELTRUN e-Business Research Center. Evangelia holds an International MBA from Athens University of Economics and Business and a B.Sc. in Economic Science from the University of Piraeus. Evangelia has participated in several European and National projects concerning issues of e-business and e-government in several sectors, while she is a Rapporteur for the Greek forum for Interactive Digital TV. In parallel, Evangelia is employed as an Executive of the National Bank of Greece S.A. and as a Designated Instructor for the Program of Continuous Training Institute. Previously, she has worked as a Financial and Insurance Consultant in the ASPIS Group and as an Assistant in the Public Relation Office of the General Secretariat of Information Systems of the Greek Ministry of Finance.

Durga Shankar Mishra, B. Tech in Electrical Engineering, Post Graduate Diploma in Human Resource Management, MBA in International Business, and Post Graduate Diploma in Governance, Democratization, and Public Policy, is an officer of Indian Administrative Service, and is posted as

Principal Secretary to Hon'ble Chief Minister, Principal Secretary, Urban Development and Principal Secretary-Cum-Director General, Civil Aviation in State Government of Uttar Pradesh in India, at present. His administrative experience spans over 27 years serving India's largest state and federal governments. He has published several articles/research papers in Indian and foreign journals; has participated in many national/ international seminars/ conferences; and has delivered lectures in training institutions and universities. He has been a visiting faculty at the University of Guelph, Canada.

Fadzliwati Mohiddin, BA Management Studies (Universiti Brunei Darussalam); MBA (Lancaster University, UK); PhD Information Systems (Curtin University of Technology, Western Australia), is a lecturer at the Faculty of Business, Economics, and Policy Studies, UBD. She lectures in ICT, Management of Information Systems, and Business Statistics, and is also the examiner to MBA's theses. Her current research interest includes Information Systems Success, Knowledge Management, and E-Government. Currently, she is involved with several ICT projects, which include Knowledge Management Systems and E-Learning Systems for the Ministry of Education.

María de Miguel Molina has a Law Degree from Universitat de València and a PhD on Management from the Universitat Politècnica de València – UPV (Spain). Currently, she is Associate Professor at the Management Department of the UPV. She is Lecturer in different subjects, such as Computer Ethics and Law and Public Management. She is the Director of an MBA at the Business Administration and Management Faculty of the UPV (joined international degree with Ansbach University, Germany). Her research field centers on the relation between regulation and organizational social responsibility and has several publications in international journals, mainly related to self-regulation in different sectors as mobile telephony, Social Network Sites (SNS), public administration, and environmental protection. She is a member of the European Business Ethics Network (EBEN) and of the DerechoTICs Network.

Rebecca Levy Orelli, BA Business Economics, University of Bologna, Italy, Ph.D. Management of the Public Sector, University of Salerno, Italy, visiting scholar at London School of Economics, Accounting Department, UK, Lecturer in Business Economics, Department of Management, University of Bologna, Italy, is affiliated with EBEN. Her current research interests are new public management and public services changes in local governments, management accounting, and management control in the public sector.

Emanuele Padovani BA Business Economics, University of Bologna, Italy, Ph.D. Business Administration, University of Ferrara, Italy, Associate Professor of Management Control and Auditing in Public Sector Organizations in the Faculty of Economics and Department of Management at the University of Bologna, Forli Campus, Italy. Has been invited as guest lecturer at the undergraduate, graduate, and executive levels in different Italian universities and abroad (University of Valencia, Spain; Methodist University of Sao Paulo, Brazil; Aarhus Business School, Denmark; Michigan State University, East Lansing, MI; University of Washington, Seattle, WA). His research interests are public management with specific reference to management control systems, performance measurement for management and auditing, benchmarking, and management of outsourcing.

Jong Sou Park received the M.S. degree in Electrical and Computer Engineering from North Carolina State University in 1986, and he received his Ph.D. in Computer Engineering from The Pennsylvania State University in 1994. From 1994 – 1996, he worked as an Assistant Professor at The Pennsylvania State University in the Computer Engineering Department, and he was President of the KSEA Central PA Chapter. He is currently a full Professor in Computer Engineering Department, Korea Aerospace University. His main research interests are information security, embedded system, e-government, and hardware design. He has published many papers in international journals and in conferences. He is a member of IEEE and IEICE, and he is an executive board member of the Korea Institute of Information Security and Cryptology and Korea Information Assurance Society.

Manish Pokharel obtained his Bachelor of Engineering in Computer Science from Karnataka University, India. He received his Master of Engineering in Software System from Birla Institute of Technology and Science, Pilani, India. He is currently a Ph.D. student in Korea Aerospace University, since September 2007. His research interests are in e-government, enterprise architecture, software technology, fault tolerance, and cloud computing. He was involved in making e-government master plan in Costa Rica as an International Consultant. He has published various research papers in international conferences and in journals.

Mohammad Habibur Rahman received his Ph.D. from the University of Wales in UK. Currently serving as Senior Lecturer in the Faculty of Business, Economics, and Policy Studies at the University Brunei Darussalam, Dr. Rahman previously taught graduate and undergraduate students in the areas of public administration, political science, and development studies in universities in Bangladesh (Dhaka and National), Canada (Lakehead), and Fiji Islands (South Pacific). He was a Senior Fulbright Scholar at Syracuse University in USA and a Visiting Fellow at York University in Canada. His current teaching and research interests include public policy, governance, public sector reform, e-governance, public sector human resource management, and urban governance. Dr. Rahman has served UNDP, World Bank, Asian Development Bank, USAID, NORAD, JICA, Ford Foundation, and Asia Foundation on several governance and development projects in Bangladesh, India, Sri Lanka, and Nepal, as well as conducted independent research in USA, Canada, Fiji, and Brunei.

Leila Sadeghi, Ph.D., is an Assistant Professor with the Department of Educational Leadership in the Nathan Weiss Graduate College, Kean University. Dr. Sadeghi's research interests include Web 2.0 and social media, utilizing new media in the classroom, and citizen participation via technology. She has written numerous articles, chapters, and blogs on these topics, and regularly provides training sessions for state and local public administrators on the effective utilization of social media in government.

Mehdi Sagheb-Tehrani is an Associate Professor of CIS/MIS at the College of Business, Technology, and Communication at Bemidji State University (BSU), USA. Dr. Tehrani teaches System Analysis, Computer Business Application, E-Commerce and Web Development, and Corporate Information Systems, as well as Structured Application Development. Before joining BSU, he taught Business Network Systems Management, Virtual Business, Management Information Systems, Expert Systems, and Advanced Topics at the graduate level, and Technology Management and Systems Analysis, Project Management, Internet Applications, Introduction to Computers, C++, Visual Basic, and COBOL programming at the undergraduate level. He received his PhD in Informatics (old name, Information and Computer Science) from Lund University, Sweden, in 1993. He has published over 48 papers in various international journals and proceedings, and he has also published a book, *Management of IT*. He has publications in *International Journal of Applied Systematic Studies*, *International Journal of Management in Education*, *The Journal of Knowledge Engineering*, *ACM/ Computer and Society*, *ACM/SIGSOFT*, *ACM/SIGART*, *Journal of AICOM*, *IEEE-IRI*, and *The IS Education Journal*. He was an IT Manager as well as a Consultant in a number of organizations, including IRISL, Cutting Tools Manufacturing, NI Register Organization, IDP Company (formerly IBM), and others.

Noore Alam Siddiquee is Senior Lecturer at Department of Politics and Public Policy, School of Social and Policy Studies, Flinders University. He holds PhD in Public Administration from the University of Manchester, UK. In the past, he taught in the University of Dhaka, International Islamic University, Malaysia, and Universiti Brunei Darussalam for over 15 years. Between 2001 and 2004, he served as Associate Professor and Head of the Department of Political Science, International Islamic University, Malaysia. He has published widely in internationally refereed journals. His current research interests include public sector management and reform, e-government, local governance, and development focusing on South and Southeast Asian countries. He has also undertaken consultancies for government, NGOs, and international donors.

Carlos Ripoll Soler is Industrial Engineer by Universitat Politècnica de València (Spain). Currently, he is Director of the UPV International Campus of Excellence (VLC/CAMPUS). He is Vice President of the Euro-Latin American Network for Continuing Education and Secretary of the Working Group for Continuing Engineering Education at the European Society for Engineering Education. He has coordinated several European projects, and he is also a frequent speaker at various international forums about higher education management and has publications on the field of continuing education management. He has been Director of the Marketing Department at the Lifelong Learning Centre from UPV, with a high specialization in the areas of marketing 2.0 and on the development of Customer Relationship Management (CRM) systems applied to continuing education.

Zdeněk Struska received his MSc (Ing.) in Strategy Management from the University of West Bohemia, Pilsen, in 2004, and Ph.D. degree in Information Management from the Czech University of Life Sciences, Prague, in 2008. His present professional position is a Consultant in the area of Technology Integration in Deloitte Advisory, Czech Republic (since 2005), and Assistant Professor of Information Management, Department of Information Engineering, Faculty of Economics and Management, Czech University of Life Sciences in Prague (2007-2009).

Roberto Santana Tapia currently works as Information Analyst at the Ministry of Security and Justice, The Netherlands. He received his PhD in Information Systems (December 2009) after completing his thesis: "Assessing Business-IT Alignment in Networked Organizations." He also received a MSc degree in Information Technology Management from the Tecnológico de Monterrey (ITESM, Mexico) in 2002, and a MSc degree in Information Systems from the Universidad de Tamaulipas (UAT, Mexico) in 1999. He has also five years experience working as a System Analyst and an IT Advisor in the government and education sectors in Mexico. His main research interests are business-IT alignment, inter-organizational collaborations, and e-government.

Pascal van Eck received his MSc degree in Computer Science from Vrije Universiteit Amsterdam in 1995. From 1995 until January 2000, he worked as a Research Assistant at the Artificial Intelligence Department of Vrije Universiteit. In 2001, he successfully defended his PhD thesis on a formal, compositional semantic structure for the dynamics of multi-agent systems. Starting February 2000, he works at the Information Systems Group of the Department of Computer Science, University of Twente, as an Assistant Professor. His research interests include IT governance and alignment of business and IT architectures at the enterprise level.

Jiří Vaníček received his MSc in Mathematics from Charles University in Prague, his PhD in Computer Science from the Czech Technical University in Prague, is a member of the Czechoslovak Academy of Sciences, and is a Professor in Information Management at the Czech University of Life Sciences in Prague. He was a Head of the Software Research Division in the Czechoslovak Research Institute for Computers (1965-1989), Director of the Government Institute for Informatics in Education (1990-1994), Associate Professor for Computer Science at the Czech Technical University in Prague (1994-1997), Professor for Information Management in Department of Information Engineering (deputy head), Faculty of Economics and Management at the Czech University of Life Sciences in Prague (1997-present), and has served as President of the Czech National Technical Standardization Committee for Information Technology since 1990. His research is oriented to mathematical backgrounds of computer science.

Marcus Vogt holds a Master of Information Technology, a Bachelor of Arts in Business Administration, a Diploma in Information Systems, and has recently submitted his PhD thesis in the area of IT-Governance and Emergency Management at Bond University, Australia. Before his academic career, Marcus Vogt has worked as an IT Manager and Business Consultant for several years. He is currently employed as a Research Assistant and Lecturer at Heilbronn University in Germany, where he is involved in multiple European Union funded research projects. In addition, he is working as a Visiting Lecturer for IT-Management at the Baden-Wuerttemberg Cooperative State University. His research interests lie in the application of IT-Governance methods in public administrations and emergency management organizations, as well as the development of domain specific information systems.

Ivan Vrana received his MSc, PhD, and DrSc degrees in Radio Electronics from the Czech Technical University in Prague. Since 1990, he has been a Full Professor and Head of the Department of Information Engineering, Faculty of Economics and Management, at the Czech University of Life Sciences in Prague. His research is on systems engineering and projecting of information systems. Recently, his research priority is group decision making in conditions of ambiguity. He is a Founder and a President of the EUNIS-CZ (European University Systems Organization in Czech Republic).

Roel Wieringa is Chair of Information Systems at the University of Twente, The Netherlands. His research interests include value-based requirements engineering, business process modeling, conceptual modeling, and research methodology for requirements engineering. He is Scientific Director of the School for Information and Knowledge Systems (SIKS) that provides advanced education to all Dutch Ph.D. students in information and knowledge systems. He has been Associate Editor in Chief of *IEEE Software* for the area of requirements engineering, and serves on the board of editors of the *Requirements Engineering Journal* and of the *Journal of Software and Systems Modelling*.

Index

A

Access to Information (A2I) 220, 232, 323
Administrative Office of the Courts (AOC) 136, 323
Administrative Staff College of India (ASCI) 237, 323
Advanced Research Projects Agency (ARPA) 3, 323
Advocacy Coalition Framework (ACF) 250, 323
Agénce France Presse (AFP) 252, 323
AmericaSpeaks 116
Analytical Hierarchy Process (AHP) 110, 323
Annual Development Programme (ADP) 221, 323
Applied Information Economics (AIE) 101, 103, 323
ARPANET 4, 13, 323
Artificial Intelligence (AI) 51, 58, 63, 323
Awami League 218-219, 323
awareness for citizen 66, 323
awareness for decision makers 66, 323
Awareness, Interest, Desire, Action (AIDA) 285

B

Balanced Score Card (BSC) 101-102, 323
Bangladesh Association for Software and Information Services (BASIS) 220, 323
Bangladesh Bank (BB) 224, 323
Bangladesh Bureau of Educational Information and Statistics (BANBEIS) 221, 323
Bangladesh Bureau of Statistics (BBS) 225, 323
Bangladesh Computer Council (BCC) 220, 323
Bangladesh Computer Samity (BCS) 220, 323
Bangladesh Telecommunications Regulatory Commission (BTRC) 220, 323
base measure 168, 323
Brain Drain 68
British Broadcasting Corporation (BBC) 252, 323
Business Case (BC) 101, 323
Business-IT alignment (B-ITa) 132, 323
Business Process Reengineering (BPR) 239, 323
Business Value of IT (BVIT) 101-102, 104, 323

C

Cable News Network (CNN) 252, 323
Capability Maturity Model (CMM) 97, 323
Capability Maturity Model Integration (CMMI) 97, 323
capacity building 71-72, 104, 116, 128, 231, 323
Central Procurement Technical Unit (CPTU) 228, 323
Citizen Attention Service System (CASS) 140, 323
Civil Service Reform Programme (CSRP) 247, 323
Commercial Internet Exchange (CIX) 4, 323
Common Industry Format (CIF) 184, 194, 323
Community Clubs (CCs) 207, 213, 323
Consumer Protection Association (CPA) 253, 323
Control Objectives for Information and Related Technology (COBIT) 97, 323
corruption entrepreneurs 240
Credit Information Bureau (CIB) 224, 323
Critical Success Factors (CSF) 99, 323
Cross-Governmental Partnership (CGP) 132, 323
Customer Redressal Efficiency System (CRES) 238, 323
Cyberdemocracy 77, 323

D

data quality 175
Deed-Writers (DWs) 237, 323
Department of Administrative Reforms, Pensions, and Grievances (DARPG) 240, 323
Department of Information Services (DIS) 136, 323
Department of Licensing (DOL) 136, 323
Derived measure 168, 180, 323
Dhaka Electricity Supply Authority (DESA) 222, 323

E

Earned Value Analysis (EVA) 100, 323
E-government Action Plan 198, 323

Electronic Birth Registration Information System (EBRIS) 223, 323
emerging presence level 227, 324
enhanced ROI (EROI) 103, 324
E-Savvyness 266-267, 324

F

Fault Tolerance 171, 324
Fordist 15
Frequently Asked Questions (FAQs) 210, 324
Full Online Availability (FOA) 199, 201, 324
Functional suitability 170-171, 181, 324

G

Gartner Hype Cycles 187, 324
General Purpose Principles (GPPs) 2
General Purpose Technology (GPT) 2
governability 40
Government Integrated Financial Management Information System (GIFMIS) 249, 324
Government to Business (G2B) 25, 27, 324
Government to Citizens (G2C) 25, 27, 324
Government to Government (G2G) 28, 324

H

Hewlett Packard's ITSM Reference Model (HPITSM) 98, 324
Hong Kong and Shanghai Banking Corporation (HSBC) 213, 324
Human Resource (HR) 211, 324
Hyderabad Metropolitan Water Supply and Sewage Board (HMWSSB) 238, 324

I

i20 Government Action plan 266, 324
Indian Administrative Service (IAS) 234, 324
Indian Institute of Management, Ahmadabad (IIMA) 238, 324
Information and Communication Technology for Accelerated Development (ICT4AD) 245
Institut Technologi Brunei (ITB) 213, 324
Intelligent Traffic Systems (ITSs) 59, 324
Internal Rate of Return (IRR) 100, 324
International Electromechanical Committee (IEC) 163
International Financial Organization (IFIs) 246, 324
International Media Relations Bureau (IMRB) 252, 324
Internet Service Providers (ISPs) 220, 225, 324
IT Infrastructure Library (ITIL) 97, 324

J

Jhansi Jan Suvidha Kendra (JJSK) 238, 324
Justice INformation Data EXchange (JINDEX) 135, 324

K

Key Goal Indicators (KGI) 99, 324
Key Performance Indicators (KPI) 99, 324
Kondratieff cycle 6

L

Language Instruction for Newcomers to Canada (LINC) 142, 324
Large Tax-Payer Units (LTUs) 222, 324
Liquid Government (L-Government) 51
Local Area Networks (LAN) 235, 324

M

Magic quadrant 186, 189-190, 324
Management Information System (MIS) 273, 324
measurement function 168-169, 324
Metro Customer Care (MCC) 238, 324
Micro-blogging 84-85, 324
microprocessor 11
Microsoft Operations Framework (MOF) 98, 324
Ministries, Departments and Agencies (MDAs) 249, 324
Ministry for Housing, Spatial Planning and Environmental Management (VROM) 141
Ministry of Citizenship and Immigration (MCI) 143, 324
Ministry of Religious Affairs (MRA) 222, 324
Moore's Law 11

N

National ICT Task Force (NTF) 220, 324
National Institutional Reform Programme (NIRP) 247, 324
Net Present Value (NPV) 93, 100, 324
New Patriotic Party (NPP) 247, 324
New Public Governance (NPG) 76, 196, 324
New Public Management (NPM) 39, 49, 76, 196, 247, 324
No Objection Certificate (NOC) 222, 324

O

Office of the Press Secretary (OPS) 251, 324
Ontario's Women's Directorate (OWD) 143, 324
Open Directive 118
Open Government 118
Operating Systems (OSs) 138, 325
Organization for Economic Cooperation and Development (OECD) 43, 51, 325
Overall Online Sophistication (OOS) 199, 201, 325
Overseas Citizenship of India (OCI) 239, 325

P

Participatory Rapid Appraisals (PRA) 240, 325
Payback Period (PBP) 93, 100, 325
Personal Digital Assistants (PDAs) 55, 325
Persons of Indian Origin (PIOs) 239, 325
Power Development Board (PDB) 222, 325
Prime Minister's Office (PMO) 220, 325
Process Reference Model for IT (PRM-IT) 98, 325
Project Management Body of Knowledge (PMBOK) 97, 325
Public Administration (PUMA) 43, 325
Public Internet Access Points (PIAP) 31, 325
public-private partnership (PPP) 198, 231, 240, 325
Public Procurement Rules (PPR) 228, 325

Q

quality in operation 166, 325
quality in use 166, 168, 170-175, 183-185, 194, 325
quality subcharacteristics 169

R

Rajshahi City Corporation (RCC) 223, 325
Relationship Marketing 78
Resource Utilization 171, 325
Return On Investment (ROI) 93, 98, 100, 325
Rights, Tenancy, and Crops (RTC) 237, 325
Risk Management Process 72-74, 325

S

Scalability 281
Search Engine Optimization (SEO) 85, 325
Semantic Accuracy 176, 325
Settlement Sector Client Administration Network (SSCAN) 143, 325
Shared Interoperability Framework (SIF) 154, 325
Simple, Moral, Accountable, Responsive, and Transparent governance (SMART) 43, 325
Snellen's approach 77
Social media 121
Social Network Sites (SNS) 83, 325
Social System Theory 32-33, 325
Software Engineering Institute (SEI) 97, 325
Software Quality Requirements and Evaluation (SQuaRE) 164
Statewide Electronic Collision & Ticket Online Records application (SECTOR) 136, 325
Structural Adjustment Institutional Support (SAIS) 247, 325
Structural Adjustment Programmes (SAPs) 246, 325
Sustainable development 280
Syntactical Accuracy 176, 325

T

Tax Identification Number (TIN) 222, 325
Time Behavior 171, 325
Total Economic Impact (TEI) 101-102, 325
Total Value of Opportunity (TVO) 101-102, 325

U

Union Information Service Centres (UICS) 224, 325
Unique Grievance Number (UGN) 238, 325
Universal Service Obligation (USO) 3, 325
US Citizenship and Immigration (USCIS) 32, 325
User Centricity (UC) 199, 325

V

Value Added Tax (VAT) 27, 325
Vehicle Information and Communication Systems (VICS) 59, 325

W

Washington State Department of Transportation (WSDOT) 136, 325
Water and Sewerage Authority (WASA) 222, 325
Web 2.0 117
Wide Area Networks (WAN) 235, 325
Wireless Application Protocol (WAP) 55, 325
World Trade Organization (WTO) 41, 325
World Wide Web Consortium (W3C) 31, 325

CPSIA information can be obtained at www.ICGtesting.com
Printed in the USA
BVOW030621130912

300086BV00001B/1/P